Green Transition and AI

Utopia or Fool's Paradise

Harald Knudsen
University of Agder – School of Business and Law
Kristiansand, Norway

CRC Press
Taylor & Francis Group
Boca Raton London New York

CRC Press is an imprint of the
Taylor & Francis Group, an **informa** business
A SCIENCE PUBLISHERS BOOK

First edition published 2026
by CRC Press
2385 NW Executive Center Drive, Suite 320, Boca Raton FL 33431

and by CRC Press
4 Park Square, Milton Park, Abingdon, Oxon, OX14 4RN

CRC Press is an imprint of Taylor & Francis Group, LLC

Library of Congress Cataloging-in-Publication Data (applied for)

ISBN: 978-1-032-84480-0 (hbk)
ISBN: 978-1-032-85578-3 (pbk)
ISBN: 978-1-003-51881-5 (ebk)

DOI: 10.1201/9781003518815

Typeset in Times New Roman
by Prime Publishing Services

Foreword

By professor emeritus Jonny Holbek

In the flood of scientific publications about the "green transition", there is an occasional book that may catch the reader's attention, because it is different. This is such a book. It is not only different because it presents a 'meeting' between "green transition" and "AI" (artificial intelligence), but primarily because it moves the reader into a new knowledge area of great importance for our planet, and it does so with deep understanding.

The introductory part of the book raises questions of high significance about the green transition: Why is such a "transition" important? What are the main problems and why aren't they being solved? These are questions frequently asked, and Harald Knudsen attends to them with scholarly and frequently interdisciplinary answers, generally within the frameworks of political economy, science and ethics.

When the author of this book was awarded a festschrift in honor of his contributions to academia, he was referred to as "the manifold" academic – pointing to his wide scope of interests, talents, and qualities. I personally learned to know Professor Harald Knudsen in a cooperative work context over many years, teaching strategic management, innovation and change. During our cooperation, my "manifold" colleague, among a great number of activities developed his "leadership jazz" concept, using live jazz to convey key lessons of leadership, creativity, teamwork, and energizing rhythms. First presented to the business community in Norway, it was later part of more than a hundred leadership sessions with internationally leading firms, such as BMW, Standard Chartered Bank, IBM at the Egrove Park venue of Oxford University.

His "manifold" career also has involved significant travelling, spanning a sabbatical year on internalization and cultural studies in Hawaii, and a long chain of post-glasnost visits to Russia and Eastern Europe. He was able to observe the variety of transition paths at close hold – transitions to democracy, market economies, liberty and the rule of law, while also discerning the later collapse of democracy and rule of law in Russia, Belarus and the former Soviet republics. Closer to home, Harald Knudsen has spent a lifetime as college administrator, researcher, lecturer and doctoral advisor.

Not all academics would risk their careers by straying from their specialized fields. But Knudsen is the kind of person who can take a "bird's eye" perspective and seek answers to the "wicked problems" behind all the tensions, bringing us that much closer to deep answers and the potential of hope. As a scholar and lecturer, he has been loved by students for his ability to make difficult subjects understandable. That is also the objective of the present book. The wide scope of his approach in this book is a tribute to the complexity and interactivity of the issues.

It is all about the present and the future of people and all living things on planet Earth. Since the Last Glacial Period 12000 years ago, throughout the Holocene epoch, people and animals enjoyed a mild and stable climate, ideal for a flourishing civilization. With the coming of the Anthropocene, both climate and the natural environment are being threatened by man-made means, causing entirely new threats to arise and new answers to be sought. It is not just about climate and the quest for green energy. It is also about nature depletion and the extinction of species, about environmental toxins and all kinds of pollution. And, as is clear from the title of the book, it is increasingly about human ingenuity and the use of artificial intelligence (AI) in dealing with the issues.

Since the concept of "Anthropocene" was first introduced, a strong and informative discourse about man-made impact on the Earth's natural systems has followed. Whereas the term was finally rejected as a geological epoch name by the International Commission on Stratigraphy (ICS) and the International Union of Geological Sciences (IUGS) in 2024, non-geologists still find the term useful.

Increasing political tension, and a compelling challenge in Knudsen's book, is linked to the future of AI, robots and automation, and the lure of a "singularity" with the development of AGI (artificial general intelligence). Accelerated AI development opens for immense breakthroughs in most areas, and also for a successful green transition – for climate and energy, nature and ecosystems, and for control with new chemical entities, plastics and pollution. Breakthroughs will also be seen in areas as diverse as medicine and finance, social sustainability, business and public governance, fusion nuclear technology and military operations – for good or for bad.

AI has been seen as a tool for mankind, to make life easier, healthier and more interesting. But with the anticipated coming of an AGI "singularity" and high-level AI agency, there is also the chance that "unaligned AGI", potentially serving non-human interests and values, will be running much of what goes on in the world. Supercomputers or super-robots will be accessing intelligence levels and possessing communicative capacities widely beyond the capacity of any human being. Most likely, AI development will overall be for the common good. But it also comes with a significant risk. Knudsen refers to Toby Ord who (in "The Precipice") has estimated the chance of collapse due to unaligned artificial intelligence at one in 10, greater than all other threats added together, and bringing the totality of anthropogenic risks of existential catastrophe to one in six – in Knudsen's terms equivalent to playing Russian Roulette with the Planet.

As for environmental threats, large anthropogenic imprints have increasingly been destructive. Climate change, biomass losses and (plastic and chemical) pollution have accelerated over the last 50 years. While climate change, as a natural phenomenon, has always existed, the present man-made interference causes temperature rise at a faster and more persistent pace, threatening both human societies and ecosystem with collapse while challenging our capacity to come up with non-fossil energy solutions and broad-scale mitigation.

Important terms in the present text include the notion of an "environmental burden" – signifying the overall gap between the present overshoot in the use of natural resources and the Earth's carrying capacity. It is also a burden loaded disproportionally by the world's wealthiest nations and individuals, and it is a burden to be carried disproportionally by the world's poor. The main ethical challenge linked to the green transition, mentioned in innumerable reports and restated regularly at international venues, is that poor nations and poor people generally contribute very little to climate change and ecological collapse, but they are at the same time the main victims of deterioration. It is about an inequality that is also likely to be prolonged in an age of AI, where poor nations are at a disadvantage both in the development of programs, and adaptation and application for practical purposes.

Another important term is Knudsen's use of the word "stickiness". It is mainly about the varieties of energy and feedstock uses of fossil fuels (including steel and concrete, plastics and ammonia) that add to the environmental burden and cause the green transition to be so slow. The "stickiness" is also about how business as usual (BAU) drives lock-ins to all kinds of business and governance systems, making transformational change more difficult.

Pointing to a contradictory fact about globalization, Knudsen identifies a "detour to development". It is about how poor people, in order to gain access to what they need for basic life satisfaction must produce (luxurious) things that rich people in world markets will buy, but which they really don't need – again adding to the environmental burden.

As an alternative to Yuval Noah Harari's prediction of a class of "useless" and "irrelevant" labor following AI-driven productivity gains and joblessness, Knudsen invites us to envision a "derivative sector". It is a sector to be supported by guaranteed minimum income schemes, turning job losses into a space for future human flourishing. Productivity rises will cause mass unemployment unless accompanied by strong economic growth. Unless there is a full (absolute) decoupling between economic growth and additions to the environmental burden, however, full employment will only be achieved at the expense of the environment. Given "limits to growth", AI-driven productivity gains will necessitate the rise of a new (derivative) sector, outside of the AI-impacted commercial sector.

Knudsen builds on expert advice from several holds. For successful change, he refers to John Kotter, who has underscored the salience of "sense of urgency" in organizational change processes. When climate change and nature depletion turn from latency to manifest damage, a "sense of urgency" or "sense of crisis" may arise and motivate concerted, solidary action and agency for the environment and for controlled AI development. Of course, urgency may also be a cause of greed and competition for resources.

For the future of the political economy, Knudsen refers to Albert Hirschman's concept of "possibilism", suggesting that political-economic change may take many forms and occur in many parts of the system. Given enormous societal, technological and economic changes, to be accelerated by AI, we shall need a release from the "exoskeleton" or "straight-jacket" holds of the present capitalist order. Knudsen foresees what in cross-country skiing and cycling sports is referred to as a "breakaway" – or several "breakaways" from present lock-ins. Possibilism is a political-economic framework that asks for renewal of rules for economic activity, better suited to deal with our present predicaments.

For philosophical inspiration, Knudsen borrows the notion of "Reductio ad absurdum" from Gottfried Wilhelm von Leibniz. Logical arguments may sometimes be strengthened by asking "what if"-type questions at the extreme. Knudsen asks "what if" nearly all jobs were taken over by robots? What would happen to the demand curves of economic models? Who will buy the products of automated production if most people do not have job incomes? Might capital owners, in order to save capitalism, be driven to distribute guaranteed minimum incomes and inheritance benefits to the needy? Or would the "useless" and "irrelevant" surplus labor simply be left to die along with the capitalist system? Alternatively, Knudsen's suggestion of a "derivative sector" will offer both meaningful and environmentally friendly voluntary engagement, to include arts and hobbies, philosophy and science, sports, spiritual engagement and social activity.

For ethical backing, Knudsen refers to Immanuel Kant and his "categorical imperative", concluding that the business class consumption on board (Kenneth Boulding's) "Spaceship Earth" is simply unethical. In Kantian terms, and basing the argument on premium class consumption levels, if everybody acted "according to that maxim whereby you can at the same time will that it should become a universal law", the Earth environmental system would simply break down. Global "overshoot day" consumption levels are already near the double of planetary ecological tolerance. The Kantian "universal law" suggests that we all ask ourselves what the result would be if our habits of production and consumption were taken over by all. The ethical challenge and the categorical imperative are also there, for billionaires, top income earners and top wealth holders to contemplate and for policy makers to respect. The present consumption levels of the wealthiest, and in particular the prospect for future multiplication of consumption levels is only feasible to the extent that the global lower half of income earners are sure not to realize their aspirations.

In the last part of the book, Harald Knudsen introduces three metaphors, or three lines of reasoning about change and hope. The "war-economy" metaphor is about public and private investments and mega-projects for the green transition – bringing change from above. The "world music" metaphor is about learning and innovation – bringing change bottom up and facilitating Hirshman-style possibilism. Finally, Knudsen traces the "dugnad" metaphor to the (Norwegian) tradition of cooperative work based on mutuality and reciprocity. The dugnad concept can also be traced to practices of "commoning" and similar approaches around the world. It is all within a frame of hope and transformational change. For a final anchoring of hope, Knudsen reminds us about the power of life as such – life sustaining life, about the good will of most people, about the incredible viability of all living species, the richness of human innovation and agency in history, and the prospects for ingenuity in the future, also underscoring the force of hope – of hope driving hope.

Preface

Sometimes, writing a book or a report is just a matter of putting down in writing what is already clear in your mind. At other times, it is about extended searches to find out about things, writing as you move on, rewriting as you find new sources or discover ever more complexities and alternatives to what you have already written. The present book is of the latter kind. It is in a way a parallel to the emergence factor in strategic management. We know that not all planned strategies are implemented, and some of what is implemented had better been left out. The world is constantly changing, and some plans turn out to be irrelevant or badly conceived. So also with writing a book. Some facts and ideas turn out to be irrelevant or misconceived, and while writing, new information turns up all the time, new technologies are being launched, and economic and political contingencies are shifting.

My engagement in writing about the green transition stems from a deep-seated concern with nature. I have always enjoyed outdoor activities, and I have loved animals all my life. I have enjoyed the wonders of nature – at sea and in the forests and mountains of my home country, Norway. For more than 50 years I have been a member of Friends of the Earth Norway ("*Naturvernforbundet*"). And in teaching business strategy and international leadership at undergraduate and graduate levels, I have expressed concern about the main environmental and ethical aspects of my field. There are issues of leadership and responsibility, about development and global injustice, and about environmental depletion and the legitimacy of our extractive growth models. It is also about corruption and political-economic issues related to greed and power-grabs, and about private and collective interest collisions. Still, the main content of my work career has been framed within the main orientation of my professional calling – to teach presumably sound managerial advice based on presumably sound organizational and market research.

All the time, however, I have been questioning the presumed soundness of it all. The business school and business culture standards of soundness, expressive of what is otherwise known as "Business as usual" (BAU), comes with long lists of reservations. Anchored in neo-classical economics and legitimatized in endless books, articles and conference papers, the important contingencies related to economic models are not always made clear to students and to the public, and they are extremely important in the face of environmental degradation and geopolitical controversy.

With increasing uneasiness, I have recognized that the basic contradiction between unlimited material economic growth on the one hand and the planetary limitations of a "Spaceship Earth" on the other is gradually closing in on the global society. I have observed that the "Spaceship Earth" metaphor, so eloquently exposed by Kenneth Ewart Boulding, was hardly ever recognized in the business school literature. It was soon followed by Jay Forrester's (of MIT) World Dynamics model, and the publication of the "The limits to growth" report to the Club of Rome in 1972 by Donella and Dennis Meadows, Jørgen Randers and William W. Behrens III. The latter model basically predicted tensions and collapses in the growth-based production/consumption systems by the mid-21st century. I found it embarrassing that already before the year 2000 colleagues at home and abroad made fun of the "limits" predictions that had not yet materialized. Economics is not a fully rational science, and "economic man" is not always right. Prediction results for the 2050s should not be judged by statistics from the 1990s.

During the latter years of my academic career, reports about nature depletion, animal extinctions, biomass loss, pollution and toxicity have started to evoke increasing concern. In particular, there has been an increase in the number and quality of reports on climate change and extreme weather incidents, nature depletion and the spread of environmental toxins. In parallel with warnings, and with policy statements and international agreements to deal with climate change, however, there have also been heavy investments in denial literature and doubt-propaganda, with strong financial backing by the fossil fuel industries. Also highly visible has been a growing gap between environmental and climatic mitigation ambitions on the one hand and the slowness of real-life mitigation on the other.

After reaching retirement, I felt a strong urge to learn more about environmental threats and scientific predictions. Signs of climate change and environmental depletion became increasingly visible. Peak oil and peak fossil energy dates were approaching, and geopolitical rivalry over minerals and material resources was on the rise. It was time for re-awakening to the "limits to growth" hypothesis and to the "Spaceship Earth" challenge to economic inequality. I decided to at least examine the main messages and the science behind it all.

The *emeritus* position at the University of Agder, School of Business and Law, offered me, for the first time, freedom and time for work across disciplines and with subject matters outside of my previous teaching domains. At the time of retirement, I started doing my new homework, collecting reports and taking notes, mainly about climate change. Among environmental threat areas, there is no doubt that climate over time emerged as a main concern. As I continued with my inquiry, however, it became increasingly clear that the green transition is not just about climate and green energy. Especially with the 2019 IPBES-report (Intergovernmental Science-Policy Platform on Biodiversity and Ecosystem Services) it was clear that man-made nature depletion is a threat of – in many ways – equal concern to that of climate change, as are also the threats caused by environmental toxins and the spread of new chemical entities.

When invited to write a book about the green transition and AI (artificial intelligence), I felt on the one hand positively challenged to systematize and make sense of my notes and broaden my scope. On the other hand, I also felt a bit overwhelmed. Thousands of reports have been written about the greenhouse effect alone. Adding other elements of the green transition, such as bio-losses, energy gaps, and environmental toxins would contribute to the complexity of it all. Also, as my work on environmental threats progressed, we were all witnessing the AI-revolution of the 2020s. We had for decades witnessed the enormous progress of digital science and machine learning. But the introduction of AI chat-programs and AI solutions to a wide array of analytical and technical problems in a short time reached proportions qualifying (in Thomas S. Kuhn's terminology) as more than a "scientific revolution" – more like a societal transformation of immense historical significance. And while recognizing the velocity and momentum of AI developments, and its potential for good and bad in relation to the green transition, adding AI to the issue of the green transition also would mean adding complexity to my inquiry. There was also the question of what to do with the rest of the threat-areas – such as natural and manmade pandemics, international and civil wars, violence and terrorism, threats of nuclear war and natural disasters – all potentially influenced by man-made environmental and digital developments.

Writing a book about a rapidly changing theme is a bit like a marksmanship competition with moving targets. While we are constantly informed about climate developments as moving targets, there are also some major leaps in the positioning of targets. As a case in point, in August 2025 (after the proofs for this book were already in place), "The Lancet Countdown on health and plastics" was published, flagging plastics as "grave, growing and under-recognized danger" to both human health and the health of the planet. Plastic pollution is found at the top of Mount Everest and at the bottom of oceans. More than 16,000 chemicals are involved, and based on only three of the chemicals in 38 countries – PBDE, BPA and DEHP, the costs to health was estimated at $ 1.5 trillion a year. Most plastic is single-use, and most are (contrary to industry and petrostate lobbyist claims) chemically complex entities that cannot be recycled, allowing the authors of the report to state that "the world cannot recycle its way out of the plastic pollution crisis." Plastic pollutants are found in blood and brains, breast milk and placentas, semen and bone marrow, most damaging to foetuses, infants and young children, but also linked to strokes and heart attacks among adults. While much was known beforehand about the downsides of plastics, microplastics and nano-plastics, and also about the climate-impact of production, use and incineration, new updates in all the areas of the green transition add to the movement of the target, in my case adding challenge to the writer. At the stage of writing this preface, ever new reports about climate changes, extreme weather, and environmental depletion are coming in. Also, ideas for mitigation and adaptation are constantly in the pipeline.

The challenge of writing about a moving target is especially pronounced when writing about AI developments. Presently, Google's DeepMind division is about to launch the Genie 3 "world model", seen as a next step towards artificial general intelligence (AGI), allowing AI systems to interact convincingly with the world and taking on human-like capabilities for navigation and agency. In an interview with The Guardian (August 4th, 2025) head of DeepMind, Demis Hassabis, suggests that change in our AI future will be "10 times bigger than the Industrial Revolution – and maybe 10 times faster', anticipating human-like AGI to be available in five to ten years, significantly earlier than most predictions. We still live in a pre-AGI civilization, after which "nothing may ever be the same again" – whether apocalyptic or utopian. It opens for radical abundance and for all kinds of advances, but also for deepfakes, and for large-scale energy consumption and productivity-driven job losses.

The term "Anthropocene" was first informally coined by biologist Eugene F. Stoermer in the 1980s, later popularized by meteorologist Paul J. Crutzen. Without doubt, meaningful discourses have been conducted about responsibilities, guilt, impact, and corrective action linked to the term – also after formal rejection by the International Union of Geological Sciences (IUGS) in 2024. Recent AI developments suggest that maybe time will soon be ripe for still another epoch identification, however. With coming generations of supercomputers enabling AGI, whether in five years or in a longer timespan, and with computer and robot agency and superior managerial abilities, computers will increasingly come to the forefront of all human activity. Humans may still be "showmasters", administering elections and holding offices, serving on boards and holding patents. But policies and decisions will emerge from the robots behind. At some stage, time may even be ripe for a robot president. What used to be seen as a "tool" may soon become the new "master". That also will be the case in relation to the environment and the green transition. At the stage of AGI deployment and AGI agency, our "Anthropocene" concern with man-made environmental degradation may shift to a "Technoscene" (or "AGIscene") concern with machine-made and machine-orchestrated environmental impact or ecological collapse.

In the present book I seek to answer three main questions – (1) What is the problem - with a focus on the main environmental threats? (2) Why isn't it solved - why aren't the threats being dealt with in time? And (3) Is there still hope – what are our main bases for solutions and hope? Each question also comes with an AI juxtaposition – what AI may contribute to threats and resolve, to speeding or impeding mitigation, and to hopes for both mankind and wildlife.

We are facing an enormous "possibility space" – containing opportunities and problem areas, analytical approaches, discourses, predictions, and possible conclusions, representing endless complexity. It is a "space" that is wider and more exciting than any environment facing human beings before in history, but also riskier, less predictive, and more threatening than ever before. More than enough food is being grown on the planet. With ethical and sensible sharing, "Spaceship Earth" has enough resources for all and for lasting peace. Still, more people than ever before, especially the young, are living with anxiety and hopelessness. Investments in rearmament on earth and hopes for the colonization of Mars does not seem to offer any comfort.

It is a "possibility space" of vast complexity and interactivity. Thousands of books and articles deal with selected problem areas, in specialized disciplines, studying specific threats and quandaries. Dealing with the complexity and interactivity of the full possibility space is not possible – not in one book, not in many books. We may not ever, even with AI, gain a full overview of the dynamics of all relevant discipline areas and interactions behind the green transition and the impact of AI. On the other hand, leaving the job of integration of knowledge and overview formation to the readers would mean giving up on the author part of the job. For me, the invitation to write about greening and AI in one book was an invitation to expand my homework and enact a discourse about the wider "possibility space".

For academic teaching and research, specialization is mandatory. For dealing with the many threats and challenges facing us, thinking across disciplines is equally mandatory. Studying cross-discipline issues also means to recognize the complexity and interactivity of the threat areas. It means that some in-depth knowledge of the various threat areas has to be sacrificed at the altar of overview and "bird's view" understanding. It is a sacrifice built into the very title of this book, covering several disciplines linked to the green transition, and adding the complexities of AI developments. I do not pretend to bring new original insight about climate models, green energy or the technicalities of AI developments. My hope is to bring a better understanding and a more integrated view of the issues, of the forces of development and the grounds for hopes for concerted action, solidarity and mitigative agency.

In my thinking and writing about ongoing developments and outlooks for the future, the dichotomy of utopia and dystopia regularly has come to mind. The extremes of outcomes delimit the possibility space, spanning grand optimism and grand pessimism, signifying both ultimate threats and ultimate prosperity for all.

From my Norse heritage, the "*Edda*" epic is a compilation of tales from early Viking times, put in writing for the first time around 1220 by the Icelandic bard Snorre Sturlason. It is a narrative about the Norns, the old Norse spinners of fate, where a *völva* (a female seer) predicts the coming of *Ragnarok* at the end of times. Fate will bring natural disasters, with the prelude of a *fimbulvetr* – a Great Winter, with biting winds, endless snow, and the sun failing to shine – lasting the full length of three normal winters, with no summers in between. After the *fimbulvetr* there will be the burning of the world and a final submersion of society into an underwater world, causing the fighting of all against all, and a battle among the gods, with the main Norse mythological heroes Odin, Thor, Týr, Freyr, Heimdal, and Loki all perishing. It is about natural destruction, the wrath of gods, and man-made rivalry and catastrophe. The burning of the world is a reminder and admonition of present-day progressive heat-waves and extreme weather. The submersion of society in the ocean is a reminder of present-day heatwaves and the melting polar ice, glaciers and icecaps. To the early Vikings, Ragnarok was final – a prophecy of what was to come, encapsulating their world in time, space and spirit.

At the other end of the optimism-pessimism scale is the notion of utopia. Sir Thomas More, England's "Man for All Seasons" Lord Chancellor, in 1516 wrote the book "Utopia" about a dreamland across the Atlantic Ocean. It was his main written work, a two-part book of socio-political satire. His overall vision of utopia may not appeal to well-to-do citizens of modern societies, but we may still notice certain improving ideas. While in England at the time, human lives and wealth were wasted on wars and fruitless endeavors, in Utopia there were no wars. Enemies were captured instead of being killed. While in England enclosures of common land produced poverty, starvation and crime, in Utopia private property was abolished, and there were no locks on doors. While in England criminal codes were written in ways that made crime worse and more pervasive, in Utopia there was no crime and no lawyers. Laws were made simple enough for everybody to understand, and they all knew right from wrong and acted accordingly. Life was designed for simplicity. They all learnt how to grow food and produce meals, and they all had to work, but only six hours a day. Sir Thomas Moore's Utopia is a reminder of potentials – more than a recipe for the good society.

Beyond Ragnarokian nightmares and Utopian dreams, we have strong reasons to hang on to hope. Right here on Earth. Now! First of all, our planet is blessed with an abundance of life – with finite resources but infinite forms of life. And life breeds life. It is an immense source of hope. Secondly, there are billions of people on the planet, and while many are misled, most people are of a good nature, the friendliest of survivors. Thirdly, we have "trees of green, and red roses too" blooming everywhere, making nature our closest ally. Fourthly, we have a rich history of people finding solutions and building good "commoning" societies, fostering "world music" and endless innovativeness. Fifthly, we have expectations for science and technology to deliver solutions for a future of good living and prosperity for all. Finally, we have a capacity for turning hope into a self-fulfilling prophecy. The gift of hope helps us make what we hope for come true.

Writing a book of considerable dept and volume is most of all a lonesome job, necessitating hours of lonely work behind a computer screen. But it is also an interactive activity. Several people have been part of that interaction.

Many years ago, my oldest daughter, Solveig, noticed that in textbooks and lectures, economic growth was generally presented as an undisputable good, and she asked me why. This book is my brief answer.

Most importantly, my wife has been patient enough to let me postpone a great number of janitor jobs around the house. She had hoped that with retirement, her husband would have endless time for fixing things, painting the house, repairing stuff, getting rid of stuff and tidying the place. She has been very patient, and also loving and trusting enough to support my engagements, even where the outcome was uncertain.

My old friend and colleague, Jonny Holbek, has been kind enough to read and offer feedback on most of my draft chapters and outlines. He has been endlessly patient and supportive, frequently bringing in new information relevant to my work. He has also offered to write a foreword. Another friend and colleague, Arild Sæther, was generous enough to review parts of the manuscript covering economics and analytical elements, while two younger colleagues, Bjørn-Tore Flåten and Rotem Shneor helped out with corrections and ideas for improvement of early chapters. Hans Seland, friend, neighbor and mountain-hiking comrade, read the initial draft and offered ideas for organizing the material and finding a suitable style of writing. While all of these have made valuable contributions, remaining shortcomings and errors are surely my own responsibility.

The interactive element also includes responses from audiences attending lectures and talks, including "jazz, nature and climate" performances in a variety of venues. It includes discussions among band members for eliciting melodies and "demos" to underline the green transition message. The interactive part also includes open-hearted exchanges in my workout group ("Friends of Tomm") meeting twice a week and in a "Men's group" coming together once a month. Both groups represent varieties of experience and disciplines, and they are both arenas for frank talks, useful for calling beliefs into question while adding support and challenges to our undertakings for the last 40 years.

Harald Knudsen

Content

Introduction

In 1992, 1,700 prominent scientists signed the "World Scientists' Warning to Humanity", under the auspices of the Union of Concerned Scientists. The group, which included most then-living Nobel laureates in natural sciences, specifically warned against depletion of the ozone layer in the atmosphere, reduced freshwater resources, threats to marine life, reduced forest areas, extinction of species, reduced biodiversity, climate change, and overpopulation. The main argument was that humans caused the ecosystems to decay, and that we were quickly approaching the limit of what the biosphere can withstand before irreversible and irreparable damage occurs. Measures were strongly recommended to limit population growth, reduce emissions of greenhouse gases (GHG), phase out the use of fossil fuels, and reduce deforestation.

25 years after the original message, a new warning came in 2017, this time signed by 15,000 experts in the natural sciences. Their report pointed out, among others, that fresh water supplies have continued to decrease at the same pace as before. After a long period of increasing catches, the world's fish population has declined. The destruction of nature has intensified, and the climate situation has become more threatening. Only in one area did the signatories find a marked improvement, namely reduced emissions of ozone-depleting gases – thanks to the global agreement from Montreal, 1987. In all other areas the situation in 2017 was either as bad or markedly worse than in the first report.

We have increased the consumption of fossil fuels. We also have increased meat consumption and agricultural processes which now account for a large part of the GHG emissions and habitat destruction. Global deforestation in tropical regions has continued, and where reforestation has taken place, it has mainly come at the sacrifice of traditional, complex ecosystems. Endangered animal and plant species and habitats, and the entire biological diversity are more vulnerable today than 25 years ago, and we are spreading more chemicals and environmental toxins than ever before (Ripple et al., 2017). Of course, the message has been repeated and underpinned in endless reports, in IPCC climate reports and IPBES environment reports, and in scientific research reports by the thousands. In 2022, another update was presented (Ripple et al., 2022), stating that we are at "code red" on planet Earth, facing dramatic climate emergencies.

Since 1992 there has been a near 40 percent increase in GHG emissions, causing a temperature rise predicted at about 3°C by 2100, higher than for the last three million years, contributing also to further environmental and wilderness depletion, to more of extreme weather, and even societal collapses (Kemp et al., 2022). Among the many answers to the reports have been policies for carbon taxation and green energy support, but also an increase of subsidies for fossil fuel production and for some of the wealthiest individuals on the planet, who are behind the GHG emissions. Also, on the side of injustice, while the global south represents less than eight percent of historical GHG emissions, the same countries represent 99 percent of deaths related to climate change, extreme weather and heatwaves.

Extended threats

Saving the world is not just about climate. A virtual Nobel Prize "summit" was held for former Nobel laureates, prominent researchers, and politicians 26–28 April 2021. Over three days, the participants discussed the future of the planet. I was pleased to notice that the founding document for the summit did not focus exclusively on climate changes, but took stock of four areas of threats (Folke et al., 2021). The four areas include broadly three environmental threat areas – the destruction of nature (diversity loss, extinctions, loss of habitats and wilderness); climate changes and tipping points (global heating and extreme weather, green energy needs); and the spread of environmental toxins (in air, water, the ground and food, from new chemical entities, plastics and microplastics).

In addition to environmental threats are threats to societal cohesion and agency linked to societal disintegration and the absence of social sustainability (economic inequality, poverty and exclusion, famines, failure to reach sustainability goals, and the threats of dystopic societietal development). Still, on top of environmental and social threats, are what many people probably think of as the worst threats – those of nuclear war, devastating pandemics, or dystopian political developments.

Referring to environmental threats and to "saving the world", we may think of extinction of animal species and even of humanity. At the extreme we may think of a breakdown of human civilization, or some other level of global catastrophe. Or we may have some other, tempered level of threat in mind. Reminding us of the short history of homo sapiens (200,000 years), and important former transitions – in agriculture (10,000 years ago), science (500 years ago) and industrialization (200 years ago), Toby Ord, in his book "The Precipice" (2020) has challenged us to visualize the human blooming potential for great many years into the future – and the threats to such a future - "threats to humanity's long-term potential", what we may think of as an ultimate threat.

The present work is concerned about ultimate threats but is mainly focused on the mix of improving and threatening elements that will determine less extreme, tempered threats. It is about threats to humanity and civilization, and to animals and ecosystems, with a focus on the hardships and advances, calamities and successes, mitigation attempts and political-economic battles that lie ahead – that ultimately may lead to a fundamental, ultimate loss of human potential, but most likely will not. The text is mainly about threats that are tempered by human action and inaction, but not necessarily at an ultimate level.

The messages from environmental scientific experts are quite clear. They come with strong warnings of risks, and we may ask critically why societies are so reluctant to respond with radical mitigation. All of us have to take side – believing or disbelieving the scientific reports.

We live at a time of mixed signals. News media daily bring depressing news – about tragedies, accidents, conflicts, and trouble from all over the Earth. At the same time, we are also regularly reminded of enormous progress across the board. In the Monty Python film "The Life of Brian", Eric Idle invited us to "always look on the bright side of life". Some among us have taken the appeal closer to heart than the rest of us.

On the sunny side

The main narrative on the sunny side is that scientific progress and technological innovation has brought product and process invention, productivity and economic growth. Economic growth is seen as a guarantee for both wealth and employment, and for overall well-being and happy living. It follows that the best course for further development is to stick to business-as-usual – BAU, and not unnecessarily rock the boat. BAU generally refers to the continuation of social and economic practices and structures, not to the exclusion of change, but with change limited to gradual and cumulative adjustments, not including disruptive or revolutionary interference (Ollinaho, 2023). There seem to be an overall pattern of sunny side perspectives favored by BAU-adherents, often positive to green transition, but not to dramatic policy changes demanded by climate- and environmental scientists and activist (Hausfather and Peters, 2020).

There are strong prophetic, industrial, and entrepreneurial voices asking for more positive news and more talk about the sunny side of life, about progress and development, security, happiness, and affluence. After all, while the agricultural revolution 10,000 years ago, brought us a worse diet, harder work, more broken backs, more of infectious diseases and large-scale wars, it also brought accelerating development, starting with husbandry, cultivation and breeding of crops and animals, concentration of more people on less land, draught animals for extra energy and transportation, the building of cities and states, invention of tools and products, development of technology and culture, communication in writing, and logical thinking in mathematical form.

The enlightenment and scientific revolutions of the 18th century brought us theory and understanding of nature and the universe, new platforms for technology and invention, and the ability to grasp earthly changes and opportunities for progress. The industrial revolution brought large scale transformation of production and consumption, facilitating a better life for millions and ultimately billions of people, extending global life expectancy to 72 years, from 20–30 in the wake of the agricultural revolution, reducing severe poverty and hunger from threatening 19 out of 20 people to now one in ten, and dramatically reducing infant mortality rates. With economic progress, the number of people with access to schooling and public health support improved dramatically. And moral improvements have also been detected – for gender equality, children's rights, lgbtq+ (lesbian, gay, bisexual, transgender, queer and more) rights, poverty alleviation, interracial tolerance, immigrant integration, and religious tolerance. Not bad at all.

In support of the sunny side, the Canadian-American psychologist, Harvard professor, Steven Arthur Pinker, is best known for the books "The Better Angels of Our Human Nature" and "Enlightenment Now" (Pinker, 2011, 2018). An important message is that modern humans tend to overestimate the violence-threats of our own time. We overestimate the danger of terrorism and violent crime, while overlooking the greater violence in historical societies. The world according to Pinker has become a much safer place! For some time, it might seem that he was right. By 2025, the number of supporters seem to have declined.

Another champion of positive feedback and sunny-side views was the Swedish medical doctor, statistician and social scientist, Hans Rosling (died in 2017). Rosling was a brilliant educator and emissary, broadcasting that most people have a distorted and generally far too dark picture of present developments in the world. In 2012, he was named one of the world's most influential people, later to be known for the book "Factfulness" (Rosling, 2023 [2018]), in which he highlights ten common "thinking errors" that he believed most of us - both scholars, Nobel Prize winners and lay people – are guilty of. To progress so far, we just have to add artificial intelligence, and utopia will be right in front – fully checked for "factfulness".

Absent from the sunny-side

Rosling was correct in disclosing our faulty tendency to exaggerate the tempered threats of terror, plane crashes or local accidents, and to underestimate developmental achievements. But Rosling had his own delusions, looking at prosperity without measuring what threatens prosperity, applauding economic growth without considering resource depletion and environmental costs, cheering about increased freshwater access without worrying about the emptying of aquifers, and generally disregarding problems of climate, extinctions, ecosystems collapse, and environmental toxins. He generally overlooked the more critical questions of whether the achieved progress, and the need for grand-scale celebration, has been based on a faulty ground of nature destruction and resource depletion. He also overlooked the impact of advanced AI – for good and for bad.

What Pinker, Rosling and many like-minded people can be criticized for is that they underestimate the shadows of the future, and that their rather one-sided praise is taken as support of problem denial and BAU at a time when big threats are real, and we need to re-consider the BAU-position. We may appreciate progress, but not without a side-view to problems hiding in the shadows of life.

There are still think tanks, political parties, business operators and commentators rejecting the main reports about climate and the environment, and warning signals about unrestrained AI development. For people who distrust science, it is perhaps more convincing to instead rely on business data. According to a survey by leading French insurance-multinational (AXA, 2024), sampling 3,000 experts from 50 countries, and 20,000 individuals representing the general public from 15 countries, climate change was ranked as risk number one among globally top 10 emerging risks areas. 77 percent of respondents claimed to feel vulnerable to climate change in their everyday life, up from 73% in 2023. Next on the list of 10 were geopolitical instability (ranked as a top risk by 16% of experts, compared to 9% in 2023), Cyber security (closely linked to geopolitical instability and to the capabilities of AI and dependency on large providers), AI and big data, social tensions and movements, natural resources and biodiversity, energy risks, new security risks and terrorism, pandemics and infectious disease, and finally financial stability risks. Also, it was disclosed that 93% of the experts believed that crises represent "an increasingly significant impact on our daily lives", that 88% agreed that "misinformation has serious consequences for the functioning of political institutions", and that 91% "stressed the importance of the role of insurers" in protecting people against emerging risks. With increasing, manifest damage from climate change and environmental depletion, insurance companies are in the front-line for the unobtrusive estimation of impact.

An overall observation was that the risk areas are not isolated, to be counted one by one, but dynamically inter-connected, potentially reinforcing each other. The report highlighted "the growing complexity of managing risks, making decisions and tackling challenges in a world marked by rapid technological advances and increasingly interconnected risks, from geopolitics to security to misinformation". While we may not want to stir unnecessary anxiety, it would be dishonest to just ignore the signals from the shadows, in this case from a main, global business player like AXA.

Among the top 10 risk areas, three are central to the green transition (climate change, resources and biodiversity, and energy risks). Two are about digitalization and AI (cyber security, AI and big data). Three are about societal and political-economic issues (geopolitical instability, social tensions and movements) – all to be covered in the present book. Finally, new security risks and terrorism, pandemics and infectious disease, and financial stability risks are not main themes of the present texts, but they are all part of the backdrop.

In the shadows

Despite much progress, there is a deep shadow embracing the world community. Reports and news about climate changes, biological losses and extinctions, environmental toxicity and societal disintegration have become increasingly serious and increasingly frequent. Jean-Jacques Rousseau, the most read European author of the 18th century, paved the way for skepticism of progress, criticism of civilization, subversion of moral tradition and respect for nature. Claiming that "Man is born free, and everywhere he is in chains", he also inspired revolutionary, political-economic reforms. Rachel Carson's "Silent Spring" from 1962 produced warnings about DDT, environmental toxins and extinctions and served to initiate and inspire an American, soon to be global, environmental movement. The Limits to Growth report to the Club of Rome warned about a future breakdown of industrial societies and the coming collapse of economic growth models a hundred years hence (Meadows et al., 1972). The report inspired estimation of annual ecological overshoot trajectories, wildlife statistics, ecosystem breakdowns and environmental toxins impacts. Added to the original wilderness concerns and concerns about food and pollution, climate reports from the 1980s, and international negotiations and agreements about greenhouse gas (GHG) emissions served to expose the reality of big environmental threats in the shadows of modernization, technological progress, and economic growth.

Still, despite warnings and reports, depressingly little has been achieved in terms of practical improvements. A recent report from a group of climate science insiders (Ripple et al., 2024), reviewing six IPCC (Intergovernmental Panel of Climate Change) reports, 28 COP (Conference of the Parties) meetings, hundreds of other reports, and tens of thousands of scientific papers concluded that "the world has made only very minor headway on climate change". And as we shall see, beyond the threats of climate change, not much more headway has been made in other areas of environmental concern. Some of the barriers to change are technological, where technical solutions need to be developed for a green transition. Some are cultural or political, where people resist change and what they perceive as threats to an otherwise good life and to historical patterns of development. And some barriers are political-economic, anchored in inefficient market systems, faulty political governance, meaningless property rights, the powerlessness of the poor (who are the main victims of environmental and climatic breakdowns) and the incredible power of the wealthiest 1 percent and 0,1 percent (who are the main destroyers of climate and nature and the main consumers of resources).

Depressing news

For well-being to prevail in a society, social and economic security is a number one priority. Over the last 50 years, depression and anxiety have reached epidemic proportions in modern societies, causing great damage and pain, with causes being largely unexplained and symptoms being badly treated by modern medicine. Topping the trend, however, researchers are now finding a quite dramatic increase of "Eco-Anxiety" around the world, especially among the young. According to a global survey (Pearson, 2024), more than half of people in the 16–25 age group struggled with sadness, felt anxious or powerless, or had other negative emotions related to climate change. Some are anxious about the future, based on news and rumors. Others suffer from direct, traumatic experience – of forest fires, droughts, heat waves, hurricanes, or flooding. Some are ready to give up, after ploughing and sowing, again and again, just to see their harvests destroyed by drought, flood, wind or heat.

And there is more to come. Artificial intelligence (AI) may turn out to be what most among us have hoped for – but it may also turn out to pose the biggest threats of all. The backdrop is a world of everlasting improvement. We were to live longer, become more knowledgeable and experience more lasting happiness in a world of security and prosperity, also expecting miracles from AI. We were promised a rose garden, a prolonged life in prosperity and good health, a playful existence on the sunny side, with everlasting (pill-assisted) happiness and divine-level (AI-assisted) knowledge.

We have come to realize instead that life on the shadow side is a life amid conflicts, wars and refugee migration, corruption and kleptocracy, economic inequality, fake news and hidden lies, increasing food prices and newborn poverty, with dire prognoses for climate and the natural environment. A main problem with life in the shadows is that much of negative information tends to be invisible or difficult to spot, sometimes emerging so slowly that we do not even notice, and most of the time clouded by outright denial of problems.

There has been an enormous increase in wealth over the last few decades, and millions of people have been lifted out of extreme poverty and starvation (the sunny side). The proportion of people living in extreme poverty has decreased dramatically. However, in absolute terms there are as many extremely poor and hungry people today as there were 30 years ago, with the numbers actually increasing over the last few years. There has been an exceptional increase of wealth, but not a corresponding increase in general prosperity and welfare – too much of the new wealth is concentrated in far too few hands. And life in cities tends to be more costly than life in self-sufficient farm communities: Modern large-scale urbanization in poor countries tend to present an inflated picture of improvement when more people pass the poverty threshold of $ 2 per person per day – not sufficient for a decent living in a high-cost city.

Child mortality, low life expectancy and "poverty diseases" (mainly due to hunger, poor sanitation, cramped housing, uncleanliness, dangerous, tiring work, and miserable social conditions) have been dramatically reversed. But "poverty diseases" have to a significant extent been replaced by "prosperity diseases". Cancer, cardiovascular disease, diabetes II, arthritis, Alzheimer's, auto-immunity diseases, and diseases linked to lifestyle, obesity, ultra-processed food, poor nutrition, and the spread of environmental toxins have in many cases taken over. These are issues to which we shall return in Chapter 3.

At times there are eruptions of protest and critical mass movements, as in Seattle against globalization in 1999 or at COP meetings where the fossil fuel industry mobilizes lobbyists to undermine climate action. There has as yet been no real large-scale mobilization for climate and the environment, or for stronger regulations of AI and protection against possible AI-caused unemployment. The dominant trend is still for small scale AI regulations, small scale carbon taxes, small scale promotion of green energy and the like – all within the political domain of BAU.

Breaking with BAU

Several times in the following chapters, I shall refer to Business As Usual (BAU). The term has traditionally been used about business organizations sticking to normal modes of operation as opposed to radically changed modes. More relevant to our topic is the metaphorical use related to environmental threats, where BAU is referred to as the overall political-economic practices predominant in modern societies, practices which may or may not allow for mitigation and an effective (as opposed to greenwashing type) green transition. Among others, Ronan Connolly and colleagues refer to BAU "conditions", in their study of mitigation policies and merits (2020). In some studies, BAU is also referred to as a baseline concept of emissions levels (Fei and Shuang-Qing, 2012). In the present text it is the "conditions" version that is relevant. The exact nature of these conditions is seldom specified. In the present text it is the predominant political-economic ingredients for keeping up BAU that are referred to. These will be discussed in detail in several chapters.

Behind the dozens of issues related to environmental threats, social sustainability and AI in this book, there is a major issue lurching in the back. Can we deal with such threats within the political-economic BAU format, or do we have to seek metamorphose level societal transformation, and if so, how? What do we need to do in order to regain hope?

The present text rests on two basic premises. First of all, it is based on the premise that the natural sciences, and warnings about environmental threats, are to be taken seriously. As a social scientist I cannot discuss the quality of climate models, dispute the soundness of biomass estimation or the toxicity of new chemical entities and environmental toxins. Generally, the conclusions reached in scientific studies tend to be unanimous and sound. But as a lay person in the natural science, I can only follow what Robert Kl. Merton in his address "Science and the Social Order" (1938) suggested, namely to trust the scientists.

For the disbeliever, there is additional confirmations about man-made climate disruption and extreme weather consequences. Already in Chapter 1 we shall note warnings from the insurance sector about in some location steeply increasing payments for climate-caused damage and repairs, causing premiums to rise and in some cases making insurance commercially impossible. With an insurance breakdown, huge sectors of finance will collapse. This is not about empty threats by academics and activists. It is about calls from the heart of capitalist institutions, warning that failure to mitigate will threaten even the capitalist system as such.

Also, for the disbeliever, daily news about extreme weather and climate disasters from around the world is a constant reminder that change is underway. In my home on the South-coast of Norway, I live in one of the world's most stable climates. Still, as a senior I notice marked changes from my childhood seasons. And on the evening news, I receive constant reminders that other areas of the globe are increasingly exposed to weather extremes.

In Chapter 8 we shall observe the diversity of experiential reactions to manifest environmental damage, adding "proof" of ongoing climate change to that of the climate researchers. Accelerating manifest damage from climate changes and from environmental depletion bear testimony of urgency and crisis beyond scientific models and insurance statistics, all too easy for all of us to experience and at least for some of us to prove beyond doubt that the climate experts are right.

In Chapter 14 we shall encounter another "proof" of seriousness. Following the news in December 2023, that Mark Zuckerberg was constructing a 5,000 square foot bunker on the Island of Kauai in Hawaii, for protection against a possible catastrophic "event" (due to climate change or nuclear war or anything in between), the bunkers construction business in the US took off. In 2022, another tech billionaire, Peter Thiel (of Paypal) had similar plans for a "doomsday bunker" in New Zealand thwarted by a local council. As we shall see in Chapter 14, people of means take the coming of an "event" for serious – confirming the seriousness of the present situation. While making enormous profits from businesses that in part are adding to the likelihood of an "event" and to the seriousness of the outcome, more and more billionaires are seeking escape for themselves and their families – ready to "flee from the coming wrath". In the times of ocean sail ship navigation, it was a sign of doomsday when the rats were seen to abandon ship. At our time it is a similar sign of trouble, but also of testimony of climate research reliability, when tech billionaires are abandoning our civilization for personal safety reasons.

As for AI developments and threats to human civilization, we have no corresponding "panels" and mathematical prognoses. What I can present, instead, is a number of historical premises and speculations about the future – related to greening contributions, and to scientific developments, societal threats, and long-term issues of employment and engagement.

Secondly, my thinking is essentially ethical, both for interpretation of consequences and for norms of mitigation and damage restriction. At heart is the individual and collective good, for people and animals and everything living. While mainly concerned with consequentialism, at one point my reasoning turns Kantian and deontological: In Chapter 11, I review some implications of the "Spaceship Earth" metaphor for the circular economy and for unrestricted BAU growth. Immanuel Kant's categorical moral imperative from his "Groundwork of the Metaphysics of Morals" states that you shall "Act only according to that maxim whereby you can at the same time will that it should become a universal law". I cannot will that the consumption patterns of the wealthiest societies on Earth were to become universal. Planetary ecosystems would collapse. In fact it wouldn't be possible. The present economic model of the wealthiest is only possible as long as the bottom half of the world population is effectively prevented from attaining the same level. That is not healthy and it is not moral, and I feel obliged to search for alternative, more equitable modes of living.

An outline

The purpose of this book is to summarize a series of threats to modern societies, and the political-economic challenges caused by environmental threats, by social in-sustainability, and by AI-robotization. It is not just about climate. A main contradiction facing us is the one between, on the one hand, two hundred years of economic and social progress, promising further improvement and, on the other hand, threats of a critical tipping point in life conditions on this Earth.

Literature about environmental and digital developments is abundantly available. My own contribution is a response to a common call: There is so much going on, so many threats, so much uncertainty.... Why? And where can we find some kind of overview? Where does it take us? I do not hope to bring original insight into a better interpretation of climate models or green energy delivery. But at least I have tried to come up with a better understanding of the connecting lines between threat areas and arenas calling for agency, of causal impacts, lock-ins and areas of possibilism.

We may use the metaphor of a grand party, a real feast for humanity. Perhaps the party will continue forever, with more and more of everything, everybody living longer, being happier, and endowed with more and more intelligence. On the other hand, perhaps the party will soon be over. We have helped ourselves to the best of goodies, to the cheapest, highest quality and easiest available resources. We have partied without restraint, but the bottles are soon emptied, and the feast will soon be over. We will return home with a hangover, longing for a glass of cold, clean water and resigning to living within constraints, cultivating true values and austerity, and supporting political-economic amendments compatible with sobriety.

This book takes the view that at least we are in for some dramatic changes and – in the end – for societal transformations to make the green transition real. The hangover element comes in many forms, for many reasons and with different degrees of severity. And while productivity improvement and economic growth is by most seen as a buffer against crisis, my position is that "coupled economic growth" (economic growth coming with an increasing environmental burden) will escalate the threats rather than help us deal with them – calling for transformative, societal change and decoupling of economic growth.

The text has three main parts. In Part One, Chapters 1–7, a set of threats and challenges is introduced. The question is asked: What is the problem? In Part Two, Chapters 8–13, political-economic and social concerns are further examined. Here the question is: Why isn't is solved? In Part Three, Chapters 14–16, some main solutions and options for hope conclude the work. The main question is: Is there hope?

A visual overview is shown in Figure 1. The numbering in the figure refers to chapter numbers in the book and to relevant chapter titles. The intent of the figure is to highlight the dynamics of the presentation, starting in Part One (Chapters 1–7) with the main environmental threats and the challenges of a green energy transition, summarized as the totality of the "environmental burden". To the environmental burden, Artificial Intelligence (AI) is added – both for positive solutions to the "wicked problems" of the environment and for the opposite possibility of generating more threats and political challenges that threaten life on the planet. AI is predicted to be likely the greatest societal game-changer of all time. Only time will show exactly how pervasive and in what direction.

Part Two includes six Chapters (8–13) dealing with denials, delays, and economic and technological realities linked to environmental action and inaction. Environmental disruptions and damages, so far at a stage of prognoses and latency, are increasingly becoming manifest, encouraging a sense of urgency, resolve and action. Manifest damage does not always lead to mitigative concerted and solidary action, however. Societal disintegration and political-economic straight-jackets are hurdles that needs to be overcome in order for there to be effective mitigative action. In turn, mitigative action needs to include a full decoupling of economic activity and economic growth from negative environmental impact. Also, in order for resolve, economic redistribution is needed intra-nationally and internationally, for truly concerted action to prevail. Adding to the political-economic challenges, extraordinary future productivity gains from AI, robotization and potentially AGI (artificial general intelligence), can only serve the interest of full employment under conditions of extraordinary economic growth, which under conditions of decoupling (or de-growth) may be extremely demanding, suggesting the emergence of an otherwise "useless" or "irrelevant" class – and better, a "derivative sector" sustained by a guaranteed minimum income scheme.

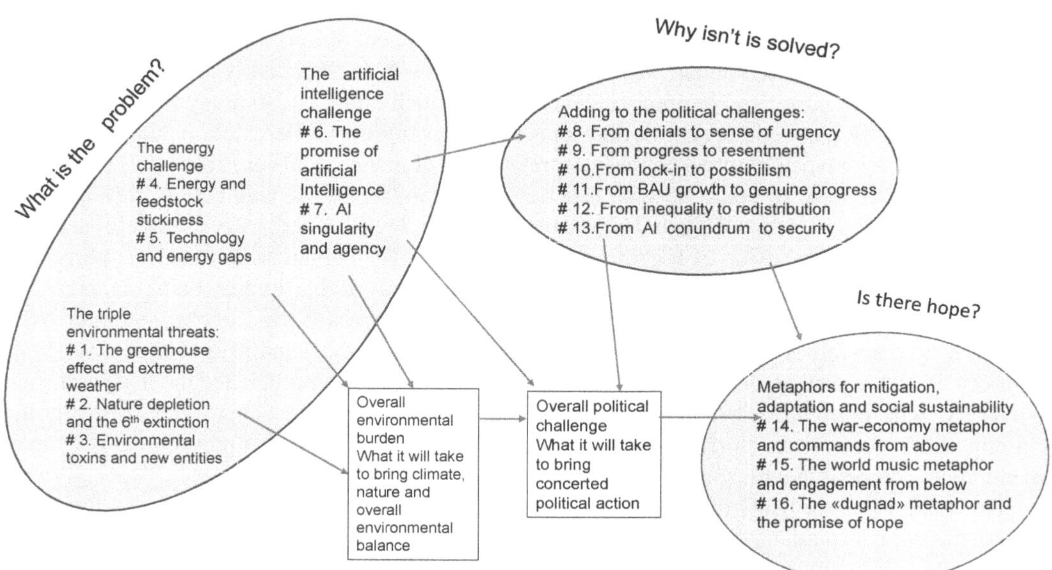

Figure 1: Overview.

In Part Three (Chapters 14–16), three metaphors introduce three different angles of environmental action and hope. The war-economy metaphor refers to top-down participation and mega-projects for the environment. The world music metaphor points to the innumerable ways people around the world may come up with green solutions. And finally, the "*dugnad*" metaphor is introduced to rethink the governance of work life in a future of an increased "derivative sector", for creative, compassionate and meaningful lives.

Figure 1 provides an overview of the full chapter outline, serving offer a preview to readers. While most readers may prefer to read chapter by chapter in the numbered order, it is fully possible to choose your own ordering of the chapters.

PART ONE— WHAT IS THE PROBLEM?

In Part One the question asked is: What is the problem? What is the problem with the Green Transition and with Artificial Intelligence (AI) – hoped to assist in curing the problem?

The green transition is about man-made climate change, but also about nature depletion and species extinctions, and about pollution and environmental toxins – forming a set of triple environmental threats. The transition also includes, perhaps as the most demanding, finding solutions to green energy transition, presenting us with the total environmental burden.

For dealing with the environmental burden, there are hopes that new technologies, fronted by Artificial Intelligence (AI), will come to play a major role. At the same time, AI development, in addition to the promise of ingenious enhancement of human life and brilliant solutions to environmental problems, also comes with its own set of ultimate and tempered threats to mankind and the Earth, presenting us with the overall political challenge.

In dealing with the overall political challenge, there is a fundamental question to be answered, whether to take the scientific warnings about environmental threats and the expert advices about AI regulations seriously. The position of the present text is to listen to the nearly hundred percent of scientists who warn about the triple environmental threats and to the voices of AI insiders who warn about the consequences of billionaire-driven, laissez-faire style AI unchained growth

Chapter 1

The Greenhouse Effect and Extreme Weather

The message is crystal clear, not only from scientists, but from widespread direct experience of extreme weather. Temperature rise, climate changes, and nature depletion need to be taken for a fact. We do not have ultimate proof that the changes are entirely man-made, but there is a direct correlation between a man-made CO_2 and other climate-gases' rise in the atmosphere and the rise of earthly temperature, as predicted by greenhouse theory since the mid-1800s. And there is no alternative explanation for the temperature rise, which makes the greenhouse effect a strong scientific case for causality of extreme weather events.

Climate change is not the only environmental concern, however. To the threats of climate change, nature depletion and environmental toxins and pollution have to be added. Authoritative sources tell us that the direct destruction of nature is as serious as the threat of climate change, also that the spread of plastics, new chemical entities and environmental toxins and pollutions are again as serious as nature depletion and the sixth mass extinction of living species. The present chapter presents in condensed form the climate element of the environmental threats.

Climate changes for real

Climate change is not just about record temperatures, but increasingly about extreme weather and damage. In 2010 there was a record flood in Pakistan, with 20 million dispersed people, and in Russia a record heatwave with 56,000 people dead; in 2011, the southwestern US had the worst-ever drought, and the Missisippi had the worst floods in near 100 years; in 2012 the artic summer ice sheet melted at a pace suggesting an ice-free North Pole within decades, Hurricane Sandy gave New Yorkers a green awakening, and Australia experienced the hottest-ever summer with record bush-fires; In 2013 the super-typhon Haiyan, the strongest ever measured, hit the Philippines, causing 10,000 casualties; and in 2014 the beginning of collapse of the West Antarctic ice sheet was announced.

Those were early signals of a transition from latent to manifest damage, and they have continued at an accelerating pace, with in 2015 an extraordinary artic heat wave, floods in South America and the US Midwest, drought in South Africa and Ethiopia causing food and water insecurity to 40 million people, record cyclones in Yemen and the Philippines, deadly heat waves in Pakistan, India, Iran and the Middle East. And it goes on and on, year by year, with hurricanes, flooding, droughts, heat waves, and gradually with impact on the global ocean current systems and ecosystems, fish and animal migration patterns and survival potential. In 2023, there was record-breaking heat across most of Asia, (the dry) Libya had destructive floods, there was heavy snow in Los Angeles, a strong cyclone in Southeast Africa, sandstorm in Beijing, cyclone Mocha in Myanmar, spring heatwaves in Australia and a category five Typhon Mawar hitting Guam. 2023 was the warmest year ever recorded, and 2024 for the first time broke the 1.5°C target set by the Paris Agreement for 2050.

DOI: 10.1201/9781003518815-2

Climate change is not just about global temperature getting a little bit higher every year, but about temperature causing greater evaporation, and greater humidity holding more water in the air, combined with higher temperatures causing more turbulence, stronger winds, more of both rain and drought, floods and heat waves, forest fires and flooding. The 2020 South Asian floods resulted from extraordinary heavy monsoon rains and caused more than a hundred billion US $ damage, most of it in India, but with severe human and economic cost also to Afghanistan, Bangladesh, Nepal, Pakistan, and Sri Lanka. In Pakistan, from mid-June till October 2022, a new record flooding caused 1,739 casualties and record economic damage, resulting from heavy monsoon rain and melting glaciers following a severe heat wave. One third of the land was under water, displacing 33 million people – half of them children, destroying critical infrastructure, including 30,000 schools, 2,000 health institutions and hospitals, and 4,300 water systems. To the scale of damage must be added that while the global south (poor countries) account for only eight percent of historical GHG emissions, these are the countries suffering most from climate change – with poor people overall counting for 99 percent of resulting deaths.

Beyond human suffering, climate change is a great environmental threat. The "Great Dying", some 252 million years ago, was the largest mass-extinction in history. Caused by massive volcanic eruptions in Siberia, followed by an extreme temperature rise, 90 percent of all life was wiped out. Ocean animals were unable to breathe, and some 96 percent of marine species died. Today, the Earth is heating even faster, extinction is proceeding at a higher rate, and ecosystem collapse is a real threat. At the time of the "Great Dying" it took millions of years for life forms and species to multiply again.

Grades of threats

Threats come in many grades and shades. At the most extreme, we may imagine a severe event where all life on Earth is wiped out, a total dying, bringing the world back to time zero, nearly as it was 4.5 billion years ago, or at least to the time of land-based plant and animal creation, the Ordovician period more than 400 million years ago. We may think of this as a total threat, very unlikely to occur.

In everyday speech we tend to use the words threats and risk interchangeably. For short, we may think of a threat as a negative event to which we are vulnerable, whereas risk is the chance (probability) that the negative event will occur. Thus, a somewhat smaller than a total threat, but at greater risk, is what Toby Ord (2020) describes as "threats to humanity's long-term potential" – a situation where humans may survive, but where we are at least back to the birth of the sapiens some 200,000 years ago, with a very uncertain future and no definite prospects for cultural, intellectual or technological advancement. I shall refer to this as an ultimate threat, according to Ord with a 1 in 6 chance of occurring sometime the next hundred years (to be further explained in Chapter 7).

Threats in the present writing are mainly about something still more probable and closer to mind, but less destructive. I refer to this as a tempered threat. Tempered environmental threats are high-risk, high-probability, human-caused threats, almost sure to occur unless we dramatically change production and consumption patterns, or unless near-miraculous technology innovation changes everything. And even miraculous technology innovation may be a threat – whether in the form of unaligned AI or in the form of climate-saving energy that stimulates nature-depleting economic growth.

The threat of dramatic climate change is caused mainly by emissions of CO_2 and other greenhouse gases (GHG), and by the problems of phasing out fossil fuels as chief energy sources. For a long time, the global average temperature has been about 15°C, but for long periods of Earth's history over the last 540 million years, it has been higher, up to 25°C. But human-like beings have only been here for a couple of million years, and the sapiens genus only for 200,000 years. For the last 800,000 years, atmospheric CO_2 density has varied between 180 and 280 ppm (parts per million) – only since the Second World War has it taken off, to pass the 400 ppm mark, with correspondingly greater temperature rise. And already in the 2020s, the temperature has risen to a higher level than ever before in the history of the sapiens. Summer temperature in 2023 was the highest in 180.000 years and the annual average of 2024 represents a new recorded high, surpassing the (Paris) target of a 1.5°C rise since the beginning of industrialization. And temperature changes make a difference.

The Greenland ice sheet, or "ice cap", is an enormous body of ice, covering 1,710,000 square kilometers, or 80 percent of the Greenland territory. The average thickness is 1.5 km (1500 meters) – and at the deepest it measures over 3 km (3,000 meters). For survivability and agriculture there is a huge difference between bare ground and 3 km of ice. If we look back to the latest ice age, the global mean temperature 20,000 years ago, when temperature was at its lowest, was about 6°C cooler than today. At the time the ice started receding in Northern Europe around 12,000 years ago, the temperature was about 5°C cooler than today – and the average ice thickness was more than 1 km, about the same as in Greenland today.

The data tells us that temperature varieties bring about a near total change of environments. The five-degree difference between an ice-age environment and today's temperate environment tells us that another five-degree difference, for a warmer climate, would bring enormous changes of landscapes and natural environments.

Early warnings

Before considering man-made greenhouse effects, we may remind ourselves of the miraculous work of the atmosphere. The atmosphere is like a shield, with a protective effect equivalent to a 4.5 meter thick layer of concrete, which absorbs cosmic rays and ultraviolet radiation and repels many kinds of particles from space. Even raindrops would cut right through our bodies if the atmosphere didn't cushion the blows and protect us. Among the different layers in the atmosphere, it is only near the ground level (up to 5,000 – 5,500 meters) that humans can live – at least for an extended period of time. This is where we find the entire biosphere – the only space in the entire universe known to support life, at a mean temperature around 15°C and an oxygen level of some 21 percent.

The greenhouse effect was first identified by the French mathematician and physicist Joseph Fourier in 1824. He observed that when the sun shone on a box with a glass lid, it would be warmer inside the box than outside. Also, the French physicist Claude Pouillet and the English mathematician and geologist William Hopkins contributed to the early, basic insight, that heat from the sun had an easier time penetrating the atmosphere (on the way in), while the return radiation from the Earth (on the way out) had more difficulty getting through. The early discoveries suggested that while glass can create a greenhouse effect, there must also be something in the air, in the atmosphere, that has a similar effect.

The American scientist and feminist, Eunice Newton Foote, was in 1856 the first to demonstrate that a change in the amount of carbon dioxide in the atmosphere led to a change in temperature. The Irish physicist, John Tydall, was soon to follow. He was not aware of Foote's work but became known in Great Brittain in the 1850s for his studies of diamagnetism (what makes magnets repel each other) and later studies of infrared radiation and the physical properties of air, demonstrating in 1859 to the Royal Society of Great Brittain how the content of CO_2 in the air led to a greenhouse effect.

The Swedish chemist Svante Arrhenius, an expert on Ice Age research, was in 1896 the first to explain the relationship between water vapor changes and the large fluctuations between historical warm and cold periods and variations in extreme weather. He was also the first to estimate the temperature consequences of CO_2 in the atmosphere, concluding that a 50 percent increase in the CO_2 content would lead to a global warming of more than 3 °C, while a doubling would lead to a temperature increase of 5–6°C – not very far off what we observe today.

Still later, the Canadian amateur scientist and steam engine engineer Guy Callendar, in a 1938 article in the Royal Meteorological Society journal, proved that warming was indeed taking place, and that increased CO_2 emissions were an important cause (Anderson et al., 2016; Callendar, 1961). However, neither for Callendar nor for the other early climate watchers, global heating was a cause of alarm. In the Northern hemisphere, a bit higher temperature was generally seen as an advantage – for people and crops.

From the 1950s onwards, however, several scientists, especially physicists and oceanographers in America, became alarmed about possible negative consequences, and in the 1970s and 1980s, a number of reports and statements were produced. Physicists dominated the professional circles for a long time, but eventually geologists and meteorologists became more active. For a long time, however, there was little political attention to the greenhouse effect and man-made warming. In the widely published "Limits to Growth" report from 1972, climate was not an issue. It was only when researchers later that decade began to write reports and give talks about the dangers of climate change, and the need for mitigation, regulations, and bans on emissions of GHGs, that politicians and lay people were alarmed. It was also at this time that skeptics, along with newly established conservative "think tanks", with solid financial support from the fossil fuel industry, and from super-rich families, became active on the denial side.

Precise measurements and serious reports

Since 1990, the UN climate panel has produces six major reports about climate change. International agreements have been entered into, and policy statements have flourished. However, emission of greenhouse gases (GHGs) has continued to grow, and very few experts believe that the Paris targets from 2015, of a maximum temperature rise of 1.5°C by 2050 – or an ultimate maximum of 2.0°C, will be met. Already, the temperature rise in Europe is around 2.5°C, and the global average (including atmosphere over the oceans) has already passed the 1.5°C target.

If we calculate the sum of CO_2 emissions and CO_2 equivalents of other GHGs, we burn globally the equals of an Olympic-sized swimming pool filled with oil, or 300,000 liters per second. The level of CO_2 in the atmosphere has increased from approximately 280 PPM (parts per million) before the industrial age to a record 427 PPM in 2024.

The oceans have so far absorbed some 93 per cent of the heat energy from fossil fuel combustion and GHG emissions. It takes enormous amounts of energy to heat the world's oceans. The Chinese atmospheric physicist, Lijing Cheng, lead author of the United Nations - IPCC Special Report for Ocean and Cryosphere, has calculated that in the last 25 years the oceans have absorbed heat energy equivalent to 3.6 billion Hiroshima atomic bombs, equivalent to four such "fire-bombs" per second (Croeser, 2020).

For a period of time since the early 1990s, the atmospheric temperature rise was lower than what many researchers had expected. According to a report by researchers at Columbia University and the University of Exeter, an ice-free Arctic may have been delayed by 15 years because of another factor than CO_2 emissions. In order to save the ozone layer, the 1987 Montreal Protocol put an end to the use of the extremely powerful chlorofluorocarbons in refrigerators, spray cans, etc. (England and Polvani, 2023). However, chlorofluorocarbons were not only a threat to the ozone layer, but also a GHG, 10,000 times more powerful than CO_2, and able to stay in the atmosphere for up to 100 years. The Montreal Protocol was enough to impact not only the ozone depletion threat, but also the greenhouse effect, and significantly delay global heating in addition to saving us from being burnt alive by profligate UV-radiation.

Beyond this delay, the IPCC reports have become increasingly precise and authoritative, confirming beyond any doubt that present climate changes are real, and they are man-made (IPCC, 2023). Reliable temperature figures are available from the World Meteorological Organization (WMO - UN body for weather, climate and water), based on data sets collected regularly by leading institutions, from thousands of records on land and in the sea, on ships and buoys and in the air from aircraft and satellites. The main institutions for calculating global and regional averages are the United States National Oceanic and Atmospheric Administration (NOAA); NASA's Goddard

Institute for Space Studies (NASA GISS); the United Kingdom's Met Office Hadley Center; the University of East Anglia's Climatic Research Unit (HadCRUT); the European Center for Medium Range Weather Forecasts and the Copernicus Climate Change Service; and the Japan Meteorological Agency (JMA). When the data are analyzed, millions of meteorological observations and ocean research data are used. There is today a full scientific consensus about the basics of temperature rise and a consequential increase of extreme weather incidents, while of course a great number of finer points are being further discussed.

To the measurement of temperature rise, we need to be aware of regional varieties and trends beyond the global average. It has for a long time been seen that the temperatures in the Arctic are rising at a pace of three times the global average, with Svalbard topping the global trend, with accelerated snow and ice melting and significant disturbance of local ecosystems, fauna, and global climate systems. A still more significant trend is that the temperature rise over dry land exceeds the rise over the oceans by a factor of some 50 percent – meaning that a global average temperature rise of 2.0°C will bring an average temperature rise of 3.0°C where we all live. The air over the oceans heats up more slowly, mainly because the humidity has a cooling effect. The humidity over land has been decreasing in recent years, which further enhances the warming over land compared to ocean areas (Byrne and O'Gorman, 2018).

To that may again be added that places with a lot of sun and drought will have a significantly higher rise, in addition to problems of the "humidity islands" of big cities making high temperatures feel extremely hot. In fact, modern cities can be compared to baking ovens, building up heat and humidity and preserving it for long periods of time. Since the first observation of the "urban heat island" (UHI) effect was made in London by Luke Howard (1833), the particular heating patterns of cities has been widely studied, also including studies of heat and humidity combinations (Huang and Song, 2023). With nearly 70 percent of the world population expected to be living in cities by 2050, and with outlooks for global heating exceeding the Paris targets by a wide margin, urban heating challenges will intensify. Much can be achieved by allowing for the greening of cities, and even more from learning from termites how to build self-air-conditioning housing and business structures (Lüscher, 1961), also learning from Persian history how to build "windcatcher" structures in hot and dry climates (Sirror, 2024). Locally, standard air-conditioning can handle increasing heat challenges for those who can afford it, but air conditioning takes significant energy and it also directly add to the heat of the surrounding environment (Salamanca et al., 2014) – adding to the overall environmental burden.

More extreme weather means that deadly heat waves, massive forest fires, floods, storms and hurricanes will come more often and hit much harder than today. Where an area today has 116 heat wave days a year (of more than 35°C, a normal, long summer as in Arizona, USA), 179 to 229 such heat-days may be expected with a temperature rise of 3°C, also with temperature peaks significantly higher than today. The world is likely to exceed both the Paris target of a 1.5°C rise towards 2050 and the 21st century maximum 2.0°C target. We are steering towards a global land-based temperature rise of at least 3–4°C, with dire consequences for nature and people, and with prospects for passing important tipping points leading on to a new, self-reinforcing planetary trajectory change, even with the possibility of irrevocable collapse.

A climate-skeptic argument has been that temperature variations on Earth are due to variations in radiation from the Sun. In order for sun radiation to be the cause of temperature rise in our time, the temperature in the upper layers of the atmosphere would have to show a greater degree of warming than the lower layers. But that is not the case. It is mainly the lower layers of the atmosphere that are now getting warmer, which suggests that the changes are coming from earthly activity.

Settlement and migration

A 2020 survey published by the US National Academy of Sciences (Xu et al., 2020) looked at what could happen to settlement and migration as a result of global temperature rise. The authors started with identifying historical "human climate niches". These are areas on Earth where people in the past and in our time have chosen to settle. Most people in the world have historically settled in areas with an annual average temperature of 11–15°C, which is slightly cooler than the global average.

Of the Earth's more than 7 billion people, today fewer than 25 million live in the warmest areas, most of them in the Sahara, in areas where the average annual temperature is more than 29°C. If the rise in temperature is not brought under control, a third of the Earth's population, or some 3,5 billion people in 2070, may end up either migrating or living in areas with temperatures at this level – areas that currently covers only 0.8 percent of the Earth's land areas. With unmitigated temperature rises such temperatures will affect large areas of Africa, India, the Middle East, Central- and South America, South-East Asia and Australia. Of course, if billions of people are forced to migrate, neither excessive emigration nor immigration is ideal for building stable societies.

Many of the areas that are today among the warmest will also see the highest temperature rises in the future. Some of today's cold areas will become more pleasant, but the main effect of global warming will be that huge areas that until today are seen as climatically ideal, such as Southern Europe, North America and Northern Asia, will belong in the warmest layer, while areas which today belong to the warmest of inhabited areas, will become more or less uninhabitable and also unsuitable for food production. Many types of crops and livestock are best adapted to temperature levels below 15 degrees Celsius, with higher survival rates, less disease and larger yields than in warmer areas. Also, the overall strongest economic development potential has been in the colder part of the implied "comfort zone" of 11–15°C. Both physical and mental work become heavier in the heat - and the need for rest larger (Burke et al., 2015). Heat waves not only affect physical health and work performance, but also mood and mental health, cognitive capacity, social (and antisocial) behavior and the danger of conflicts and war. Also, large migrations and large variations in settlements have historically been linked to temperature changes and climate (Büntgen et al., 2016; Zhang et al., 2011). Unless there is a marked change of mentality and migration policies, in the future climate changes will be linked also to conflict and violence.

Emitters and emissions – nations and people

There are four types of gases that together make up 97 percent of man-made emissions of GHGs. Most important is carbon dioxide (CO_2) from fossil fuels and industrial processes (65 percent of the greenhouse effect) and from deforestation and agriculture (11 percent). Next on the list is methane gas, mainly from meat production (16 percent), and increasingly from thawing of tundra areas and cold ocean-bottoms. Then follows nitrogen oxides from car engines and over-fertilization in agriculture (6 per cent), and finally chlorofluorocarbon compounds (3 percent). The methane gas is approximately 30 times more powerful than CO_2, but it weathers after only 6–10 years in the air, which means that the fastest way to bring down the greenhouse gas (GHG) level is to drastically reduce methane emissions, mainly by reducing the amount of meat in standard diets.

If we convert all types of GHG emissions into CO_2 equivalents, 25 per cent originates from power plants and heat production, 24 per cent from agriculture and food production, 21 per cent from other industrial production, 14 per cent from transportation, 6 per cent from construction and housing, and 10 percent from other uses of energy. Among lessons we can glean from such figures is that private automobiles (at about 9 percent) have received much attention relative to impact. Also, we may note that the energy needs of the growing data sector is often overlooked (passing the emissions from aircraft, at two percent, on its way toward 8-10 percent of the future total).

Looking at emissions by country, China alone accounts for 30 per cent of the world's CO_2 emissions, while the USA (with significantly higher per capita emissions) accounts for 15 per cent, the EU 9 per cent, India 7 per cent, Russia 5 per cent, Japan 4 per cent, and the rest of the world 30 per cent.

A political-economic concern mandates that we also look at emissions per capita, an issue to which we shall return in Chapter 12 on the political economy of inequality. According to figures from "The Global Carbon Project" (GCP) the global average annual emission of CO_2 per capita in 2020 is 4.5 tons (Friedlingstein et al., 2020). Among countries, Qatar is at the top with 37.1 per capita tons, while all 21 countries with the lowest emissions are in Africa, with Congo (0.03 tons), Somalia and the Central African Republic at the bottom. Among the most populous countries are India at 1.8 tons and China at 7.4 tons. Among the richest countries, the USA is at the top with 14.2 tons – more than twice that of the EU with 5.8 tons. Within the EU, Luxembourg is at the top (13.1 tons) and Latvia at the bottom (3.6 tons). If we include other GHGs than CO_2, the situation is fairly similar to that for CO_2, with a global total average of 6.5 tons of CO2e (CO_2 equivalents) in 2018.

We may also look at per capita figures within countries. The contribution of the wealthiest people to planet destruction is massive. The relationship is almost linear and positive: The richer the person (or the country), the greater the damage to the environment, to nature, climate, and sustainability. According to a report from Nature Climate Change (Schöngart et al., 2025) two thirds of global warming is attributable to the wealthiest 10 percent of the global population – 6.5 times the average, and one fifth of warming is attributable to the wealthiest one percent – 20 times the average. The same pattern holds for contribution to extreme weather events, 1-in-100 years extreme heat events, and Amazon droughts. On top of this, we have the influences from the very super-richest 0.1 percent (Paddison, 2021). Overpopulation is not a major risk factor – the overpopulation of wealthy people is! The group of the wealthiest 10 percent of the global population includes about 50 percent of the gainfully employed populations of the wealthy parts of the world, suggesting that there is something fundamentally wrong with modern lifestyles, with consumption and investment patterns, and with economic distribution – to which we shall return several times in this book.

The economic differences between countries explain much of the emissions differences. The richest countries also score highest on "Earth Overshoot Day" measurements and on emission tonnage. But the income differences within the countries are even more decisive. In addition to other arguments for equalizing income and wealth, climate change is good case. The wealth and consumption patterns of the very rich contribute disproportionally to climate change and the wrecking of the environment.

Reaching tipping points and beyond

Until now, we have been able to say that "it is not too late to do something - climate change is man-made, and it is within human reach and competence to reverse it". But climate change is not endlessly reversible. At the stage of tipping points, some change may be reversible, some may be irreversible but stable, and some may be irreversible and self-reinforcing.

The idea of "tipping points" comes from physics, where there are plenty of examples, such as water-change at the point of boiling, or a weighing scale tipping at a certain weight. The principle has also been observed in economics (stock market crashes, "bank runs", cracks in housing markets), social sciences (revolution, social movements, neighborhood deterioration) and medicine (epidemics). Malcolm Gladwell made the principle known to a wide audience in his book "The Tipping Point: How Little Things Can Make a Big Difference" (2006) with examples of sudden changes in markets, societies and ways of living.

Climate at the precipice

Related to climate, it was NASA climate scientist James Hansen, who in a presentation in 2005 initiated a new trend in climate research, where the tipping point problem more and more came to take a center stage (Hansen, 2005). The questions of whether and when one or more tipping points in climate change will occur have become a core issue in recent climate research. Also, in relation to nature conservation, the question of tipping points concerns both whether and when ecosystems break down and species are threatened with extinction.

We can imagine tipping points in three varieties, or three degrees of seriousness. Reversible tipping points are the most innocent. Boiling water can be cooled down again. It is reversible. When, on the other hand, damage in some area is irreparable, for example when an animal species has died out, or a piece of land has been blasted away to make room for a modern highway, it is generally irreversible.

Similarly for climate. Local climate change caused by deforestation may be largely reversed by the planting of new trees, making it locally reversible. New trees bring shadow and humidity. If on the other hand a large area of a tropical rainforest has been stripped, burnt or otherwise dried out and changed into a savannah, it would be very difficult to reverse the process – in an extreme case such as the Amazon, it might for practical reasons be considered irreversible. Similarly, for local ecosystems. New trees may have the same capacity of carbon storage as the ones cut down, but they do not automatically reproduce the ecosystems of the previous virgin forest, indicative of irreversible natural damage.

The most dangerous variant of a climate tipping point is when it is both irreversible and self-reinforcing – in systems terms called "a tipping point with positive feedback". This is already underway in artic regions. Emissions of greenhouse gases and the greenhouse effect leads to warmer water and air, particularly in artic regions, causing the artic icecap to melt. Hi-reflection white snow and ice give way to black, open sea, which in turn leads to less reflection of sun radiation (the Albedo effect), which again intensifies the warming. The process becomes self-boosting. The same happens when permafrost areas thaw, which leads to increased emissions of CO_2 and methane gas from the ground, which in turn reinforces the greenhouse effect and give rise to temperature increases, which in turn increases thawing and emissions in a vicious, self-reinforcing circle. A further example is that the heating of the ocean floor will cause a release of methane, stored as icy clathrates (like gaseous snowballs) at the bottom of the oceans, and causing further carbon release in the process. Nobody knows what quantities of clathrates are hidden on the ocean floor – estimates range from emission potentials of twice what has been emitted into the atmosphere so far, to several times that level. As for methane stockpiles in permafrost areas, the quantity may be 250 times the quantity presently in the atmosphere. Self-reinforcing tipping points are very difficult to do anything about, once they are started. That is why climate mitigation the next few years is imperative.

How many climate tipping points do we have? Firstly, we have countless local tipping points, some of them relatively easy to isolate and reverse – others more likely to add to global tipping points. Among the biggest global tipping points – which hold the capacity to bring about dramatic changes in global climate conditions, are the aforementioned Albedo effect (radiation effect) of the black ocean surfaces of de-iced polar regions, the thawing of permafrost areas in Siberia and Northern Canada, and of methane ("clathrates") at the ocean floor. Other instances of global impact include change in monsoon winds in Africa and Asia, change in the El Ninjo ocean current in the Pacific Ocean, degradation of tropical rainforests in the Amazon, changes in the Gulf Stream (the Atlantic Meridional Overturning Circulation - AMOC), destruction of coral reefs, melting of glaciers - especially in Greenland, the Himalaya, and the Antarctic. Not all of these are equally self-reinforcing, but they all have the capacity to impact the climate trajectory of the planet. Also, for every tipping point that is being passed, the costs of mitigation, adaptation, and repair will increase, and the greater the geopolitical and economic conflict potential.

Temperature rise and a new ice-age

The threats of changes in the Gulf Stream (AMOC), and related ocean currents have received attention for a long time. Recently the warnings have become more vocal (Boers, 2021; Caesar et al., 2021; Michel et al., 2022). In particular, the article "Warning of a forthcoming collapse of the Atlantic meridional overturning circulation" (Ditlevsen and Ditlevsen, 2023), predicting imminent collapse sometimes in the 2025–2029 period, stirred great concern. Already at its weakest point in the last thousand years (Caesar et al., 2021), the normal circulation of the AMOC brings warm water northwards in the Atlantic ocean. As the water streams north, it heats up the surrounding land mass (and makes living in Brittain and Scandinavia quite comfortable), but the current itself cools down, and the cooling together with salt content causes the current to sink towards the bottom as it approaches the North Pole. From there it turns southward for a new circulation with a cooling effect of tropical areas.

There are long-term variations of speed and spread in the large ocean currents, that may influence present developments. But the chain of events that may cause a collapse soon, starts with the greenhouse effect and the gradual warming of land and sea – more so in the North than for the global average. Heating causes the Grenland icecap to melt, at a rate of 250 billion metric tons a year – at an accelerating pace. The outflow of fresh water into the ocean causes desalination, which together with warming slows down the normal circulation, at some point potentially bringing the current to a halt (the tipping point, and the start of a new trajectory). Although we cannot say exactly what the consequence of an AMOC collapse will be, it may quite likely be a dramatic cooling, at a scale of 5 – 10°C over Northern Europe and North America's east coast, and accellerated heating, with more of both heat waves, droughts and floods in the South Western USA, Southern Europe and global tropical areas. Also, a collapse would add to sea level rise, and reduce the carbon storage capacity of the oceans, impacting marine ecosystems.

The ICCP (United Nations Intergovernmental Panel on Climate Change) has estimated the chance of AMOC collapse before the end of the century at a "medium" probability. Some researchers are reluctant to predict collapse at such a time schedule, while others warn that standard climate models may have underestimated the danger of near future collapse. Much depend on reaching the carbon zero target by 2050 – even if some tipping points may occur before.

Jet-streams topping ocean streams

While ocean streams are generally slow-moving and slowly changing, jet streams are generally less predictive, and researchers are trying to figure out whether global temperature rise is causing the jet streams to become more erratic. There are four jet streams, which are air currents flowing west to east around the globe, two in the northern hemisphere and two in the southern hemisphere. Nearest to the poles are the polar jets, strongly impacted by the tension between polar icy air and temperate zone warmer air. Near to the equator are the subtropical jets, navigating the tension between temperate air and the hot air of the tropics. The jet streams are behind the "normal" weather patterns of regions, but also a cause of extreme weather events. The greenhouse effect brings a temperature rise that again brings more humidity into the atmosphere, which in turn brings varieties of extremes of rainfall and flooding, heatwaves and cold-waves, storms and tornados, all generally of increased intensity, duration and spatial extent – reflecting changes and erratic behavior of the jet winds. The torrential downpours in the Valencia region in Spain in early November, 2024, killing 231 people and causing enormous damage was a result of erratic jet wind behavior as are most similar events around the globe.

Jet streams tend to be erratic at the outset, and the enormous variation of jet stream behaviors in altitude and latitude, also seasonally and over time, makes it difficult to trace the direct consequence of climate change. In particular, there is the increased occurrence of "atmospheric blocks", often caused by a weakening of higher atmosphere winds (Woollings, 2020) . As the jet slows, it causes weather systems to offer cold spells or hot spells, intense rain for days on end, or lasting drought – as in North-west America in June 2021, and in Siberia 2020. In Europe, summer heat wave blocking seem to be a main trend.

Runaway departure from orbit

Even at the stage of self-reinforcing tipping points, and even with dramatic rises in sea-water level, and in the level, frequency, spread, and longevity of extreme weather, humankind and strong elements of civilization may survive and support humanity's long-term potential. The main concern of the present text is with the potential damage and human and animal pain caused by climate change at a time of approaching tipping points and severe temperature rise.

But there is also an ultimate threat that climate may come totally and ultimately out of control. With an unmitigated, runaway greenhouse effect, higher temperature would cause more water to evaporate, reinforcing the greenhouse effect and speeding up the heating process through positive feedback, causing amplification towards a deadly ending. The chance is rated at 1 to 1,000 by Toby Ord – which is not high, but still worth a warning. There is also a chance of beyond-tipping points damage approaching a runaway situation, without necessarily causing irrevocable collapse, extinction, or permanent ruin.

A full-scale runaway greenhouse effect is not seen as possible within the present century, but our actions the coming decades may put us on a trajectory that opens for such a possibility – even a deadly "lock in" situation that irrevocably brings permanent ruin of humanity, civilization, and bio-diversity. The chance of such an occurrence is small, but the consequence would be infinitely negative.

All of the greenhouse gases work as a shield or a blanket, trapping heat in the Earth system, preventing heat from escaping into space. But water vapor is also a greenhouse gas, and warm air can hold more vapor than cool air, which again means that a planetary, runaway "super-tipping-point" may be reached, where global warming causes a further vapor build-up, and the build-up causes more warming, causing more build-up .... in a self-reinforcing, "positive feedback" manner.

The possibility of a runaway greenhouse has been studied since the early 1990s (Nakajima et al., 1992), and the main question of possibility has still not been settled. There is a limit to how much the Earth can heat up, however. The not very sobering fact is that warming will cease at around 1400 Kelvin, or 1126,85°C – too hot for jogging! At that temperature level radiation of heat away from the Earth will be in the near-infrared zone, and water vapor will cease to function as a greenhouse gas. Also, at that temperature level, all planetary life will be extinct, and the oceans will be fully evaporated. Such a burn-out most likely is what happened to Venus, the hottest planet in the solar system, and the one closest to the sun, billions of years ago. It may also happen to Earth, but not for man-made causes, and not for another two billion years, due to increased solar luminosity (Goldblatt and Watson, 2012).

Near all evidence indicate that a man-made runaway greenhouse effect is strongly unlikely. More likely, we would run out of oil and coal before being able to heat up the planet for a burn-out take-off. And after all, we have long lists of candidate technologies for cooling. Even so, we have no experiential data for studying the dynamics of radiation and cloud physics at relevant temperature levels. Neither are there any full-blown models of all possible tipping-points interacting to produce un-anticipated outcomes. And, as mentioned, even if self-reinforced, vapor-induced warming does not cause a runaway effect, it can turn out to be extremely bad for humans and most life forms on Earth. More likely than a runaway situation is a gradual stabilization of temperature at some point in the process, called a moist greenhouse effect (Kasting, 1988; Popp et al., 2016). Even such an outcome may be a great catastrophe, with deadly temperature rises and the drying out of water resources.

Compared to previous periods of extreme warming, the present temperature rise goes much faster, and nobody can know for sure what mitigative steps may be taken, nor what the consequences will be. Increased heating the last 50 years may seem like a slow rate over a long time, but in a more relevant historical perspective it is indicative of near instant change – it goes incredible fast! Mitigation must come now!

Extreme weather consequences

The most obvious and direct consequence of global heating and climate change are the direct health consequences of heat itself and the increase of damage from extreme weather events. Dying from heat stroke is one of the most painful and traumatic ways to die, with organs simply collapsing one by one, as a direct result of exposure to heat. But heat waves also cause a gradual deterioration of health (Wong, 2023), even accelerating aging (Ledford, 2024).

People who have to work outdoors during heatwaves are the main victims. City dwellers without air-condition are equally vulnerable. Also, animals with fur will find it harder to sweat, and they are more prone to heat stroke than humans. They are also more dependent on being able to run and move to get food or to avoid becoming food for human and animal carnivores. They also need long north-south and east-west corridors for migration away from heat traps,

Damage to nature people and property

Heat and extreme weather will have unforeseen consequences for nature, for vulnerable forests and habitats, for vulnerable animals, insects, and plant species, and of course for people exposed to all kinds of extreme weather – hurricanes, floods, draughts, heat-waves, and wildfires. Extreme rainfall and floods lead to destruction of property and contamination of drinking water, and extreme weather will lead to more destruction of roads, electricity supply and other infrastructure, buildings, and homes – adding to the costs of living.

With higher temperature and increased evaporation, weather events are becoming more extreme, heat waves are more frequent and hotter, and they are lasting longer. Dry areas are becoming drier, and drought periods are more frequent and longer lasting. Hurricanes are more powerful and more frequent – the same are flooding disasters and forest fires. Also, the time it takes for a strong wind to reach hurricane or tropical cyclones level is significantly reduced, and the frequency of such "intensifications" has been drastically increased over the past 20 years as compared to 40 years ago (Garner, 2023). Food production may increase in some areas, but much larger areas are becoming less suitable for agriculture. New diseases, some of them infectious and pandemic, will arise. Warmer and more acidic seas will cause fish deaths and the end of many coral reefs. Extreme weather will increase the chance of "compounding disasters", which means that destruction occurs in many ways at the same time or in rapid succession, such as several hurricanes coming one after the other and often together with extreme rain and floods. One million biological species are currently threatened with extinction. With a global temperature rise of three degrees, a quarter of the Earth's species could become extinct or be on the way to extinction.

Area risk and insurance premiums

Rising sea water is sometimes presented as a main threat caused by climate change. It is certainly true that coastlines will change as a result of rising sea levels, but it will happen over a very long time, initially making small island communities and coastal residents the most vulnerable, also making nearby building entrepreneurs offering high elevation building sites the main beneficiaries. In a longer time-span, large metropolitan areas and whole countries may be exposed to rising waters. Coastline living and business has been popular around the world.

For more immediate threats, torrential rain and flooding, hurricanes and tornados, and extreme heat are the main climate threats to both humans, animals and ecosystems. And topping the combination of high temperatures, droughts and wind are wildfires of exceptional intensity. The common denominator of all this is increased vulnerability, pain, anxiety and loss. It is about loss of property, but also loss of lives and livelihood, welfare and security. And beyond losses to humans, it includes the loss of animal lives of all kinds – to which we shall return in the next chapter.

As for loss of property, the insurance business is in the forefront to appreciate the losses from changing climate impact. The wildfires in the greater Los Angeles in January 2025 were estimated to cost insurers some $30 billion. Some insurance companies had pulled out of the state before the fire – based on previous losses. And premiums have soared in the last few years for those who chose to stay. From an average of 653 property destructions per year in 2004–2013, the number has risen to 5,669 in the 2014–2023 period. But it is not just about California.

Around the world the insurance business is in a strategic squeeze, either for rising the premiums for those most directly exposed – not only to wildfires, but also to flooding or hurricanes, making it very difficult for owners to obtain needed insurance at all, or on the other hand seek some kind of public regulation and compensation for handling insurance for those most directly at risk. An additional problem is that while the sector traditionally has based risk estimates on historical records, emerging climate changes tend to make historical data less reliable for estimation of future losses. The general trend is for more extreme weather and more damage, but it is quite difficult to adapt the general picture to into local projections for insurance premium estimation.

Recent research from the US, covering all counties, show that those in the top fifth of climate risk had property insurance premium increase of 22 percent, as compared to an overall average of 13 percent (Keys and Mulder, 2024). As premiums gets more expensive, even to the extent that insurance for many becomes unaffordable, property values will depreciate, at some point potentially threatening a general financial crisis. The case of Florida highlights the issue. Among the ten US cities where house prices dropped the most that year, four were located in Florida. Keys and Mulder in their report include a map of climate risk across the US, providing visual evidence of insurance premium rises, demonstrating that the south-east (hurricanes and flooding), the mid-west (tornados) and south-west (wildfires) are worst hit. In an interview with Oliver Milman of the Guardian (2024) Ben Keys expressed the message quite clearly: "You can deny climate change for whatever motivations you have but when insurance is going up because you live in a risky area, that's hard to deny".

Keys also compared premium rises to the canary in the climate coalmine – now hitting households' pocketbooks. It is a strong signal, but still narrowed down to property damage, and to risks of wildfires, storms and flooding. To this we may add what is most often not insured, including human and animal health and life, the intrinsic value of ecosystems and nature, and the health consequences of extreme heat and heatwaves.

In an interview with Damian Carrington of the Guardian (2025), Günther Thallinger, on the board of Allianz SE, one of the world's biggest insurance companies, and former CEO of Allianz Investment Management, expressed concern that "The climate crisis is on track to destroy capitalism". Arguing that "with the vast cost of extreme weather impacts" the insurers will not be abler to cover damages caused by extreme weather. Without insurance, the financial sector will not be able to finance mortgages and property investments. Much of the finance sector may collapse.

Beyond insurance and housing, the escalating climate crisis is already driving inflation. According to studies by The Potsdam Institute for Climate Impact Research and the European Central Bank (Kotz et al., 2024) climate change is increasingly driving inflation, especially food prices. Also, climate-related inflation may overall add an extra percent to annual inflation towards 2035 (Blyth and Fraccaroli, 2025). Droughts and crop failures are bound to increase with more extreme temperatures, but extreme weather of all kinds adds to the general pressure on prices for repairs, adaptation and mitigation. On a spaceship earth planet of limited natural resources, persistent increase of exploitation is bound to put pressure on prices. Of course "Drill baby, drill" policies do not alleviate climate problems.

Higher temperatures will also make tipping points kick in, intensifying the warming and adding to incidents of extreme weather, in a self-reinforcing process, with unforeseeable consequences. And with higher temperatures and more extreme weather, we may have reason to expect more erratic human behaviors and an increase of crime, including behaviors that may undermine democracy and social order. On the other hand, with more manifest climate damage, there will be less space for denial and more reason for both mitigative and adaptive action – to be examined in Chapter 8. This does not all have to happen, however. It all depends on engagement for mitigative action, including energy issues and mobilization for cooling and stabilization, to which we shall return in Chapters 4 and 5.

CHAPTER 2

Nature Depletion and the 6th Extinction

The destruction of nature by humans has been going on for centuries, with the hunting down of species, burning of fields, deforestation, exploitation and toxification. But destruction has been intensified with industrialization, and in particular, with the high-consumption economies of the post-World War II era. Around 73 percent of global biomass has been lost during the last 60 years, habitats and wilderness areas have been continuously attacked, more than a million species of all kinds are threatened with extinction, the oceans are getting warmer and more acidic, coral reefs are decaying, and nearly a quarter of freshwater animals are threatened with extinction. Of great concern is also the spread of invasive species, mainly in the economically advanced regions, causing frequent damage to eco-systems.

A fundamental environmental concern is that efforts for the planet are going too slowly. Mitigation and damage-limiting measures come too late and too little. We generate and amplify the problems faster than we generate solutions to solve them. This applies both to climate and to the general degrading of nature and ecosystems.

The nature status report

In 2019, we received the first major report from The Intergovernmental Science-Policy Platform on Biodiversity and Ecosystem Services, IPBES. It concluded that loss of biomass and biological diversity now threatens huge ecosystems, and the consequences could be as severe as those caused by climate change.

After waiting for three years for a decisive UN meeting in Kunming, China, to set a new global standard for the protection of wilderness, in December 2022 the meeting was moved to Montreal in Canada and an agreement was reached – The Kunming-Montreal Global Biodiversity Framework. 195 countries agreed to a lasting protection of a global third of land territories and sea areas, restoration of a third of damaged ecosystems and a doubling of resources for protection of nature. It was about time - wildlife and nature have waited a long time for protection, and they are still waiting for concrete steps to be taken.

The long run

Physicists have estimated the age of the universe at 13.8 billion years, the age of the Earth at 4.5 billion years, and the age of single-celled life in the ocean at around 3.7 billion years - perhaps more. About 3.5 billion years ago, cyanobacteria arose, capable of photosynthesis and oxygen production. After another billion years, oxygen production had become large enough to threaten what was left of anaerobic life forms - all underwater. For such life forms, which hated oxygen, it

DOI: 10.1201/9781003518815-3

was a disaster, but almost three billion years later, oxygen was essential to near all other life forms and land-based plants, animals, and in the end humans.

After the first signs of life in the sea, more than 3 billion years would pass before fish life in the sea and animal life on land emerged. Less than 400 million years ago, some of the species in the sea began to adapt to a life on dry land, setting off the final stage of global evolution, with the emergence of all kinds of land-based life forms, many of which we see today, and many who later died out.

We are now in the midst of The Sixth Extinction of species and life forms. The first major natural destruction, the Silurian extinction, occurred nearly 450 million years ago, and it only affected species in the sea - life on land did not exist. Close to 90 percent of all species in the ocean may have died out, presumably due to a severe cooling event, followed by warming, with sea level rise and lack of oxygen, perhaps due to volcanic eruptions. Later, we have had four dying-out events, with varying levels of extinctions, before the sixth that we are now entering. Only the sixth extinction is the result of human activity. The fifth occurred 65 million years ago, caused by a giant meteorite or asteroid that landed in Chicxulub, Mexico. It is the only one with an extraterrestrial cause. 70 percent of the species on Earth disappeared, including the dinosaurs.

A little more than two million years ago, the first human-like creatures appeared somewhere in Africa - *homo erectus*. Unlike other creatures, humanins walked upright, and they soon learned to use fire. Over time, several different species of hominin creatures appeared, among them the Neanderthals. Homo sapiens appeared late in development. The common great-grandmother of all presently living humans gave rise to the sapiens genus somewhere in East Africa a little over 200,000 years ago.

The birth of an ecological serial killer

Approximately 70,000 years ago, the sapiens somehow went through a "cognitive revolution" that enabled them to form complex societies, develop technology and communicate about abstract things. On their way from East Africa to all corners of the Earth, enriched cognitive capacity also allowed them to dominate and exterminate other species, including some 10 different branches of hominin creatures. Both the early hunter-gatherer societies and the subsequent agricultural societies show a picture of homo sapiens as entrepreneur, discoverer, colonizer, and exterminator.

With the role of colonizer came also a role that the historian Yuval Noah Harari in his book about "Sapiens" calls an "ecological serial killer" (2014). Harari also refers to the period before the agricultural revolution some 12,000 years ago as "an original welfare society". The hunters and gatherers lived quite comfortable lives - better than the later farmers, workers, clerks and financial experts. They were taller and stronger, and more knowledgeable than their farmer descendants. They worked less than we do today, lived interesting lives and they did little boring housework. As Harari writes, in good times they mostly did what modern people at best can do on holiday – hunting, paddling, fishing, going for walks and chatting with friends. It is not clear that ever more economic growth, with new technologies, new financial instruments and modern lifestyles always will contribute to the betterment of life!

The latter statement needs some modification, however. The hunter/forager version of holiday life in prehistoric time does not account for the enormous varieties of life and work at the time. Perhaps best explained by David Graeber and David Wengrow in "The Dawn of everything" (2021), there is no linear stage-line of pre-historic progress, from hunter-gatherer socies of dispersed bands, to foragers, to farming, to complex societies and modern civilization, the final stage led by Europeans. The massive Göbekli Tepo constructions of southeast Turkey, uncovered in the 1990s, was built by a pre-farming society, and impressive public works from around much of the world, similarly contradict the stagist thinking of traditional historians and anthropologists. Looking back, pre-history points to an enormous variety of lifeforms and political organizations. As we shall see in later chapters, the discovery of variety in the past is a wonderful position for seeking variety of possibilities in the future.

Our early forefathers had access to possibilities for alternative ecological adaptation. What they lacked, however, was any kind of overview and feedback that could tell them about the ultimate consequences of their hunting practices. For homo sapiens, with newly acquired cognitive capacity, there was little to prevent them from helping themselves to food and territory. With expertise in boat building, geographical curiosity, and well-developed organizational and navigation skills from around 50,000 BC, they were ready for discovery and "island hopping", and they made their way from Africa and the Middle East to distant places like Indonesia and Australia. They took steps to the top of the food pyramid, in the process becoming the most benevolent but also the most dangerous species in the history of the planet.

In Africa, they had adapted to existing ecosystems and existing wildlife. In Australia and elsewhere they engaged in colonization, they practiced both adaptation and change. In Australia, some 50,000 years ago they encountered a two-meter kangaroo weighing 200 kg, a marsupial lion the size of today's tigers, koala bears much larger than today's cozy, stuffed-like animals, birds that could not fly but were twice the size of modern ostriches, dragon-like eagles and five-meter-long snakes, and finally a bear-like diprotodont weighing two and a half tons. Apart from the birds and reptiles, all these Australian animals were marsupials, with their young in a pouch on their stomachs – a phenomenon unknown elsewhere. Within a few thousand years, 23 of 24 species weighing 50 kg or more were extinct. We cannot prove that homo sapiens defeated them all. Some have suggested that climate change might be to blame, but there is little evidence of major climate change in Australia during the period. Moreover, in Australia a pattern of conquest and hunting was set which was repeated elsewhere, indeed across most of the globe. The serial killer was well underway - to Asia and the Americas 16.000 years ago.

The result was worst for the large land mammals. Before the original spread of sapiens beyond Africa, some 70,000 years ago (the start of "colonization"), there were approximately 200 genera of large land mammals over 50 kg in the world. At the start of the agricultural revolution 10,000–12,000 years ago, only about 100 remained. It happened before people had learned to grow grain, before they invented wheel and iron tools, and before they could express themselves in writing and start making guns.

If to the extinction of the large animals we add the extinction of at least ten other hominin species and the destruction of fauna on a large scale, as in Australia and North and South America, Harari says that the first wave of the sapiens' colonization was one of the biggest and fastest ecological disasters to hit the animal kingdom, and homo sapiens emerged as most dangerous species in nature, with a world record for driving plant and animal species to extinction. For the early sapiens, however, it was not possible to see the consequences of their own actions. There were no satellite images, no intergovernmental panels, and no modern mathematical modelling techniques. The sapiens were looking for food and for protection against dangerous animals. They had little opportunity to understand that the last hunting trip or the last fight against an attacking animal also could mean the extinction of a race or a species.

Modern man, on the other hand, knows what is going on, and we still take part in development and growth processes that break down ecosystems, remove habitats and livelihoods for a million species, spread environmental toxins and plastic litter over large parts of the Earth, and produce greenhouse gases that lead to climate change that will threaten the livelihoods of both animals and humans.

Shocking numbers

It was a shock to many when headlines about the IPBES report in May 2019 told us that the ongoing destruction of nature in the world was at least as big a challenge for humanity as climate change. A headline in New Scientist stated bluntly that "Destruction of nature is as big a threat to humanity as climate change" (Le Page, 2019).

Around the time of the IPBES report, the WWF published the "Living Planet Report 2020" (from the World Wildlife Fund – WWF, in collaboration with the Zoological Society of London - ZSL). The report was based on the Living Planet Index (LPI), which is a database that shows the development within more than 20,000 habitat areas. It showed that in 2020, one million species were threatened with extinction, also that three quarters of all land areas on Earth and two thirds of the oceans are undergoing significant change due to human activity. In 2023 a new report was released, doubling the estimate to two million species threatened with extinction (Hochkirch, 2023). There are 8–10 million species in the world, more than seven million described as animal species, and more than 90 percent of them invertebrates. But only a little more than two million have been properly described – half of them insects.

Most people are aware of the key role of pollinating insects for food production. Fewer are aware of their ingenuity overall. Anne Sverdrup-Thygesen has a fun summary in the introduction to her book about "Extraordinary Insects: The Fabulous, Indispensable Creatures Who Run Our World" (2020). Long before humans were present on Earth, insects were busy with farming and husbandry. Termites cultivated mushrooms for food. Ants kept aphids as milking cows. Wasps were the first to produce paper from cellulose. Werefly larvae made nets to capture other animals for food – millions of years before humans made fishing-nets. Insects also solved advanced mathematical problems of aero-dynamics and navigation millions of years before humans. The longest insect is found in China, at 62.4 cm, and the smallest, at 0.16 mm, is too small to be seen, still able to sit down and enjoy life at the short end of a human hair.

In October 2024, the World Wildlife Fund's "Living Planet Report" (WWF, 2024) reported a 73 percent decline in the number of mammals, birds, amphibians, reptiles and fish over a 50 years period (1970–2020). The report also warned about approaching tipping points due to habitat destruction and climate change, with dire prospects for humanity. The index tracks almost 35,000 vertebrate populations of 5,495 species. Worst is the decline in freshwater populations, at 85 percent. Terrestrial decline is measured to 69 percent and marine decline to 56 percent. Geographically, the decline is greatest in Latin America and the Caribbean, at 95 percent, followed by Africa at 76 percent. The lower figure for Europe is mainly due to the sad fact that the extinction started earlier – meaning that much of the job of extinction was already done.

The main factors behind the decline are loss of habitats due to expansion of agriculture, cities, and infrastructure, commercialization of forests (at the expense of primeval forests and complex ecosystems) and generally less space for wilderness and habitats. The spread of invasive species, disease, and climate change add to the problems. The report tells us about dramatic losses already incurred, stock reductions and extinction threats, and tipping points ahead – classified by WWF-US President Carter Roberts as a "slow motion catastrophe."

Almost half of the habitable land areas, and three quarters of administered fresh water resources are used for agriculture and meat production. Excessive hunting and fishing, excessive production and consumption, pollution and the spread of plastic and microplastics, climate change and species migration intensify the destruction. It is also noticeable that areas belonging to indigenous people are regularly better cared for than areas managed by modern industrial states. At the same time, indigenous people's areas are under constant attack from growth-hungry political regimes, industrial agriculture, and multinational companies.

Of the total Earh surface, the oceans account for 71 percent while land areas account for 29 percent. Of these 29 percent, 71 percent is habitable land, while 10 percent is glaciers and snow-covered areas, and 19 percent is bare land and mountains. Of the 71 percent that is habitable, 50 percent is used for agriculture, 37 percent is covered by forest, and 11 percent is scrubland (Richie and Roser, 2024). Of the 50 percent that is used for agriculture, 77 percent is used for meat production and 23 percent for vegetarian food (fruit, vegetables, nuts, beans, etc.). But the shocking thing is that these 23 percent vegetarian "whole foods" account for 82 percent of the total calorie supply and 63 percent of the protein supply to people.

For ease of recall, we can say that approximately 80 percent of the areas give us only 20 percent of the food value, or that 20 per cent of the areas contribute an amazing 80 percent of the food value. A net consequence is that humans (by weight) today represent 36 percent of mammals on Earth, whereas livestock (for meat and dairy-production) represent 60 prosent – and wild animals only account for four percent (Bar-On et al., 2018). The biomass of wild mammals has fallen by 83 percent since the agricultural revolution 10–12,000 years ago. Five out of six wild mammals have disappeared.

For dealing with the loss of wilderness and natural ecosystems, three main strategies will be commented on in the following. First there is the strategy of stopping the destruction – which is the ambition of the Montreal agreement signed in 2022. Secondly, around the world, a great number of rewilding projects have been launched to restore land areas and wild animal populations to a natural (uncultivated and undestroyed) state, including the reintroduction of former species and so-called keystone breeds. Thirdly, there is also the alternative of what I call wilding – simply "leaving alone" areas that have been destroyed and let nature do the job of re-cultivation. To wilding we may also add the confluence of nature and people. Several reports on nature and wilderness have commented on the fact that the natural environment generally (not always) seem to have gained from the harmonious coexistence of indigenous cultures and nature, where indigenous people also have used small-scale mitigation to help making nature more resilient. Similar results can be found in many traditional village communities.

Stopping the destruction

1992, the UN adopted the "Convention of Biodiversity", but as late as 2020 it was clear that 80 percent of the targets had not been reached, and half of the countries did not even have a plan to reach them. In October 2010, an agreement was concluded in Japan, with updated targets for 2020 – the so-called "Aichi targets". But neither of these goals were achieved. The 2019 IPBES report was a stronger warning – it stirred considerable alarm, and in December 2022, the Montreal Agreement was signed by 193 countries.

The agreement aims to put a stop to the degradation of nature and the destruction of ecosystems. According to the agreement, by 2030, 30 per cent of the Earth's land area and sea area must be permanently protected, and 30 per cent of natural areas where the ecosystems have been weakened must be restored. Invasion of alien species must be brought under control, food waste in the world must be halved, and the use of environmental toxins, fertilizers, and pesticides must be greatly reduced. Selection of conservation areas must also be representative. Both high mountains and forests, coastal areas and inland areas, marshlands, wetlands, peat, and moss lands are to be protected. Of course, it is also important in this context that the areas must provide protection for the fauna, with both predators and grazing animals and all kinds of birdlife and complex ecosystems. The areas must also be large enough and contiguous enough to allow animals to migrate.

Unfortunately, for the 30 percent agreement, there is no international standardized method for counting and protecting wilderness areas. Even at the national level, and even in nations with strong economic and political resources, there is a shortage of the data and statistics, maps and regulatory systems necessary for the implementation of the Montreal Agreement. Much work remains to be done – and it is not proceeding very fast.

It is also part of the Montreal agreement that rich countries shall contribute to the protection of areas in poor countries through a fund that will add 200 billion dollars a year. At the same time, current subsidies for land development that run counter to the conservation goals must be scaled down – today such subsidies stand at about 1.8 trillion dollars a year, to be reduced by 500 billion dollars by 2030. The world community is investing much more in the destruction of nature than in rehabilitation and protection. An international Nature data bank – a gene bank – is also to be created, with public access to genetic information. Finally, through the agreement, the nations have committed themselves to an annual registration system, similar to what is in place for the Paris Agreement. The quality of such reports remains to be seen.

In extension of the Montreal agreement, in March 2023, a global agreement on the protection and sustainable use of international sea areas was reached under the UN Convention on the Law of the Sea – after 20 years of negotiations. Most of the Earth, 71 percent, is ocean, and areas outside of national jurisdiction account for two-thirds of the world's oceans and a large proportion of the globe's biological diversity. The oceans produce half of all the oxygen in the atmosphere, and at the same time act as the world's largest carbon store and temperature stabilizer. At sea, now only areas near the poles are little affected by fisheries, shipping, and littering. And soon an ice-free artic will attract increasing commercial activity.

Rivers and lakes also need protection. In March 2023, the Netherlands and Tajikistan organized the "UN 23 Water Conference" to promote a "Water Action Agenda" for the world's rivers and freshwater resources. Billions of people still live without access to safe, clean water and sanitation – defined as a human right. In many places, pollution and littering are the most important threats. Elsewhere, it is climate change that leads to droughts or floods or combinations of the two. The fight for water resources also creates conflicts related to development rights and power stations in many places. A river can be both a "dealmaker" and a "dealbreaker". Without radical interventions now, up to half the world's population in 2030 could be without enough clean fresh water to satisfy basic needs.

Reversing the trends

The ongoing extinction of species is referred to as The Sixth Extinction – and both the extinction and the general loss of habitats and wilderness are proceeding at a shocking pace. Few people are able to grasp just how fast.

We tend to "shift baselines" more or less continuously and subconsciously – suffering from the "shifting baselines syndrome" (SBS). The essence of the syndrome is that the thresholds for what people consider a "normal" is steadily lowered. Young people accept as normal a situation or an experience that an older person in the same area would find surprising or even shocking. This may hold for temperature and extreme weather, but also for the loss of forest birds and seabirds, flower meadows, insects and animals, trout in the lakes, summer water levels of rivers, and for the overall health status of habitats and wilderness areas. Shifting baselines means that people are less alert to environmental changes of all kinds, and the consequence is that environmental degradation is not dealt with as needed (Soga and Gaston, 2018). Perhaps the most meaningful task for museums in the future will be to remind us of old baselines, old versions of normalcy (Janes, 2022). Shifting baselines mean that mitigation tends to come too late and too little. Most people simply don't notice the severity of changes.

Saving the old forests

According to (the now deceased) eminent biologist and Harvard professor, Edward O. Wilson (2016), 85 percent of the world's biological diversity could be preserved through a "half earth" strategy. Today, only some 15 percent of the areas are protected.

Wilson was the one who came up with the "theory of island biogeography" and ideas for "biodiversity parks". Delimited remnants of wilderness, such as national parks and development-free zones, tend to incur a gradual loss of diversity – the areas being too small and too tightly surrounded by civilization. With permanent protection and the return to the wilderness state of one third of the Earth's habitable areas, we may be able to save somewhere between 60 and 80 percent of the biological diversity.

Halting the loss of biomass and biological diversity is within reach if the Montreal protocol is duly implemented, but it means that there will be new and more narrow boundaries for land exploitation – for all kinds of infrastructure and living, and for mining and agriculture, windmills, hunting, logging and other human activity. In particular, we have noticed that meat production, claiming near 80 percent of agricultural areas, contribute only 20 percent of food nutritional value. It has been estimated that to regain ecological balance, global meat consumption must decrease by 75 percent. Industrial meat production is a "protein factory in reverse", part of a "nutrition factory in reverse" (Lappé, 2011, 2013), where in the case of beef you may have one pound of meat on the plate for every 16 pounds of grain and soy fed. The reverse protein factory element is also present in chicken and other meat production, in salmon and other fish farming, and in the cultivated or "cultured meat" – at least by a factor of three. They all produce less food than what is consumed in the growing process, even when it takes place in a laboratory or in a "cultivated meat" production hall.

Many people think of reforestation as crucial for both nature conservation and climate change. As for nature conservation, it is first and foremost the preservation of primeval forests and ecosystems that are decisive – including wetlands, bogs, peat, and moss. Reforestation and "Plant a billion trees" campaigns sometimes suggest that reforestation is not only an important strategy, but near all it takes to make a green transition. Globally, deforestation represents close to 24 percent of the total emissions of CO_2, while afforestation in both North America and Europe has to some degree compensated for deforestation in the rest of the world. But for biodiversity and ecosystem maintenance, new forestry can never be more than a poor substitute for primeval forests and a temporary and limited solution to the climate problems.

There are limitations to what reforestation can contribute to climate stability. We certainly can›t continue to expand the forests forever, and global reforestation initiatives (like the Plant a Billion Trees campaign) run up against ownership interests, production of food and objects, housing, and nature conservation. It should also be remembered that trees store carbon in nature, but they can also do so for hundreds of years in the form of building materials – if well preserved (as among others in 900 years old Norwegian stave churches). Using wood for buildings and recirculation you can eat your cake and have it too!

Re-wilding to old baselines

Rewilding is about returning areas of lost biodiversity, destroyed ecosystems and poisoned soil to their previous natural state, often to "baselines" far earlier than the industrial revolution - essentially by re-introducing "keystone species" and letting nature do the main work itself. Rewilding refers to the ecological restoration of an area, with the removal of traces of human activity, and the re-introduction of important animal species such as bears and wolves, or large, herbivorous animals that maintain plains and grasslands, and birds of prey such as eagles or vultures. By re-introducing 1–5 key species in areas ready for rewilding, close to 20 percent of the land masses may come to qualify as wilderness (Watson et al., 2018).

Rewilding brings traditional nature conservation a long step forward from the "nature protectionism" of the 1950s–1970s, which focused on limiting pollution and saving endangered species – which is of course still highly relevant. But rewilding brings forth an expanded and more hopeful perspective (Blythe and Jepson, 2020; Spiering, 2006) restoring what is already lost. The main idea is that the reintroduction of certain strategic species, often the largest predators, will lead to an ecological flourishing in the entire area. A well-publicized example is from the Yellowstone National Park in the United States. After wolves were extirpated in 1926, the park entered a long period of ecological decline. After 14 Canadian gray wolves were re-introduced into the park in 1995, the entire area enjoyed a new period of ecological flourishing (interesting videos are available on YouTube at among others https://www.youtube.com/watch?v=8rZzHkpyPkc).

Another early example, referred to as the Oostvaardersplassen experiment, or OVP experiment, took place in the Netherlands, just three miles from Amsterdam – demonstrating that rewilding can also apply to smaller, near-urban areas. In 1991, the area, formerly planned for industrial development, was turned into a nature reserve. Geese had already moved in. Deer and roe deer were brought in from Scotland and from reserves in the Netherlands. Konik ponies, descendants from the early Tarpan horse breed, were brought in from Poland. The animals are not identical to the original breed, but still able to fulfill their original function in the ecosystem.

Other examples are jaguars in Argentina, where their reintroduction was given a role similar to that of wolves in Yellowstone. From Russia we have the Pleistocene Park in Chersky in Siberia, known for demonstrating the key role of predators in the fauna. A prerequisite for the predator's place is that there is ample access to prey, which in turn presupposes the existence of extensive plains where the prey can graze.

On the two islands Ile aux Aigrettes and Round Island, both belonging to Mauritania, rewilding required that invading species first had to be eradicated, such as rats and cats, rabbits and wild goats. But following the elimination of these, there was a flourishing of scrub which also involved new invasive species. The solution was to bring back giant tortoises from the nearby Seychelles as a replacement for the islands' original tortoise species, which had become extinct. The turtles quickly wiped out invasive plant and grass, making room for indigenous species – which had a natural protection from being eaten by the turtles. Numerous such examples from all around can be cited.

There is a growing, international rewilding movement, creating new optimism within nature conservation movements in contrast to doomsday thinking. The first "World Rewilding Day" was celebrated on March 20, 2021. From Britain, one of the world's most densely populated countries, there were reports about a whole series of projects, with water buffalo in Suffolk, pine trees and dwarf birch in the Scottish Highlands, lynx roars in Northumberland, eagles in Dundreggan for the first time in 40 years, salmon, birds and animals in the Lake District.

In general, a key strategy of rewilding is the deployment of old breeds. When not possible, animals of a related breed may be brought in to fulfill a similar role in the ecosystem. Another principle is that even though people may feel offended that the predators feed on prey, and many animals struggle to make it through the winter, people have to accept that in the wilderness animals are being hunted down by other animals, and they often suffer from cold or starvation.

From Yellowstone we got the "CCC" concept, which stands for "cores, corridors, and carnivores". Existing reserves, parks, open spaces and protected wilderness areas are often too small and too isolated for species to survive. "Core" means that the protected areas must be large enough to be sustainable. "Corridors" means that the areas also must be contiguous, with space for larger animals to roam and wander. Finally, "carnivores" means that predators also have a place in the wilderness. We also have the term "keystone species" for species that have a disproportionate impact, and "flagship species" – signaling species that in a particular way symbolizes or highlights the uniqueness of the system (for example, koalas and kangaroos in Australia).

From re-wilding to wilding

Rewilding means to restore ecosystems, returning them to a previous baseline – thought of as a natural ideal. But many areas are already so destroyed that rewilding becomes near impossible. Still, there is hope. Areas may be abandoned, polluted, poisoned, filled with junk or armory – or even exposed to nuclear radiation. Still, nature may create completely new wonders, even without human assistance. The Scottish journalist and author, Cal Flyn, spent two years seeking out such places all over the world, and reported in her book "Islands of abandonment" on remarkable blooming and interventions by nature itself (2022).

Sometimes, we have to set aside the idea of a definite historical baseline, and simply leave it to the forces of nature and what is left of ecosystems to take over and generate something, often a mix of old and new. For myself, I think of this as an alternative to man-orchestrated "rewilding", and I have come to call it "wilding". With wilding, we are not necessarily thinking of any specific ideal of nature for an area, or a baseline, or a specific past era as norm-setting for how an area should look and what organisms should populate it. Under wilding, everything is left to the variety of species that are available in whatever is left of the ecosystems.

The backdrop to Flyn's work is that large parts of land in the world are now more or less fallow. In many parts of the world, birth rates are falling, people are moving to cities and towns, and agriculture is producing more food in smaller areas. Enormous areas are simply abandoned – areas that are often characterized by previous over-exploitation, poisoning, rusty iron, concrete, and asphalt.

Cal Flyn reports first from Inchkeith, one of the Forth islands in Scotland, with large, abandoned fortifications, and with tunnels, bunkers, and gun emplacements from the Second World War. Before the war, there was only one species of seabird left on the island, the eider. But after the island was abandoned by the military, it has once again become a nesting place for dozens of bird species. With increased bird life, living conditions have also improved for other species. For those who are concerned with the beauty of nature, old fortifications and life in tunnels may not be a big deal. But for those who hunt for livable ecosystems, nature's ability to recycle and develop complex ecosystems in the middle of ruins can be all the more inspiring.

Flyn further reports from the "Five Sisters" in West Lothian, a huge oil drilling area in Scotland from the 1860s, where large mountains of slag are left, not suitable for habitation and not inviting walking tours. When the area was examined ecologically in 2004, researchers and the public were surprised that the area, in spite of ugliness and toxins, was teeming with animals, birds and biological diversity.

Similar discoveries were made in bombed out areas of London after the Second World War. When people were forced to stay away, the areas became greener. The same effect was found in Berlin, where bombed-out areas were left in a no-man's land between East and West, what later became the Natur-Park Südgelände. The area covers 18 hectares with lots of animals and vegetation, built on nature's own restoration work between abandoned train tracks and train carriages, gardens, farmland and scrub.

On the Bikini Atoll, which belongs to the Marshall Islands in the Pacific Ocean, the United States carried out a series of nuclear tests in the 1940s and 1950s, including in 1954 the Castle Bravo experiment, detonating a bomb that was 7,000 times more powerful than the one dropped on Hiroshima. The result was an 80 meter deep lagoon crater, one and a half kilometers wide. Two smaller islands were pulverized. The temperature rose to 55,000 degrees C, and all life was killed. But when an international team inspected the lagoon in 2008, they found that a lush ecosystem had meanwhile developed in the underwater crater.

Flyn also cites numerous examples of corridors and no-man's lands, where hunting and fishing has been halted as a result of war and conflict – and even where stocks have increased while the war has raged. In Estonia she found that huge areas that were cultivated during the collectivization of agriculture in 1949, after liberation from the Soviet Union in the 1990s were left fallow, and now blooming with a diversity of plants and animals, starting with grass and weeds, then shrubs, before fast-growing softwoods appeared and finally dark forest with hardwoods such as spruce.

Perhaps the most astonishing example from Flynn's long journey is from Chernobyl in Ukraine. The meltdown of the nuclear power plant occurred on 26 April 1986. The area was emptied of people. Later, a few have moved back – longing for home became stronger than the fear of radiation. The remarkable thing is that both plant and animal life have recovered in the meantime. For wilderness and wild animals, it can be more dangerous to have people in the area than a collapsed radioactive nuclear power plant.

Graded intervention

Rewilding may not always mean to go "all in", and wilding may not always mean to do nothing. There may be several steps in between. In an interesting paper, using the case of the Iberian Lynx for illustration, Falco van Hassel and Bernice Bovenkerk (2023) ask how to help wild animals cope with climate change and other environmental threats to animal life and alleviation of animal suffering. The authors suggest five main positions – do nothing, reduce other threats, provide food and shelter, conduct assisted migrating and bringing the animals into captivity for protection and breeding. At each stage, and for each position, ethical arguments can be mobilized for – or against – action.

A basic premise of differentiation is between domesticated and wild animals. Domesticated animals are directly dependent on human provision and protection. In most countries, such provision is demanded by law. Owners are responsible for food, drink, shelter and – to a varying extent – a minimum freedom of movement and outdoor exposure. Without doubt, much of modern husbandry, especially for meat production, breaks with rules, and if not with the rules, at least with sensible animal ethics, including rights not generally enacted in law.

As for wild animals, there is no such general legal-ethical framework of responsibility. There are rules about hunting, etc. But in wilderness, animals are on their own, in principle leaving both carnivores and herbivores to their own destiny and for people not to interfere. In the Anthropocene, however, mankind is dramatically intruding on wild animals' territories, with increasing responsibility for animals' well-being as an ethical consequence. Man-made climate change and the spread of environmental toxins have a dramatic impact on wild animals well-being, making it a serious ethical concern.

In the case of the Iberian lynx, droughts, floods, and forest fires are climate-caused threats, but exaggerated hunting, reduced habitats, and a dramatic reduction in the number of rabbits, the main food source for the lynx, are also major causes of vulnerability. For the next decades, increased droughts as a result of climate change, is seen as a main threat of extinction.

The "doing nothing" position does not imply a BAU attitude. Rather it means to leave wilderness and wild animals to wilding. It means to stop or reduce all hunting and fishing, habitat destruction and pollution, and protect what is there for preservation and conservation.

The more the ecological systems are reduced and destroyed by human activity, and the greater the expected payoffs from some level of rewilding intervention, the stronger are the animal-ethical arguments for intervention. On the other hand, the greater the damage to environments, the greater is the chance of species extinction, and the smaller the chance that species similar to those lost will come back to restore ecosystems – unless we actively intervene.

A first step up from "doing nothing" is to "reduce other threats". It means to limit forestry, road-building and all forms of infrastructure, and to gain control with illegal hunting. Also, of great political sensitivity around the world is the separation between cattle, especially sheep and goats, and predators (wolves, lynx, wolverines). One step higher up is the direct provision of food and shelter for the protection of prioritized species, or at a still higher level, assisted migration or colonization. This raises difficult questions. As for shelter or feeding, in the case of the lynx, would it be morally defensible to breed rabbits to feed the lynx? What about the suffering of the lynx that otherwise would go starving? What about the suffering and loss of life of the rabbits who would be killed for lynx food? Many ethical arguments can be marshaled.

Generally, it seems that the arguments in favor of rewilding also support the limited defense of threatened species. Special provisions for the protection of threatened species may be used, especially for reintroduction of "keystone species". But beyond that, it is normally up to the generative capacities of the respective ecosystems to balance the power and numbering of species and individuals. It means to accept that some animals kill other animals for food and that some animals are weaker than others, more susceptible to die from cold or heat, thirst and hunger, as prey or hunter. Wild animals cannot be protected permanently, but they may have to be rescued from disaster – whether caused by man or by forces of nature. This may also include facilitation of temporary migration and return, and in some cases even colonization, meaning permanent relocalization. For preservation, "facilitated adaptation strategies" have been suggested, employing genetical "redesign" of wild animals for greater robustness and capacity for among others "draught resilience" (Torres et al., 2023). So far, there are no vaccination posts, and no wild animals waiting in line for injections.

A battle of intelligence

In 1971, the American film "The Hellstrom Chronicle", was released, with elements of documentary, horror, and prophecy, based on a script created by David Seltzer. The main character is a fictional scientist, Dr. Nils Hellstrom, who postulates that there will be a final battle in nature between humans and insects, and that the insects will win because of superior intelligence. Intelligence is defined then, not by brain capacity, but by the ability to come up with a variety of weapons and strategies. The main Hellstrom claim was that with very short generation intervals, enormous brood sizes, and enormous diversity of weapons and strategies, multiplied by the ability to develop resistance and by the frequency of mutations, the insects will come up with new forms of attack faster than humans will be able to develop pesticides to stop them. Horror scenes in the film focus on how swarms of butterflies, grasshoppers, wasps, termites, ants, flies, and insects that none of us have seen before, attack humans. Good movie!

Thousands of insect species are today threatened with extinction, and insects may not win a final war. But from the point of view of planetary survival, the variety of life forms in nature, and the capacity of species to mutate, develop resistance, adapt and multiply if given a chance, may be our best ally for survival. It happens everywhere in nature. It may take time, and it may end up vastly different from historical baselines, but sooner or later species and varieties appear that have the ability to survive and reproduce – while not necessarily being edible by humans. With dioxin and PCBs (carbon, hydrogen and chlorine) in soil and mud, there can be no "re-wilding" of such places in the foreseeable future. The restauration of a previous ideal stage will sometimes be impossible. Instead, there will be completely new ecological formations, some simpler, uglier, more dangerous than before – others friendlier or more complex and robust. An abundance of green growth is in any case the core of nature's own strategy – our most important ally in the fight for an Earth worth preservation. Sometimes, the only requirement is for people to stay away.

To all of this, we should also add and appreciate the roles of indigenous people around the world, in upholding harmonious relations with nature, developing non-disruptive life-styles and motivating practices that reflect both physiological needs and spiritual traditions (Chanza and Musakwa, 2021). Also, we should appreciate the great role potential of rural dwellers when living in harmony with nature, but also with a gift for environmental restoration and the control of invading species.

Already, more than half of the world population now live in urban environments, and by 2050 urban living will include 70 percent of all people. This means that vast areas of land will lie fallow, ready for wilding. The American botanist, taxonomist, and ecologist Frederic Clements introduced a theory of plant succession in his book *Plant Succession: An Analysis of the Development of Vegetation* from 1916. He believed essentially that human impact was mainly a disturbing element in nature and that left to itself, nature would always over time tend to recover and develop complex ecosystems and gravitate toward a viable balance. His theories have later been challenged, and while we really should appreciate nature as our main allied in the battle for nature and wilderness preservation, we also need to be realistic about the huge potential of people – of scientists and village dwellers in caring for species and protecting the weaker links against invasions. Monocultures are not only a threat posed by industrial agriculture, but also by invasive species that are often being controlled by village dwellers and indigenous cultures.

Nature compassion

Saving animals, biomass and ecosystems is part of the green transition, and we need to remind ourselves of the enormous complexity and challenges represented by the transition. It is about more than GHG-limitations and the electrification of automobiles. The green transition is about climate, the natural environment and toxic pollution. It calls for a change of economic conduct, from extraction and exploitation to preservation and circulation, and for restoration and repairs where damage to the natural environment has been done, with implications for all sectors of the economy and most sectors of political governance.

In order for the message of the green transition to catch on we need all the best data we can have, with statistics, mathematical models, predictions and projections, and we need scientific categories and strategic vocabulary. But somehow, we also need to involve the emotional side of communication, connect to people's experiences and potential for compassion. "A picture is worth a thousand words", and sometimes a singular case may be more telling than astronomic numbers. From the height of the South-African apartheid, there was a story about how a detailed description of the torture of a dog stirred more concern among the white population than solid documentation of the torture of masses of black people.

During the "Black Summer" of Australia, lasting from September 2019 to March 2020, a series of more than 15,000 bushfires affected all states of Australia, making for a wildfire unprecedented in size and impacts, directly killing 33 people and wounding hundreds, destroying 5900 buildings, and burning an estimated 18.6 million hectares of land (an area larger than France). Initial counting suggested that more than one billion animals had died. Later investigation suggested that three billion animals might have died, including 143 million mammals, 180 million birds, 51 million frogs and a staggering 2.5 billion reptiles (Haque et al., 2021). How can we communicate about the endless sufferings of all these animals? How can such suffering be accounted for in neo-classical economic models and in political decision-making? Do we call it "negative utility"?

What is still at the forefront of my visual memory of the Black Summer wildfire, however, is not the masses of data from the evening news, or thousands of buildings on fire, but a brief video of a little koala bear surrounded by flames. There was terror in his eyes. His body was frozen in anxiety. There was smoke steaming from his fur, and he was surrounded by burning brush and a shower of sparks in the air. Then a young woman entered the scene, with soot in her face, her clothes tattered. Courageously, she picked up the little koala, embraced him and started showering him with water from her bottle while carrying him to safety. "The story of one animal may be worth more than a thousand words". For me the little koala represents the whole of three billion creatures, and the young woman embodied what the green transition is to me – the green transition made manifest.

The greatest threats to wild animals in the Anthropocene are human splitting and occupation of habitats, energy consumption causing temperatures to rise and extreme weather to be worse, more frequent and longer lasting. It is all in the reports, but we talk about "loss of habitats", when it is really about stealing and greed. We talk about "vulnerability", when what we mean is extreme pain and death. We analyze "climate effects" – and it is about is hundreds of millions of mammals and billions of other animals, dying from lack of food and water, or chased from breeding and hunting grounds. We talk about "extinction" – it is about animals in vain searching for a mate, about stocks of animals being too small and too scarce, about too many succumbing to hunting and to wildfires, floods or hurricanes, or directly suffering heat stroke. Unlike people, mammals are dressed in fur in the middle of summer, with temperatures exceeding 45°C.

CHAPTER 3

Environmental Toxins and New Entities

Beyond climate change and loss of biomass, the spread of environmental toxins in air, water, and the ground, in food, plastics and microplastics, in insecticides, cosmetics, and medicines, are slowly taking a toll. 100.000 chemicals, many of them highly toxic, coming in 350.000 varieties and innumerable combinations, pose threats to human and animal life far beyond the capacity of authorities to control. Environmental toxins are of particular danger precisely because the impact normally comes from tiny doses over long periods of time, making damage-causality difficult to prove, but hurting people and animals, insects and a variety of life-forms in infinite ways.

Saving the world means changing course in all three main areas of environmental threats, also recognizing the threats in combination. Sometimes mitigation in one area may cause disturbance in another areas – as when windmills for green energy is combated by environmentalists striving to save ecosystems and the natural environment. Sometimes mitigative action in one area will lead to improvement in another area, as when wilderness and nature protection also means preservation of carbon in the ground and trees, lessening the climate impact. On the negative side, climate change and environmental toxins both serve to reinforce the destruction of nature, while deforestation and depletion of wilderness areas cause loss of carbon storage in trees and the ground, adding to climate change.

Severely overlooked problems

In 2009, a research group linked to the Stockholm Resilience Centre, at Stockholm University, undertook to assess 9 different areas of environmental threats to the global society (Rockström et al., 2009). Rockstrom and his colleagues tried to define what they perceived as a safe "operational area" for humanity, based on people's food needs and "critical goals" for life on land and in water. The researchers looked for so-called threshold values in each threat area, including climate change, loss of biological diversity on land and in the sea, the use of nitrogen and phosphorus fertilizers, reduced ozone layer, acidification of the world's oceans, access to fresh water, habitat loss and access to agricultural land, spread of chemical pollutants and air pollution. An updated overview came in 2022 (Persson et al., 2022).

In three of the areas, the researchers found that we have already exceeded the thresholds for what is justifiable, namely climate change, loss of species diversity, and disruption of the nitrogen cycle (fertilizer). In four areas they found that we are on the verge of exceeding safe thresholds, namely the use of fresh water, loss of habitats and wilderness, acidification of the world's oceans, and the use of phosphorus fertilizers - all four cases related to food production and modern industrial agriculture.

Finally, there were two areas where the researchers simply could not find any threshold value, namely the spread of chemicals and environmental toxins, and the addition of so-called

DOI: 10.1201/9781003518815-4

"new entities" or "new substances", such as plastics and microplastics. There is today no defined "threshold values" related to such toxins and new entities, and no "Paris Agreement" type of arrangement to gain control. In December 2024, a fifth round of negotiations for a legally binding treaty on plastics pollution ended without result, only with hope for new talks next year. We have local clean-up campaigns, and systems for garbage treatment, but a great deal of environmental toxins and new entities are spread without any overall system for cleaning up or for avoiding poisoning once the substances have entered circulation.

An under-the-radar problem area

Chemicals and new substances come in countless varieties and surround us in countless ways – and their threats come on top of climate change and natural destruction and extinctions. Among chemical pollutants, the researchers count a wide range of heavy metals, pesticides, chemicals, radioactive entities, and environmental toxins. To this we may add a whole series of "new entities", a term introduced by Steffen et al. (2015).

New entities are substances which, in addition to affecting animals and humans, also may affect climate, the ozone layer, air pollution, land areas, fresh water, salt water, biological diversity, and animal and human health. New or novel entities stand for all kinds of substances that are new in a geological sense, i.e., that did not exist on Earth before humans in modern times invented them. They also include traditional toxic substances which through modern industry and chemistry are used in completely new contexts, such as mercury and lead. The annual production and spread of new entities far exceed our overall global capacity for mapping and assessing the level of danger.

Total production of chemicals has increased 50-fold since 1950, and production is again expected to triple by 2050. There are now around 350,000 different chemical substances and mixtures of chemicals in circulation (Persson et al., 2022). In the last ten years alone, 70,000 new ones have been added. 30,000 of these are produced in developing countries, often with minimal environmental legislation and control. Poor countries are used as production sites for a wide range of highly polluting chemical products, pharmaceuticals, and cosmetic products.

The control problem does not only apply to developing countries. In the joint European EU REACH system (Registration, Evaluation, Authorization and Registration of Chemicals), which is intended to control the use of chemicals in Europe, only 23,000 chemical substances were registered (as of December 2020). 12,000 of these were registered as "non-intermediates" (which means they are dangerous and important enough to require special approval) with a production volume of more than one ton a year. But of the 12,000, only 2,400 were studied more closely, and of these only 403 are subject to continuous monitoring. Elsewhere around the world, monitoring and control may be significantly slacker.

There has been no lack of initiative to gain control. Protests from workers and their organizations, from neighbors and other stakeholders have led to increased oversight and sometimes shutdowns of production and storage sites. National control bodies have intervened with a ban on the sale of chemicals for specific purposes, and there have been boycotts from consumers. But protests and administrative measures are repeatedly lagging behind the development of new ones.

Small doses over long periods of time

Ambient air pollution and household air pollution are known to cause several million premature deaths a year. In big cities, pollution also causes serious health setbacks and reduced work capacity. It has also been found that pollution has a direct impact on the frequency of Alzheimer's disease – beyond the effects of aging and average longevity (Calderón-Garcidueñas et al., 2018; Power et al., 2016). For more specific forms of pollution and damages, a study was published in May, 2023 that demonstrated a clear connection between ten different pesticides and the development of Parkinson's disease (Paul et al., 2023). Researchers from UCLA and Harvard found up to 53 different pesticides, all of which may have been implicated in predisposition to Parkinson's disease. Most of these have not previously been linked to the disease, and they are still in use. Parkinson's disease involves a gradual breakdown of nerve cells in the brain responsible for dopamine production, which affects both mental health and physical movement. Among the pesticides, the researchers looked in particular at ten that have a direct effect on dopamine production, of which four were insecticides, three were herbicides, and three were fungicides. Considering that in California alone there are nearly 14,000 different pesticides in use, with more than 1,000 active ingredients, it is interesting to note that combinations ("cocktails") of pesticides were more toxic than each individual pesticide.

There is a largely unexplained rise in the prevalence of "modernity-illnesses" – illnesses for which we have no clear causes and treatments. A study made by researchers from several leading universities in the US showed a 50 percent increase in auto-immunity and auto-immune diseases from the end of the 1980s to 2012. These are diseases caused by the body's own immune system, mistakenly attacking healthy cells. The study revealed that there was a particularly large increase of such illnesses among children and young people (Dinse et al., 2022). 41 million Americans now suffer from such diseases – often simply referred to as an "inflammation in the body", with no identified single cause, but with a broad set of interactions of many causes, including environmental toxins and a broad band of modern lifestyle environmental factors. The same problem of lack of causal proof also holds for the increasing prevalence of autism and for "new diseases" of diffuse origins, such as ME (myalgic encephalopathy – or chronic fatigue). More and more people also have problems with allergies, which are often – again – difficult to trace to specific sources and causes.

No one can know the final consequences of the chemical spread, and knowledge of all possible combinations of substances is far beyond what is testable and controllable. The main research problem is that exposure to extremely small toxic doses over long periods of time may cause illnesses, but the cause-effect relationship is almost impossible to prove.

Chronic optimists may think that the human body is so adaptable that it can handle almost anything. But the adaptation does not apply to all chemical compounds, and it takes at best a few thousand years: the modern chemical industry has existed for 150 years, and most of the new chemicals have been around for less than 40 years. The immune system does not have much of a chance to defend us – it is totally unprepared for the new type of attacks against body and mind, and it increasingly misses the mark.

Beyond auto-immunity and the increase of "modernity illnesses", it is noticeable that the number of people with depression and anxiety has multiplied over a couple of generations. Such an increase obviously has little to do with genetics and inheritance, but it is unclear to what extent chemicals and environmental toxins are to blame, as compared to changes in the socio-cultural environments and a stressful world situation. Still, a recent meta study of depression linked to occupational exposure to pesticides among agricultural workers revealed that most studies (78 percent) found a clear link (Cancino et al., 2023). Organophosphates (17 studies), herbicides (12 studies), and pyrethroids (11 studies) were found to be the most common causes, demonstrating a "clear association" with depression. Again, there is the problem of low exposure over long periods of time, and the researchers ask for more high-quality longitudinal studies in order to better control for sociocultural variables and make controlled studies of pesticide-specific biomarkers and depression. For more general purposes, such studies are also needed for the larger population to better understand depression and anxiety – as linked to environmental toxins in food, air, and water, and in many kinds of consumer products.

Diseases of prosperity

Diseases of prosperity affect large numbers of young people, and we can agree that it is a tragedy when cancer, depression and anxiety affect an increasing number of teenagers and youngsters. Concluding from a large study of cancer among people under the age of 50, British and Chinese researchers gathered that "Global incidence of early-onset cancer increased by 79.1% and the number of early-onset cancer deaths increased by 27.7% between 1990 and 2019" (Zhao et al., 2023). While cancers affect people over 50 in 9 out of 10 cases, the rise of many forms of cancer among young people is now reaching epidemic proportions in both rich and poor countries (Ugai et al., 2022). The final cause has not been identified, but 8 out of 14 cancers on the rise are linked to digestion. Ultra-processed, industrially produced food is seen as a main cause. Affluence has its costs.

There is a constant stream of research reports about the increased prevalence of (more than 80) different auto-immune diseases. Issues include autism, depression and anxiety (especially among the very young), hesitance to have children and inability to have children, further linked to gender confusion and gender identity issues. It is about diseases of diffuse origins and about the (hard to research) long-term consequences of small-scale exposure to 350.000 varieties of 100.000 chemicals, to environmental toxins, pesticides, micro-plastics and nano-plastics. In the Introduction I made a distinction between the "sunny side" of modernity and the "shadow side". Consequences of new chemical entities belong to the "shadow side" – difficult to observe and not a sign of progress.

Conferences and agreements

The first UN conference on environmental pollution was held in 1972. It resulted, among others, in the establishment of the "United Nations Environment Programme" and, in 1987 the Montreal protocol for limitation on substances that threatened the ozone layer in the atmosphere.

A new conference on environment and development was held in Rio de Janeiro in 1992. It resulted in Agenda 21, which listed 100 program areas related to poverty, nature management and the environment, voluntary organizations and economic support schemes. A separate UN Commission for Sustainable Development (CSD) was also created, and the preparation of national and local Agenda 21 plans was encouraged. But the spread of environmental toxins and chemical pollution still received little attention. Agenda 21 was later replaced by the UN Millennium Development Goals, and in 2015 by the 17 Sustainable Development Goals for 2030.

Two of the 17 sustainability goals relate to the production and control of chemicals and environmental toxins. Sustainability goal 3, on health and quality of life, states (target 3.9): "By 2030, significantly reduce the number of deaths and illness from dangerous chemicals and air, water and soil pollution and poisoning". And in goal no. 12, on sustainable production and consumption, target 12.4 reads:

"By 2020, ensure environmentally sound control of chemicals and all kinds of waste throughout the entire life cycle, in accordance with internationally agreed frameworks, and to a significant extent reduce emissions to air, water and soil with the aim of minimizing the harmful effects on human health and the environment".

We may note that the target date is set to 2020 – not 2030 as for other targets, and that specific chemical groups and damage areas are listed under the sub-goals, but also that by 2020 very little had been achieved to meet the targets. A central driving force in the system is of course the chemical industry itself, but also important users of the chemicals, such as the construction industry, agriculture, electronics, cosmetics, the pharmaceutical industry, mining, and the textile industry. Clearly, the growth within the sector and the spread of new substances in products and processes currently far exceeds what is safe and controllable.

There have been industrial initiatives for "sustainable chemistry" (Blum et al., 2017), "one world chemistry" (Matlin et al., 2016), for "circular chemistry" and generally for "cleaning up" the industry (Anastas and Eghbali, 2010). But the enormous growth of the sector, its strong integration in other sectors (food, medicine, textiles, electronics, buildings....), and its very strong financial backing, and constants calls for innovation and the development of new product lines, makes limitation and control very hard. Also, since so-called "cocktail effects" (effects of combinations of substances) are a dominant feature of chemicals use, full control is in principle impossible. Chemical compounds, which may be harmless in the first place, may become toxic in combination with other substances, and much of the mixing of content is done by innocent consumers. Damage has mainly been observed locally and regionally, but increasingly also globally, affecting species extinctions, insect eradications and reductions in biological diversity – in addition to human health.

While we may often think of air pollution stemming from automobile traffic as a main threat to health, recent research has revealed that almost 40 percent of the world's tiny particle pollution comes from ammonia from livestock manure and excessive use of fertilizers in agriculture. In some countries, with strong controls of other pollution sources (such as the UK), the percentage may rise to 60. Shifting to a vegetarian or vegan diet might save several lives a year, in addition to remedy the loss of mental capacity due to such pollution (Springmann et al., 2023).

The non-conventional conventional

There is only one true form of conventional agriculture – that of traditional organic farming. By now, however, the saturation of soil with toxic chemicals has become the new conventional. In good soil there is more variety of species below the ground than above. Actually, a recent study found that more than half of all of Earth's species live in the soil – 90 percent of the world's fungi, 85 percent of plants and more than 50 percent of bacteria, making soil the richest habitat on the planet (Anthony et al., 2023). But it can be destroyed. By the new conventional standards for soil handling, insecticidal pesticides and – not least – weed killers contribute to the destruction of countless small creatures. When the insects die, so do the small creatures that keep the soil alive, and also the birds and fish that live by eating insects.

The insect population in the world has decreased by 60 per cent in 60 years. Insects are attacked from many quarters, both in wilderness, in agriculture, in urban areas and on the highways. Insecticides affect both pollinating insects and millions of other species, both in nature and in human food production. Not all insects are equally important (or equally welcome to humans), but all have a position in one ecological context or another. Monocultures in agriculture, loss of flower meadows, clearcutting of forests, spread of artificial fertilizers and eutrophication of rivers and water, all contribute to giving insects a hard life and eventual death.

Direct spraying against insects is only a small part of the problem. Sprays aimed at weeds and fungi are in fact more important. Such pesticides are used five times as much in agriculture as the insecticides, and they affect the insects in numerous ways. To give the bees and other insects breathing space, rows of wild flowers can be planted along the fields. A remaining problem is that clouds of weed killers drift with the wind and pollute these flowers as well.

Slowly, research indicating the carciogenic consequences of glyphosate is emerging. Glyphosate is globally the most widely used weed killer (under the brand name "Roundup BioFlow" in Europe and "Ranger Pro" in the USA, from Monsanto). The substance, feared as a bee killer, by attacking bees in their digestion, has been traced to cause endocrine disrupting effects in rats and in human infants (Lesseur et al., 2021; Ward et al., 2023). Again, such research is extremely demanding, on obvious ethical grounds excluded from experimentation on humans, and bound to incur problems of cocktail effects and problems of causality detection from small doses of exposure over many years.

Added to this, there is also the material problem of diminished glyphosate effectiveness as the weeds over time tend to adapt and develop resistance. Since the introduction of glyphosate-resistant soybean and corn in 1996 and 1998, the product has come to dominate the global herbicide market, with an annual output of 8.6 billion kilograms. But as with all "silver bullet" solutions to weed control, the weeds find ways to develop resistance, in as little as 12 years, by mutation and other means – calling for either more diverse chemical treatments or for organic solutions (Landau et al., 2023).

Organic farming represents attempts to run efficient food production without the use of modern toxins and fertilizers, also with aims to improve nutritional content, food safety and taste as well as animal welfare. The price per kilo is almost always higher for organic food, but the price per essential micronutrient is regularly lower. A good overview of the issues and research results can be found in Reeve et al. (2016). The general conclusion is that organic farming produces far less toxins in the food, gives the animals a better life, takes better care of insects and flowers, is better for the soil and for overall bio-mass and bio-variety, takes better care of the organic substances in the soil that are necessary for nature conservation and sustainability over time, and provides better food quality with increased nutrient content. Only organic farming is truly sustainable, but much can be gained if "conventional" farmers can learn from "organic" that soil protection is about more than area protection and that the use of organic approaches, such as crop rotation, ground cover (for control of weeds), and catch crops (to improve soil fertility) can also be combined with conventional solutions.

It is often repeated that there will not be enough food for everyone if we base ourselves on organic farming. But according to renowned Washington State University professors John Reganold and Jonathan Wachter (2016) who compared the last 40 years of organic and conventional food production, organic farming can be both sufficient to feed a growing world population and also be profitable for farmers, as well as capable of protecting and improving the environment while being significantly safer for farm workers.

The outstanding position of plastics

Plastic is primarily a malleable, artificial (synthetic) material that can be used for countless purposes and in many different varieties. Plastic-like products, primarily celluloid (cellulose nitrate), were developed in the 1860s, initially to make artificial billiard balls that imitated ivory, later also to make film, among other things. The first synthetic polymer plastic was produced in 1907. Since then, endless inventions of new types of plastics have followed. Only recently has there also been concern about microplastics.

In many contexts, plastic is practically irreplaceable today – plastic products have somehow become too good and too cheap, and there are too many of them. The products are designed to be robust and durable. This, together with the huge increase in volume, has led to littering going too far (Hou et al., 2021).

Huge numbers

The main question related to single-use plastics has been what to do when they have done the job, and related to durable plastics, how to phase them out and deal with the leftovers. In the wake of The UN's annual World Environment Day in Paris, June 2023, FRANCE 24 produced an overview of plastic pollution (Cyrielle Cabot: "Ten statistics on global plastic addiction and its consequences", 2023 (https://www.france24.com/en/environment/20230605-ten-statistics-on-global-plastic-addiction-and-its-consequences): Globally 460 million tons of plastics are produced annually, and 8,3 billion tons have accumulated so far, most of it spread in the environment. The production figure doubled from 2000 to 2019, and is expected to nearly triple by 2060, reaching an annual production of 1,231 million tons. The lifespan of nearly two-thirds of the annual output is only 5 years or less, and much is used only once or a few times, especially for food packaging, textiles and consumer goods. Globally, only 9 percent of what is produced is recycled, meaning that presently some 350 million tons of new plastic end up as waste every year. Half of what remains goes to landfills, 19 percent is incinerated, and 22 percent discarded or burned.

When the substances break down, whether in nature, with the help of sunlight or through organized incineration, greenhouse gases are emitted. Presently, plastics contribute some 3,4 percent to global greenhouse gas emissions and climate change, which is quite significant (more than all of airline traffic). The breakdown of plastics is expected to quadruple, to 6.5 Gt (gigatons = billion metric tons of CO_2 equivalents) by 2050 (Zheng and Suh, 2019).

Between 400.000 and one million people in developing countries die annually from illnesses caused by toxic fumes from burning plastic waste. 109 million tons of plastic waste have accumulated in rivers and lakes, and 30 million tons in the oceans. Some 100.000 marine mammals die each year from plastic and debris, much from abandoned fish nets and fishing equipment. On top of this, in addition to being an enormous burden on the natural environment, plastic littering is also a huge economic waste (Garcia and Robertson, 2017).

In February 2024, The United Nations Environment Assembly (UNEA), met in Nairobi to work for the "triple planetary crisis of climate change, nature and biodiversity loss, and pollution and waste". It was the 6th meeting since being launched in 2014, and among outcomes was the initiation of planning for a plan to gain control with plastic pollution. It is not about what we otherwise would call "proactive leadership". As mentioned, the INC-5 conference in Busan, South Korea, in December 2024, ended without agreement. The intended treaty was meant to be the most significant environmental step since the 2015 Paris Agreement. It remains to be seen what will be the next step.

Microplastics

Some plastics are in the form of tiny particles – microplastics or nano plastics. Most forms of plastic products may end up as microplastics. Microplastics may originate from the breakdown of ordinary plastics such as car tires that wear against the asphalt, fleece products, clothing and textiles during wear and washing, plastic bags exposed to sunlight, or additives that leak into water or air. Also, special additives in medical and cosmetic products, polishing agents, and plastic composites among others in building materials contain microplastics. When the products break down, the particles are dispersed – and almost impossible to recover.

Microplastics are particles that are smaller than five millimeters, while nano-plastics are smaller than 0.1 micrometers (1 micrometer = 0.001 millimeter). The latter are tiny particles that cannot be seen or felt or filtered. They are found in both air and water, and are taken up by most living creatures, even plants. According to a meta study (Cox et al., 2019), the average American receives at least 50,000 microplastic particles through food and at least the same amount through air and water during a year. If mainly drinking water from plastic bottles, one may count on an additional 90,000 particles (up to 130,000 particles, compared to 4,000 in ordinary, healthy tap water). Another study found that an average liter of bottled water may contain between 110,000 and 400,000 particles of nano-plastics (Qian et al., 2024).

So far, only a few foods have been properly investigated for microplastics and nano plastics. According to a report from the World Wildlife Foundation (WWF): "No Plastic in Nature: Assessing Plastic Ingestion from Nature to People" (Advisors et al., 2019) we take in on average 5 grams of micro and nano plastic a week – the equivalent of a credit card. Happily, most of it passes through the body. But some is stored.

An important point of warning, however, comes from the observed fact that plastic waste of all sizes has a significant attraction for pollutants, which contribute to the spread of diseases, environmental toxins, and heavy metals. Micro- and nanoplastics are also found in the placenta of humans and animals, where the substances are passed on to the next generation. While researchers still know little about the specific harmful effects of the microplastics and nano-plastics, there is an emerging consensus behind warnings about future harmful effects on both people and nature (Conti et al., 2020; Gibb, 2019). A landmark study, including 312 patients who underwent carotid endarterectomy was published in the New England Journal of Medicine (Landrigan, 2024; Marfella et al., 2024). Detectable levels of microplastics and nano-plastics in patients with carotid artery disease was found in 58 percent of the patients, causing a 2.1 times higher risk of death or serious cardiovascular events. Nano-plastics has been identified in the bloodstreams of people, and they have even been observed to pass the blood-brain barrier in mice (Kopatz et al., 2023). This in fact means that nanoparticles will find their way to all organs in the body, with known damage to the cell structure – but with largely unknown overall, long-term health consequences.

Unknown consequences

The world is becoming permeated with environmental toxins and pesticides, to the detriment of nature – plants, animals, and people. A consistent historical feature of the spread is that serious controls and prohibitions are put in place only after damage has been reported. From initial acceptance to final ban often takes some 20 years or more.

While far more critical testing needs to be done before new entities are approved, there is also a need for similar thoroughgoing control of replacement products, and also of alternative uses for the banned product. The "lock-in effect" implies that when a product is banned for one purpose, the manufacturers will a) come up with a substitute (which may later turn out to be just as bad), and b) the manufacturers will find other applications for the original entity. Any product that is banned must therefore be followed up for at least 20 years to check what the manufacturers choose to do with it. Normally, they never stop producing anything – no matter how toxic and destructive.

A general problem with environmental toxins and new entities is that the quantities affecting humans and animals tend to be infinitesimally small, and harmful effects only appear after very long exposure and in "cocktails" with other entities of unknown nature. Also, as for microplastics and nano-plastics, we currently know very little about health effects to animals, to the human body, or to a newborn child (who receives breast milk blended with nano-plastics that are also magnets for environmental toxins). And when it comes to the overall amount of environmental toxins and hormone disruptors, we know very little: As mentioned, the consequences of small amounts over long periods of time are very difficult to research and control.

Affecting reproduction

An important group of new entities are so-called "endocrine-disrupting chemicals" (EDC). They are also referred to as "hormone mimics" and "estrogen-like substances", and they can have serious, disruptive effects on the reproductive ability of animals and humans, impacting also procreation and gender identity. Some types of plastics (for example plastic bottles) with special alloys, traditionally phthalates, also have such a hormone-mimicking function. Estrogen-like substances in many plastic products, and in food and the environment may undermine ordinary gender development, and more and more people experience gender confusion and identity crises.

"Most couples will need assisted reproduction in 2045" was the title of an interview with Shanna Swan, a world leading professor of environmental medicine at the Mount Sinai School of Medicine in New York (Corbyn, 2021; Levine et al., 2017; Swan and Colino, 1920). We also have reports that an increasing number of men are developing "male boops" – or gynecomastia (Deepinder and Braunstein, 2011; Swerdloff and Ng, 2023). Another warning signal is that more and more girls (also some boys) are entering puberty at the early age of 7 – which contributes to major personal and social problems. We are warned that early exposure to sex hormones (from medicines, skin creams and other hormone disruptors) may cause such disturbances (see, for example, "Precocious puberty", health information for patients from the Mayo Clinic, https://www.mayoclinic.org/diseases-conditions/precocious-puberty/symptoms-causes/syc-20351811). What often happens as a result of early puberty is that biology collides with normal intellectual and social maturation, often causing life-long tragedies of behavioral mal-adaptation and delinquency.

Added to this are issues related to reduction of sex drive, sperm quality, and testosterone levels. An article in the Guardian with the provocative title "Why have young people in Japan stopped having sex" (Haworth, 2013) caught widespread attention. Referred to as a "celibacy syndrome", the article revealed that near 50 percent of women in the 16–24 age group were not interested in sexual contact, or even despised it. More than a quarter of men in the same age group felt the same, and a third of people under 30 never dated. Similar trends have been observed elsewhere, and in several high-income societies reproduction rates are down and population size is shrinking. Population decrease may be good for the environment and for limits to growth, but according to reports in The Lancet, global (involuntary) infertility affect some 8–12 percent of couples, causing relational problems, affecting more men than women, and reflecting among others decline in human semen quality and lowered testosterone levels (Agarwal et al., 2021; Damario, 2014; Wu et al., 2022). Of course, a number of causes may be behind the trends, including lifestyle changes, and there is still a lack of research testing the relationship between sexual dysfunction or "celibacy" tendencies and environmental toxins.

Chapter 4

Energy and Feedstock "Stickiness"

A premise of this book is that models and prognoses (such as the IPCC, IPBET, WWF reports, etc.) are to be taken seriously. Only top experts in climate modelling and biological statistics can properly judge the scientific quality of relevant status markers, mathematical models, and prognoses. The rest of us will have to believe or disbelieve and wait for the manifestation of real climate, biomass and biodiversity changes (Chapter 8).

The warnings are not only about climate changes, but also about eco-system collapse and extinctions, environmental toxicity, and pollution. "Saving the world" means delivering in all threat areas (Folke et al., 2021). In the present chapter, however, the focus will be on energy and the main energy vectors that will determine degree of success of the green energy transition.

The message from both António Guterres, secretary-general of the United Nations and from an unusually agreed corps of nature-scientists, is that Business as Usual (BAU), neglect of mitigation strategies, and "wait and see" type adaptation strategies, need to be replaced by quite radical and prompt mitigation. Pope Francis in 2015 issued an encyklika, "Laudato Si' warning that climate changes were human made, and asking for ecological and lifestyle changes, austerity, and mitigation. In October 2023 he came with a second warning, "Laudate Deum", claiming that mitigation is not coming soon enough and strongly enough. The Pope asked for stronger commitment and faster economic change, condemning environmental depletion, economic inequality, excessive consumption, and climate skepticism. If austerity is to be real and large-scale, however, it is not likely to come with BAU-type economic growth and hard-core capitalism. And, if it is to be effective it must address some very serious questions about energy provision, and energy forms, use, conversion, and phase-outs.

Energy in the anthropocene

"Stickyness" refers to the quality of being sticky, which means for something to be attached to a surface, as something adhesive, viscous, or gluey. Merriam-Webster also includes a number of metaphorical uses, such as being humid, muggy, or clammy, and also disagreeable or unpleasant, difficult or problematic. In the field of marketing, "stickiness" is used as a metric for measuring user engagement with an app. In research, there is a vast area of chemical and physical studies of material adhesive stickiness, and in economics we have "price stickiness" as a measure of price rigidity for goods and labor (Sbordone, 2002; Schmitt-Grohé and Uribe, 2006). We also have the stickiness of income disposal (Mankiw and Reis, 2006), consumption growth (Carroll et al., 2011), and financialization (Sun and Gong, 2023).

For the present use, "stickiness" will refer, first of all to fossil fuels adhesiveness in the modern economy. The green transition is often talked about lightly, as a series of easy steps for GHG emission control and green energy development. Overall, the use of fossil energy sources is the

DOI: 10.1201/9781003518815-5

number one threat to climate stability, and the elimination of fossil energy sources is our number one strategy for the climate. In this chapter we shall see that fossil fuels are intimately glued to major processes in the economy, sticked to modern product choices and production systems – not easily replaceable and de-couplable from neither present consumption patterns nor from envisioned economic growth. In management texts, the word "lock-in" is used for some of the same, but while a lock-in refers to a specific method, material, or process choice, stickiness here refers to the overall variety and viscosity of material and energy uses in the economy. As will be seen, "stickiness" may also help explain the near linear relationship between economic growth, climate impact and nature depletion (Schröder and Storm, 2020).

A short energy history

For movement and work, human muscle-power was the only well-documented kinetic energy source for work until the time of agriculture, some 10,000 years ago. The sapiens may have experimented with sails for the capture of wind-power at a very early stage. But the oldest known remains of sails and sailboats are only about 7,000 years old. And while firewood was used for heating and cooking, lightening, and hunting, as an excellent form of heat and light energy, it was not convertible to kinetic energy until the end of the modern era, with steam and combustion engines.

The key to getting a good grasp of energy uses is to look at various energy sources and energy conversions. We have two ultimate sources of energy, the sun in the form of light and heat, and the Earth in the form of heat from the ground (geo-energy). The main energy conversion processes are first of all the photosynthesis effect of solar radiation, sustaining oxygen production and plant growth. By way of photosynthesis, solar energy is converted into chemical energy for all living species feeding off the plants, and further into chemical energy storage as fossil fuel (coal, oil, gas), as heat storage in plants and trees used as firewood, and into electricity using modern photovoltaic solar panels. The warming and lightening energy of sunshine, further converts into climate effects, carrying the cycle of water, air, and carbon, and allowing for wind power by use of sails and windmills. Energy from within the Earth converts from geo-energy to geothermal heat energy, and from uranium to atomic energy, via heat to electricity.

With the agricultural revolution some 10,000 years ago, people were able to acquire more energy from a smaller space by the concentration of plant growth and early domestication of animals, adding to the chemical energy supply, and facilitating more energy conversion to mechanics and movement. With the domestication of animals came the added energy supply from working animals, some 9,000 years ago – the first definite, extra-somatic, kinetic energy supply beyond human muscle power. Horses and oxen, donkeys and mules were increasingly used for field work and water pumping, and for transportation of people and goods.

After the domestication of animals followed a series of inanimate energy supplies. Sailboats were definitely in place 5,000 years ago, waterwheels 2,000 years ago, and windmills 1,000 years ago – all additions to or replacement of human muscle power, utilizing energy sources in nature: wind, water and indirectly solar.

For some reason (a huge theme discussed in the discipline of global history) Europeans, and in particular the British, succeeded more than others in entrepreneurial exploitation of wind and water energy from around the year 1,000 AD and onwards. Windmills and waterwheels were not European inventions, but they were put to more practical use here than in the East and Mid-east, and they freed human work energy from the strenuous tasks of milling, grinding, water pumping, forging, and eventually timber sawing. Also, the uses of such advanced tools stimulated learning of mechanics and material production, and supported market development and the expansion and widening of consumption patterns. Even so, Smil (2022) has estimated that by 1500 CE more than 90 percent of all mechanical energy around the world was provided by animate power, with humans and animals providing roughly equal parts. Thermal or heat energy was provided by the burning of biomass – plants (wood, straw, dung) and wood-based charcoal.

Going modern

The first step towards fossil fuel energy era started in England, with the burning of coal, surpassing biomass burning around 1620 (Smil, 2022). By 1650 coal provided two thirds of heat energy in England, reaching 75 percent in 1700 – the main foundation for early industrialization and for the first experiments with steam engines in coal mines, converting heat energy to mechanical for pumping. Even by 1800, England by a far margin exceeded Europe and the US combined in the use of coal. With James Watt's steam engine from 1774, and the Bouton and Watt steam engines (25 HP capacity) in serial production by 1800, Brittain was to maintain the lead for several decades (Smil, 2014, 2022). Elsewhere in the world, biomass still provided some 98 percent of heat energy, and human and animal muscle energy provided 90 percent of kinetic energy in farming, construction, manufacturing, and transportation.

Since medieval times, crude oil directly from the ground was used in the Mid-East. For modern, industrial usage, it was first extracted from the Drake oil well in Pennsylvania in 1859, of which is said that it gave birth to "an industry that has shaped our modern world". Oil has since then been used for combustion and lubrication and as a basic feedstock ingredient in innumerable products, including plastics.

The first production of electricity based on waterpower, was made by William Armstrong, in Northumberland, England, and it was used to power one single lamp in his art gallery. The first Edison hydro-power station started electricity production in Wisconsin in 1882, and soon hydro-electric power stations were spread in the US and Canada during the 1880s and in Europe by 1900. Experiments with geothermal electricity started in the early 1900s – without any real take-off. The first modern windmill for electricity production was built by Charles F. Brush, an American scientist, in 1887, and small turbines were used in North America before the turn of the century. The later take-off of modern wind-turbines followed the oil-crisis of the early 1970s.

The first silicon photovoltaic (PV) solar cell for direct conversion of sun radiation into electrical energy was produced by Daryl Chapin, Calvin Fuller, and Gerald Pearson of the Bell laboratories in 1954. Due to low efficiency and high costs, solar power and technology development was first affordable only for space exploration. The commercial take-off, and high efficiency solutions for solar power were available in the 1990s.

Uranium was discovered in 1789 by German chemist Martin Heinrich Klaproth. Uranium is the heaviest element naturally occurring in the universe. Plutonium is a synthetic chemical element discovered in 1940 by Glenn Seaborg and colleagues at the Berkeley Radiation Laboratory as a derivative of uranium. The first nuclear power station was put in operation at Obninsk, near Moscow, in 1954. Some countries have invested heavily in large-scale nuclear power stations, but due to high risks (and need for governmental insurance) there are still no properly commercial nuclear power stations in operations – mainly for insurance and security reasons. Advocates hope for a change, with the introduction of serial production of smaller installations starting in the 2020s and fusion based technology perhaps available by the mid-2030s.

There are also other energy forms, such as gravitational energy, elastic energy, radiant energy, and mass energy – awaiting further conversion-technology development. To the various uses of energy, fossil sources in particular, but also plant and wood sources, serve important functions in the overall manufacturing, agriculture and industrial processes, as feedstock, lubricants, and additives.

Additional hurdles

This all adds to a large list of energy sources, energy forms, energy conversions, and innumerable energy uses. And it is also a reminder that in order for there to be multiple applications, energy conversion is a key factor. For the green transition, electrification by use of wind and solar power catches much attention. For a deeper understanding of the real hurdles to the Paris targets and the net zero by 2050 target, an understanding of sources, forms, applications, and conversions provides useful information.

While BAU-geared politicians and lay people tend to speak lightly of the green transition, and while climate experts and naturalists underscore the dangers from not moving fast enough, the immensity of the task is seldom exposed, and the intricacies of energy conversions and the integration of energy in all phases of modern production and consumption is too often overlooked by policymakers. But it is these immensities and intricacies that we need to take seriously if we are to make sensible environmental policies and political-economic decisions. We are pretty much stuck in fossil energy production and consumption patterns – for food and for material comfort and survival. Fossil energy is a sticky mess, and once we are stuck, it is extremely difficult to depart and generate absolute decoupling.

In the hope of a fast green transition to green energy, policy makers often overlook a key variable in energy conversion – that of energy density. It is often forgotten that all the main fuels based on crude oil, such as gasoline, aviation kerosine, diesel and residuals, are all high-density fuels, especially well-suited for transportation of all kinds. Smil (2022, p. 26) ranks energy sources based on density (in terms of gigajoules per ton). Air-dried wood has a density score of 16; bituminous coal, 24–30; kerosene and diesel, 46. By volume (gigajoules per cubic meter), wood scores 10; high quality coal, 26; and kerosene, 38; whereas natural gas is measured in megajoules per cubic meter at 35. The latter figures indicate that the density of natural methane gas is less than one thousandth that of kerosene, meaning that to propel a modern jet plane with uncompressed natural gas instead of kerosene would take up a thousand times as much space, leaving very little allowance for passengers or cargo, and even in a compressed form natural gas takes a lot of space. As for green hydrogen to power aircraft, there are lofty plans, but quite a lot of problems to be solved before takeoff (Boretti, 2023).

Green and not so green energy

An overriding differentiation for our purposes is the one between properly green energy forms and energy applications causing disturbances to climate, nature, and health. We shall continue with some more definitions, then look at examples of conversions and applications that are part of the "stickiness" problem threatening the green transition.

"Clean energy" is generally thought of as electricity produced without the emission of greenhouse gases and other pollution. But that does not mean that it is necessarily "fully green", or entirely without "carbon costs", "footprint" or "nature costs".

Greenness of energy sources

Hydroelectricity is clean, but not entirely "green". It does not come without climate and nature footprints. The erection of dams and other infrastructure comes with massive destruction of wilderness, lakes, rivers, fish populations, wildlife habitats, and sometimes indigenous communities, and it is using massive amounts of CO_2-emitting concrete in construction. It produces "renewable" energy, but in terms of nature conservation, it is not properly sustainable – the damages done are generally irreversible.

Bio-energy may be sustainable and renewable, but mainly at the expense of food production, wilderness and habitats preservation, virgin forest and ecosystem preservation, biological diversity and traditional farming communities. In the case where American forests are sacrificed in order to serve Europe with wood pellets, a 40–50 year setback in forest carbon capture is incurred. Only a small percentage of bio-energy is properly "clean" and "green", and it seems clear that investment in bioenergy so far must have been the most unsuccessful of all forms of climate investments.

In the book "There is no Planet B" (2021), Mike Berners-Lee provided an interesting calculation. He looked at what a square meter of land in sunny California can contribute in terms of energy for transport. He compared the cultivation of grass and wheat converted into bioenergy with the effect of a one-square-meter solar panel. If you squeeze oil out of the annual growth of grass or wheat in that small area, it will support a petrol-fueled passenger car to be driven 5 English miles (8 km).

If the same bio-energy is used to charge an electric bicycle, it may travel 25 miles (40 km). Alternatively, if the space is used for a one-square meter solar panel generating electricity for the time the wheat would need to grow, it will provide enough electricity for driving a car 1,081 miles (1,750 km), or enough for an electric bicycle to pull 21,243 miles (almost 35,000 km). Admittedly, not all places have as much sun as southern California, but the tropical areas that are today being converted into plantation land for rapeseed, soy and palm oil are at least as sunny.

While biomass energy using food crops and fresh timber pellets for many reasons do not qualify as properly green energy, animal residues, sewage, municipal solid waste, industrial residues, and even some forestry residues, may qualify for green bio-fuel. So far, the volume has been too low to make a dent in the global energy supply, however, and the technologies needed for proper treatment of residues are not sufficiently developed. For such inputs, there is still a potential, however, both for heating and for electricity uses.

"Green energy" in a pure sense means energy from totally natural sources, with negligent environmental impact, both in production and consumption, and from undeletable sources. Sometimes, the term "eco-energy" is used for roughly the same. As such, properly "green energy" and "eco-energy" are hard to find.

Solar PV comes with pollution from production of panels, depletion of key natural resources, and with a need for battery functions and grid-systems that come with considerable additional environmental burdens.

Wind power meets with resistance from environmentalists and citizens who are critical of threats to wildlife, indigenous and local communities, tourist spots and ecosystems. Critics are also concerned with production costs and pollution, micro-plastic pollution, decommissioning costs and the undermining of the 1/3 Earth target from the Montreal agreement.

Still, both energy forms, wind and solar, are generally considered "green". We have to live with the fact that properly "green" and "clean" energy is hard to find, and that we have to rely on a "more-or-less" type of classification. Covering the full range of non-fossil energy forms, the term "alternative energy" is used. Covering the main green energy forms, solar PV and wind are the most promising in terms of growth, and hydro-electricity the most promising for stabilization and "battery" function, and the three of them are referred to as the WWS (water, wind, solar) of a "green" or "eco-friendly" solution.

Ambitions and realities

According to The fourth edition of the Production Gap Report, countries will produce more than double the amount of fossil fuels in 2030 than would be consistent with limiting warming to 1.5°C, and 69% more than would be consistent with the "not more than 2°C" goal (Achakulwisut et al., 2023). A radical expansion of coal production is planned for in India. Increased oil production is planned among others in the US, Canada, the United Arab Emirates, and Saudi Arabia. And Russia is planning for expansion of both coal, oil, and gas. Other exporters are adding to the challenge, each betting on an overall failure to meet the Paris targets, and each seeking a share in the incomes to be generated during the final stages of fossil energy production. Overall, the petrostates are planning for production volumes that by 2030 will be more than double what would be consistent with the Paris-target.

Through tax and depreciation benefits, and sometimes direct subsidies, governments are financially supporting policies that climate experts describe as "madness" or "insanity" – or "a climate disaster of our own making." Among 20 fossil fuel-producing nations, representing 84 percent of global CO_2 emissions in 2021, 17 have pledged to go for net zero emissions within the same time frame. There is a huge gap between words and deeds, and trillion-dollar incomes are the main cause.

Each company and each nation are being "played out" in a game where each participant believe that sacrificing fossil growth or prolongation will not lead to a general reduction of emissions – only to some other nations and companies capturing the premium (Achakulwisut et al., 2023). To this we may remind ourselves that even if all fossil sources were shut down immediately, the greenhouse effect would continue to drive temperature rise for another 15–20 years. Fossil fuel emissions do not only come with CO_2 and climate gases, but also with aerosol pollution, producing a cooling effect. When ending fossil fuel combustion, layers of aerosol will shrink faster than the layers of CO_2 in the atmosphere, driving the temperature further up before stabilizing.

The greening of electricity and of total energy use

According to Ember's fourth annual Global Electricity Review (Ember, 2023), there are two macro-scale ambitions that need to be met in order to realize the Paris agreement target of 1.5°C temperature rise (or "not more than 2.0°C"): (1) The world needs mass-scale electrification, and (2) electrification has to be based on properly green energy, while fossils are being phased out, or neutralized by use of carbon capture and storage (CCS) technology. To this we may add (3) the challenge of greening all kinds of fossil energy uses, beyond electricity, in products and production processes. For a full net zero ambition, fossil energy in all forms causing carbon leaks, has to be eradicated from all products and productions systems that are part of the global economy. Not all uses are available for electrification. At least for the near future, you cannot remove oil from the production of plastic or replace the petrochemical process (based on oil) with green electricity!

In relation to the former two targets, it is a sobering fact that even for electricity production, there has in absolute terms been an increase of fossil energy use, and in relative terms only a small decrease from 64 percent in 2000 to 61 percent in 2022.

On the optimistic side, PV (photovoltaic) solar electricity production has been quadrupled, from 1.1 percent of electricity in 2015, to 4.5 percent in 2022. Wind power has had a slower rise, from a higher initial, doubling from 3.5 percent in 2015 to 7.6 percent in 2022 – totaling 12 percent for wind and solar. Bioenergy (a controversial part of the clean energy mix) has only seen a slight rise of 0.3 percent. Nuclear energy (also a controversial part), which in year 2000 accounted for 17 percent of the world's electricity production, fell to 11 percent in 2015 and to just 9.2 percent in 2022, while hydroelectricity fell from 17 percent of the total in 2015 to 15 percent in 2022. Contributions from other forms of green energy (such as geothermal; solar mirrors; and ocean energy – currents, waves, and tides) has so far been negligible. In total, alternative and mainly green energy accounted for 39 percent of electricity production in 2022, while fossil fuels accounted for 61 percent.

Among green energy sources with minimal negative environmental impact, the geothermal solution, in combination with heat pumps and deep drilling, may pose the greatest promise. Presently, only 32 countries have geothermal power plants in operation. The US is a leading producer, followed by Indonesia, the Philippines and Türkiye. However, other countries are more reliant on geothermal power. In seven countries headed by Kenya, Iceland, and El Salvador, geothermal represents motre than 10 percent of electricity. Overall, geothermal electricity accounted for 0.34 percent of worldwide electric generation, which is still low (Gutiérrez-Negrín, 2024). But the potential is enormous. Temperature at the Earth's core is five times warmer than floating lava, and the temperature will be held for billions of years, enough to supply all the world's energy needs for whatever length of time. The challenge is to drill into extremely hard, extremely hot deep basement rock, possibly within reach with "Millimeter-Wave Directed Energy Deep Boreholes" (Woskov, 2017).

Wind and solar PV are by most experts seen as the main foundation for the emerging electrified economy. The potential of geothermal energy is still greater, but the level of development is still too small for an impact. By 2050, near 90 percent of electricity will have to be renewable in order to reach the Paris targets, and solar PV and wind will, by most estimates, have to contribute 70 percent or more. A long-term fall in nuclear power, and stagnation in the production of hydro-power and bio-energy must also be compensated, as must expected economic growth. To this must be added that not all alternative, "renewable" or "sustainable" energy is properly "green" or "clean", and that many sources of ("relatively green") energy comes with the .problem of "energy cannibalism" – when for a startup period or for the full length of service it takes more energy to produce energy than what comes out (similar to the "reversed protein factory" of meat production) – problems to which we shall return later in the present chapter.

The next 80 percent

While the record green production of 40 percent of electricity has been celebrated widely as a sign of winning for the green transition, a very sobering fact is that only 20 percent of total energy use in the world is in the form of electric power. To put it bluntly, nearly all (i.e., 80 percent) of energy use is non-electric and fossil based. This means that while solar and wind energy amount to 12 percent of the electricity package, the two sources account for only 2.5 percent of the total energy package. The non-electrified 80 percent of all energy use is nearly all based on fossil energy, and that is a really sticky part. To this must also be added the non-combustion uses of fossil energy, on which a modern economy is fully dependent.

Technology for energy conversions, for combustion and feedstock uses for all kinds of industrial and transportation uses causes "path dependence" – meaning that past decisions and technology choices makes is difficult and costly to change at a later date. There are also "lock-ins" to existing value chains, to supplier and customer networks, including contractual agreements, individualization of service or tailor-making of products, that makes switching to other technologies and other networks costly and risky. That is what the stickiness problem is all about. To the greening of near all of electricity (beyond the present 40 percent), the green transition has to account for the other 80 percent of non-electricity energy use, for the added burden of economic growth, for dealing with the "cannibalist" problem, and also for the many non-combustion uses of fossil fuels.

To get an idea of the diversity of such (non-combustion) uses, value chains, technical solutions and marketing mechanisms linked to oil, one may consider lists of hundreds or thousands of uses (in alphabetical order) and start with Adhesive Air mattresses, Ammonia, Antifreeze, Antihistamines, Antiseptics, Artificial limbs, Artificial turf, Asphalt, Aspirin, Awnings, Backpacks, Balloons, Ballpoint pens, Bandages and Beach umbrellas – ending with Toothbrushes, Toothpaste, Transparent tape, Trash bags, Truck and automobile parts, Tubing, TV cabinets, Umbrellas, Unbreakable dishes, Upholstery, Vaporizers, Vinyl flooring, Vitamin capsules, Water pipes, Wind turbine blades. The list is endless, and to a lesser extent it also points to similar stickiness related to coal and gas.

While such uses do not produce CO2 from direct combustion, the greenhouse gas from oil carbon still leaks to the atmosphere, as oil-based products gradually disintegrate and spread, adding considerably to the challenge of the net zero target.

Stickiness in time

A prime policy reference for our time is the "green transition" – and related terms of an "energy transition" and a "sustainability transition" – generally assumed to be within reach of BAU practices. The term "transition" is indicative of "stagist" thinking and the idea that change come in stages or life cycles, where the main carriers of a previous stage are phased out, being replaced with something new.

Joseph Schumpeter (1994 [1942]) is commonly credited with the term "creative destruction" – although Karl Marx had already identified processes whereby new innovations came to replace older ones, making them obsolete and causing economic disturbances in capitalist economies. The German sociologist Werner Sombart, in his book about "Krieg und Kapitalismus" from 1913, was supposedly first to introduce the term "creative destruction". The message of creative destruction is that in the history of technology, innovation constantly brings forth new solutions and practices, rendering existing products and processes outdated, effectively replacing the old with the new, adding to productivity and also to price/cost benefits and quality, bringing progress, growth and welfare to societies, while sacrificing ownership benefits and jobs from what is replaced.

Stagist thinking

The idea of a green energy transition borrows from the creative destruction theory to presume that fossil fuels will be replaced with green energy – just as wood in former times was replaced by coal, and oil lamps were replaced by incandescent lamps. In his book about "More and more and more", Jean-Baptiste Fressoz (2024) has noted how the term "age of" has been linked to "stagist" thinking, reflected in the titles of technical journals – such as *The Railway Age* (1850), *The Age of Steel* (1857), *The Iron Age* (1867), *The Gas Age* (1884), *The Petroleum Age* (1887), *The Electrical Age* (1897), *The Motor Age* (1898), *The Cement Age* (1904), and *The Coal Age* (1911) – adding up to a sum total of successful industrialization and economic progress, and rewriting history as a succession of material epochs. As noted by Fressoz, there were no technical journal named after the "age of wood" or the "age of the horse." Supposedly, we are now about to enter the "age of green energy".

No doubt, electrification and green energy are in progress. What is doubted, however, is whether the fossil fuels are being replaced. Under stagist assumption, in competitive markets, replacement (and often the bankruptcy of incumbents) is a fundamental assumption in the political-economic creative destruction literature. In the strategic management literature the notion of "disruptive innovation" (Bower and Christensen, 1995) has attracted increasing attention, advising companies about innovative (and disruptive) agency and the needs to defend against loss of market positions due to disruption. In the field of innovation management, the term "exnovation" (Kimberley, 1981) has received less acclaim, focusing more on the failure of management to phase out old instruments and practices and making room for the new – in a stage-like manner.

In direct contrast to stagist models, Fressoz points to a transition-less energy history, documenting persistence and symbiosis as opposed to transition. Obviously, at the product level, the history of life cycles includes countless cases of product maturity, decline, and death, as innovations bring forth replacement products of greater efficiency, quality and economy. Surprisingly, however, at the level of primary energies, life cycles seem to extend well beyond the oft-cited normal stages.

The stagist myth tells us that coal came to replace wood as an energy source, first in England and from there gradually around the world. Fressoz, on the other hand, documents that we are now using wood for energy three times as much as a century ago, and twice as much as before nuclear fission and hydroelectricity, and by 2019 twice as much as solar and wind power combined. As for the supposed replacement of coal by oil, hydroelectricity and nuclear power, it simply did not occur. The latter energy forms came as additions to coal – not as replacement.

The initial growth of coal took indeed place in 18th century England, spreading to Europe and America in the 19th century. But 95 percent of coal mining has occurred around the world since 1900, with the strongest growth of all time occurring globally in the 1980–2010 period – nearly all outside of Europe. During that period countries such as China, India, Taiwan, Vietnam, the Philippines, Turkey, and Indonesia, all had growth rates for coal in the tenfold range. Around the world, coal-fired power stations are by now on average of a younger age (15 years) than atomic power stations (30 years) – and often of greater efficiency. And while coal has become a pillar energy form in developing countries, even the US, Japan, and the Gulf states had significant growth in the 1980–2010 period, before falling back. Fressoz notes that America under Bush consumed four times more coal than under Roosevelt. Still more surprisingly, Fressoz mentions that Germany under Merkel consumed three times more coal than under Bismarck. Generally, Europe is seen as leading the way to coal decarbonization. But its industrial stronghold, Germany, is among the global top producers of lignite (or "brown coal"), with low heat content, used mainly for steam-electric power generation, generally considered the worst source of GHG emissions in addition to being the worst pollutant and threat to human health among energy forms. Also, Germany remains a world-leading producer of coal mining equipment, stimulating coal-based growth around the world, contradicting the ambition of a green transition.

The carbon transfer phenomenon

The essence of the transfer problem is that even when a nation like Great Brittain call a stop to all production and burning of coal, great amounts of coal may have been used to produce imported consumer and industrial goods. The case of coal is a prime reminder of what we may think of as consumption based GHG emissions or as import emissions. This is a theme of high significance for the measurement of progress towards a net-zero result around 2050 and for the understanding of the concept of environmental decoupling (to which we shall return in Chapter 11). Our present theme is stickiness in time, and the question of why fossil fuel is so difficult to replace at the global level, while many nations now claim to have reduced consumption. A relevant reason may be that a major form of use is simply not properly reported. People and politicians are simply not aware of how much is consumed.

It is a fact that several leading economies have been reporting lower GHG emissions in recent years, indicative of progress for a green transition. But it is also a fact that the same countries report overall consumption growth including huge imports from developing countries and NICs (newly industrialized countries), above all from China. What is hidden in the optimistic reports is that much of the increase of consumption in wealthy countries is based on coal and oil powered production abroad, with GHG emissions reported only in the countries of production. Consumption is reported as "clean" in the importing countries, whereas the total of global GHG emissions continues to rise, making the planet a riskier place.

An absurd version of the problem is found in the oil sector, where producers only report emissions from production, whereas the main global problem is one of consumption and end-user combustion. In the case of oil, the major portion of emissions comes from final use. In the equally absurd case of coal, the major portion of emissions comes from extraction, heating, electricity production and other uses at the stage of production. And in both cases, the main drivers of the fossil-driven economy are not reporting about the main problem.

An interesting example is that of the coal phase-out in the UK. The Ratcliffe-on-Soar mine was closed down at midnight on September 30th 2024, marking the end of coal production in Brittain, and sending signals to other countries and other energy sectors for following suit. The final closure of the mine is without doubt a significant step, both for the climate effect and as a signal to others for walking the talk. The "stickiness" problem remains, however, as long as the UK (and other wealthy countries) continue to increase the consumption of goods and services produced offshore, based on coal and other fossil fuels. Based on what he calls "coal incorporated into imports", Fressoz estimates that the UK is consuming at least 50 million tons of coal, as opposed to the nine million in official reports.

A breakthrough study of the "transfer phenomenon" was presented by Wu and Chen (2018). Using input-output methodology, covering globally 6 main uses, 188 geographical regions and 26 economic sectors, tracking the uses of coal from extraction to final applications, the authors came up with the first ever findings about the "transfer phenomenon" of "embodied" coal.

First of all, coal is a prime energy source in the global primary energy mix, providing in 2015 for 29 percent of the total energy use, 41 percent of global electricity production, and near 50 percent of global GHG emissions. At the same time, it was also an essential input in more than 40 percent of global industrial production, classified as "embodied" use. In total, embodied coal represents some 71 percent of global extraction. While about 50 percent of global coal production is used to fire electric power stations, more than 40 percent of the electricity produced by coal is used for industrial production, making the industrial sector a main indirect coal user, sustaining the embodied part of commodity imports, consumption, and GHG emissions.

Secondly, coal is likewise a prominent good in globalization, with some 70 percent of extracted coal traded internationally – either directly in raw form or indirectly in embodied form, with embodied export/import representing about three-quarters of the total in international coal trade.

Furthermore, embodied coal import is predominantly a matter of wealthy countries importing regular commodities (such as mobile phones, textiles, automobiles, food and household goods) from NICs, including China as the leading "world factory" nation. When reporting about the green transition, however, the importing, high-income nations report regularly about domestic reduction in coal use, without reporting about corresponding coal use increases in lower income, exporting nations. The climate-political pressure is put on the latter group of nations, whereas consumption in the wealthier nations is the main driver in the international system.

Mainland China is the largest overall user of coal, but also the world's leading commodity exporter, with embodied coal as a key ingredient. The Chinese per capita use of coal by households, is only a quarter of that of the US. When including embodied coal in commodities import, the US surprisingly turns out to be the world's leading "importer of coal". And even the decarbonizing UK has a higher domestic per capita consumption of coal than China – when accounting for embodied imports. Most countries are both importers and exporters of embodied coal. The US is the largest net importer, with embodied coal accounting for nearly 50 times the amount of direct coal imports. Japan is the second largest importer, followed by mainland China, Germany and South Korea. Regarding embodied exports, mainland China, the United States, Russia, Australia and Germany rank top five.

Generally, the pattern of embodied coal exports and imports (similar to embodied oil, nuclear energy, solar and wind energy) is about advanced economies imports and emerging economies exports, and it is not accounted for in climate-statistics. It is the essence of the "transfer phenomenon" of coal. Overall, the indirect trade of embodied coal amounts to more than the double of direct physical coal trade, representing a corresponding addition to the environmental burden. National climate statistics showing GHG emission-progress are generally grossly misleading due to embodied energy imports. Also, many of the factories in emerging markets that export embodied coal and other fossil fuels are owned by multinationals headquartered in the leading economies, offshoring the dirtiest operations as a built-in strategy.

Symbiotic energy

The provision of a new form of energy does not automatically lead to replacement of existing forms. There is a difference between replacement and addition, between transition and symbiosis, and failure to note the difference is definitely counter-productive, subversive of climate-saving policies and action.

While creative destruction, disruptive innovation, and exnovation and whatever stagist models appear to be relevant at the level of commodities and practices, suggestive of a more general pattern of transitions, stagist thinking does not seem to apply to the primary energy mix. While the case of coal is perhaps the most serious in terms of climate impact, Fressoz (2024) comes up with a series of historical examples. Also, small steps towards an energy transition seem to generate unfounded optimism based on among others a tendency for "recency bias" (York and Bell, 2019) – making us overlook the permanence, stubbornness, and stickiness of the full primary energy mix.

Of course, in relative terms the "incumbent" energy forms (such as wood and biomass) have tended to loose market share with the arrival of alternative energies (such as coal). But in absolute terms (the only thing that matters for the climatic greenhouse effect and for absolute decoupling), the energy incumbents have held ground and even progressed. While part of innovation history is about new things that did not work, the main story of more and more, is about the robustness of energy forms in competition, and mainly in symbiosis with the new. Like for new chemical entities that are proven to be toxic and carcinogenic, prohibition does not generally mean that production is phased out, only that the ingredient is used for some other purpose. As for prime energies, competition from new entrants does not generally lead to substitution, but rather to symbiosis and in many cases continued growth.

While Nobel laureate in Economics, William Nordhus, has claimed that the first major leap in lightening technology since Babylonean times, was the introduction of gas lamps in the late 18th century, Fressoz documents that existing oil lamps and candles, followed by stearic candles were simpler, safer, cheaper, and vastly more efficient than gas lamps based on coal. There was no phase-out stage.

The case of whale oil is a curious example. Whale oil was no big primary energy source before the coming of mineral oil, and it was not replaced by mineral oil. Coal and mineral oil did not save the whales. Instead, steam and later diesel engines caused a reduction in the costs of hunting for whales, with great efficiency improvements, greater navigational safety and greater productive capacity for the production of whale oil. Instead of "creatively destructing" the whale oil business, fossil oil caused whale oil to be competitive for a number of new applications, including margarine and (aircraft) engine lubrication, threatening extinction of whales. What saved the whales, was the engagement of activists, the change of public opinion and finally the introduction of regular prohibition in the 1970s.

Causality issues

Transitional, stagist ambitions meet with stickiness. Fossil fuels are stubbornly holding ground against the green transition, in spite of international agreements and promises. Some historical explanations are quite obvious. The global population has more than tripled since 1950, causing overall growth and demand for energy. The use of traditional energy sources is prolonged as developing countries are using more context relevant energy (Schumacher, 2011 [1973]). Prolongation is also a matter of stubborn competitiveness. Nuclear power is still not commercially competitive – mainly for security and waste management reasons, enhancing the continued competitiveness of traditional safer primary energy sources.

Probably more important than commercial competitiveness is the issue of symbiosis. As various energy forms are combined, and also used for feedstock purposes, and for combinations of both energy forms and feedstock materials, elimination becomes increasingly costly and technologically challenging. Symbiosis is basic to what is otherwise referred to as a "lock-in", or in various connections as "path dependence", "debt traps", "institutional dependency", or "maladaptation" (Goldstein et al., 2023). The essence of lock-in or path dependence is that an initial choice of technology or strategy, at the outset fully reversible, tend to make the organization, or whole societies more and more "path dependent", at later stages making it more and more difficult to depart. Added to the initial investment or strategy choice, are investments in training and competence, machinery and process capabilities, technological and marketing specialization, business systems, patents and networks. In 1967, Sweden made a quite demanding change from left side driving to driving on the right side of the road. Still, the transition was technologically successful and economically acceptable. For the UK to do the same in 2024, on the other hand, would be near impossible. The complexity and capital investment costs of modern traffic systems would simply be too great. As for company-level lock-ins, the case of Kodak has become a classic, ending with bankruptcy in 2012 after 20 years of struggling to break out of the lock-in. For a climate green transition, the case of fossil fuels is the prime example, and for environmental toxins, the stickiness of chemicals, plastics and new entities are prime examples (Blumenthal et al., 2022).

To the impact of symbiosis can also be added the "rebound effect", the "Jevons effect" or "Jevons paradox". The English economist, William Stanley Jevons, in 1865 demonstrated that technological advances that increase efficiency in the use of a resource (in his case coal), which would imply a reduced consumption of the resource in question, often contribute to increased consumption instead. Increased efficiency in the use of coal led to increased demand for coal, because coal became a more cost-effective and competitive means for heating. Similarly, overall improved energy efficiency across primary sources and use areas, would imply a reduced need for energy. But with greater efficiency, demand has instead increased. A modern living-room wood stove now converts energy in firewood to more than 80 percent heat energy. A generation ago, similar stoves would convert to some 30 percent heat. Normal economic thinking would suggest that households would now use significantly less firewood. Instead, with inflationary electricity prices people are (where relevant) switching to firewood for heating, with no creative destruction, and good outlooks for firewood forestry and stove fabrication.

To the general "Jevons effect", Fressoz (2024) add the impact of primary energy combinations and feedstock combinations. Stickiness is not just about one source of primary energy at a time producing "rebound effects". It is as much about combinations of energy forms, materials, technologies, and business practices, responding to disruptive as well as to accommodating innovations, new products and services, enabling fossil fuel users to seek new opportunities and growth rather than transition.

Finally, there is a strong emotional side to adding rather than substituting. Some people enjoy the smell of petrol, the sound of a V8 engine, or the atmosphere of a gas station, and prefer a fossil fuel automobile over an electric vehicle. And the same may hold for endless products and services that make up fossil stickiness. Charles Eisenstein (2018) has argued convincingly that the love of nature, and the threat of extinctions are more decisive for environmental action than mathematical models and climate predictions. In a similar vein, the love of a vintage automobile may be more pivotal than emission numbers in the preference for transportation.

Stickiness in space

Vaclav Smil (2022) has identified four energy- and materials-consuming sectors, deeply engrained in the modern economy, of utmost importance in sustaining modern living standards, and extremely difficult to fully electrify or decarbonize.

Based on indispensability, ubiquity, and demand, he singled out ammonia, plastics, steel, and concrete as "the four pillars of modern civilization". Common to them all are that, even if green electricity was available in great and cheap supply, a green transition will require the buildup of entirely new production systems, still to be started, and for many uses based on technologies that do not yet exist.

Starting with ammonia

In wilderness and traditional societies there has been a near-perfect recycling of organic nitrogen – essential to plant growth. Still, there was great variety in soil productivity. The justification for adding synthetic fertilizers to agriculture was to supply of more nutritional elements that were locally in short supply, thereby increasing food availability. With the "green revolution" of the 1960s came new breeds of wheat and other grains, more robust against wind and weather, faster growing, and with the added advantage of tolerating large doses of fertilizers. The result was a quite dramatic increase in harvests, producing for the first time enough basic food in less space, to supply a gigantic population increase. In addition to better crop varieties and fertilization, mechanical equipment of all kinds, irrigation, and better overall management also explains the green revolution. And of course, for all steps of the food production process, for field machinery (tractors, combines, harvester), irrigation (pumps), transportation and distribution (trucks), fossil fuels, notably oil, was a primary energy input – as it also was for the production of steel plows and all modern farm machinery, and for the provision of fertilizer and all the chemicals that came with modernization of agriculture. While the input of human labor per kilogram of wheat production in America, in the course of two centuries was reduced from 10 minutes to two seconds, steel plows and inorganic fertilizers were the main instruments, as they still are in all modernized farm communities (Smil, 2022).

Ammonia is not among the most visible targets for greening, but according to Smil (2022), the feeding of 8–10 billion people on present diets and labor input standards depends on ammonia-based nitrogenous synthetic fertilizers. Half of the world's population might be under-fed without the fruits of the Green Revolution. This may be disputed. Proponents of organic and sustainable agriculture seek other alternatives, but it is generally agreed that any transition away from ammonia dependency would take time and huge investments in training and development – it is sticky.

The main technology for ammonia production is still the Haber-Bosch process, invented in 1909, requiring huge energy use for production of hydrogen, purification of atmospheric nitrogen and production of ammonia. Ammonia production today accounts for one or two percent of global energy consumption, three percent of carbon emissions and three to five percent of natural gas consumption (Song et al., 2018).

Ammonia is in short supply in food-deprived parts of Africa, and in too-abundant supply in other parts of the world, causing unnecessary pollution, greenhouse gas emissions, and the eutrophication of lakes and rivers. The critical role of ammonia would be lessened with an alternative, less meat-intensive diet (Willett et al., 2019), and it would also be greatly relieved with lessened food waste practices, and with more of precision farming and organic and sustainable farming. Also, basing production on electrolysis for green ammonia or other green technology would help (Mohanasundaram et al., 2024). So far, however, green ammonia production is significantly more costly than the conventional. For alternative green or blue technology to be used, enormous investments would be needed to build new production facilities satisfying strict safety requirements (ammonia is both poisonous, combustible, and explosive). On the positive side, green ammonia might become an important green-energy carrier, with a higher energy density than hydrogen. Already to-day, 30 percent of ammonia is used for non-agricultural purposes.

The ammonia market is expected to grow by 5,4 percent annually towards 2030, increasing the volume from 213,9 billion tons in 2023 to 313 billion in 2030. Ammonia production is part of the chemical industry, the largest energy consumer and the third largest in terms of CO_2 emissions, following electricity, heat production, and transportation.

Adding plastics

While ammonia is a quite specific product, developed mainly for one specific industry – that of agriculture, plastics comes in innumerable varieties for a seemingly unlimited number of purposes at very low costs. For some purposes, other materials, such as wood or steel may be used, but generally only at a significantly higher cost and often with less suitable solutions. The main problem of plastics is that they are in a sense too competitive, consumed in too great volumes, with too strong prospects for growth. With popularity of use comes huge emissions of CO_2 and other greenhouse gases – both in production, in use, and decomposition, and also physical pollution of land and sea, while spreading micro-plastics and nano-plastics to enter air, water, soil, food, and animal and human bodies.

The main feedstock material in all sorts of plastics is oil. Decomposition of plastics may take 20 to 500 years, depending on choice of material, and exposure to sunlight. Ultraviolet radiation breaks down plastic molecules – emitting CO_2 in the decomposition process.

For several decades, plastics have had the strongest growth of all feedstock materials, representing seven percent of all oil consumption and emitting 4,5 percent of all greenhouse gas emissions. Previous forecasts have projected a doubling of demand towards 2050 and tripling by 2100. Recent updates are even more striking, anticipating a doubling by 2040 and tripling by 2060, as the phasing out of fossil fuel energy for most uses may cause producers to shift production to plastics (OECD, 2022). With overall GHG emissions doubling or tripling, increased production of plastics will add an extra burden on the environment, both on climate and wilderness, in the oceans and on land, threatening living species. And as we noticed in Chapter 3, there is now definite proof that the spread of microplastics and nano-plastics are threats to human and animal health. Our concern in the present chapter, however, is about the stickiness of energy.

Various ideas have been presented for reducing plastics pollution and climate impact, covering the full life-cycle of products, from electricity and feedstock provision, to production and waste disposal (Stegmann et al., 2022). Climate impact might be reduced by using bio-mass feedstock and bio-based carbon sequestration of plastic products, using landfills to bury and store the remaining carbon. More recycling would also improve the situation, and also the combination of bio-mass solutions and recycling in a "circular bioeconomy" approach. Using green electricity for production (instead of coal and gas) and biomass used as feedstock material (instead of oil), plastic waste could be returned to the ground as a carbon sink, and the remainder recycled, making a net zero target attainable by 2050 (Meys et al., 2021; Stegmann et al., 2022). Still, such a transition would take time and enormous investments, requiring strong regulations, taxation, and incentives, adding an extra burden on the greening of electricity, and needing an enormous build-up of infrastructure and logistics for waste management.

To this we may add that plans for recycling have so far been mainly an instrument of greenwashing. The report "The fraud of plastic recycling – How Big Oil and the plastics industry deceived the public for decades and caused the plastic waste crisis" by The Center for Climate Integrity (Allen et al., 2024) documents that for more than 30 years insiders of plastic production have known that recycling is not a feasible plastic waste management solution – neither technically nor economically. Still, they have kept promoting it for 40 years, telling the public that recycling is coming. Plastic is extremely difficult to recycle. Plastic products come in thousands of distinct chemical varieties that cannot be recycled together, demanding elaborate and expensive sorting systems. Large scale sorting appears to be nearly impossible. Also, plastic materials degrade for every turn of recycling, making reuse more than once or twice impractical. Therefore, single-use, disposable products have been at the heart of the industry strategy since the 1950s, while the public has been told that plastics can be disposed of by incineration or deposits in landfills.

A proper solution will also have to include micro-plastics and nano-plastics. Initial environmental concerns about plastics were focused mainly on toxicity, littering and waste – also on the deaths of sea animals and fish with their stomachs full of plastic. Recent concerns have been directed more towards GHG emissions and climate impact, but the former concerns are still of utmost importance. With the added focus on micro- and nano-plastics, all concerns are relevant, including GHG emissions. The overall hurdle at all stages and for all forms of plastics and micro-plastics, is the enormous volume, projected growth volumes, the enormous diversity of plastic-recipes, product applications and uses, the integration of plastics with near all other products – making a properly green transition exceedingly complex, and calling for scaling of infrastructure, monitoring and control orders of magnitude greater than to date (Shen et al., 2020).

Made of steel

Most modern steel is made by use of basic oxygen furnace (BOF) technologies that date back to the 14th century. Coke (made from coal) and iron ore are heated to transform ore into liquid iron by removal of oxygen, and by converting liquid iron into steel by reducing the carbon content. From there the process continues with further forming, rolling, and product making. In 2021 global production based on BOF was 1,381 billion tons, and 563 million tons were based on Electric Arc Furnace (EAF) for recycling of steel scrap. More than 3,500 varieties of steel are available, reflecting specific qualities for specific purposes. Carbon steel covers 90 percent of the market and is used for most of construction and everyday purposes. Alloy steel incorporates other metals for addition of particular qualities, such as hardness, strength, ductility (stretching ability), or elasticity. Stainless steel has added 10–20 percent chromium for protection against corrosion. And tool steel has extra tensile strength – resistance to breaking under extreme pressure, such as the grinding or cutting of other steel items, concrete, or granite.

Steel beams can carry loads 15 times greater than columns made from granite. Its tensile strength is about seven times that of aluminum, and its hardness four times greater. Aluminum melts at 660°C, whereas steel is heat resistant up to 1425–1540°C. This all means that steel is used for innumerable purposes, in extreme quantities and in nearly all environments. It is the most widely used metal, and it is also the most widely used metal for making machinery – including machinery to make other metals (Smil, 2022).

Steel products are found everywhere, from the tiniest hospital instrument parts, via bicycles and automobiles, to the largest suspension bridges and the highest skyscraper buildings. It is also a core material for ocean transportation, and for all kinds of machinery, for oil refineries and modern windmills. And while airplanes and modern high-speed trains rely on aluminum and plastics to save weight, steel is used for wheels and axels, bearings and motors, for landing gear for airplanes and tracks for trains.

Iron ore, the feedstock of steel, is abundantly available. Annual production amounts to some 2,5 billion tons, including more than 500 million tons of recycled, scrap metal. Estimated global reserves amount to 800 billion tons. Like for plastics, the extreme quantities of steel and the innumerable varieties of products and applications, is an enormous industrial strength, but in terms of the green transition a giant hurdle. Steel can by no means be "phased out by 2050". Instead, industry plans point towards a doubling of capacity by that date, mainly in developing countries, with prospects for a corresponding growth in GHG emissions. Steel production accounts for 7–9 percent of global GHG emissions – and it is said to be the world's most polluting industry. Unless greener alternatives are developed, steel production alone would be enough to sink the net zero ambition for 2050.

Highest on the list of steel-producing countries is China, with more than half of global production, followed by India, Japan, USA, Russia, South Korea and Germany. Generally, steel production is expected to stabilize or even decline in industrialized nations and increase significantly in developing nations, adding to the "transfer phenomenon" explained above.

For a green transition, initiatives for small-step energy efficiency have so far produced meager results. Carbon capture, storage and use (CCSU) has proved difficult. Hydrogen-based solutions (Vogl et al., 2018), increased use of recycling, and EAF (electric arc furnace) technology are on the list of future solutions.

So far, the steel industry and national authorities have produced promising long-term, high-aspiration targets, counting on reduced costs and general availability of green electricity for a green transition. A recent study found that medium-term CO_2 reduction attempts have not yet caused any reduction in global, coal-based steel production, and overall the industry is not on track to fulfill long-term promises (Arens et al., 2021). Adding to the overall problem, there is also a need to align the greening of steelmaking to local and national contingencies (Ledari et al., 2023). Steel is definitely a core "pillar" of modern economies. It is not yet ready for large-scale greening, and the countries with most urgent needs for green steel growth are not the ones best prepared for implementation.

A concrete turn

Steel is often used in combinations with concrete, generally with overall three times as much concrete as steel. But since concrete production is less energy intensive, total GHG emissions from the two are roughly the same – at around 7-8 percent of the total – also making the combination of the two responsible for about 15 percent of global total GHG emissions. And, as for steel, China is now a proud producer of more concrete than the rest of the world put together, and "not surprisingly, the country now has the world's most extensive system of freeways, rapid trains, and airports, as well as the largest number of giant hydro stations and new multimillion-population cities" (Smil, 2022, p. 98). The present large-scale urbanization of societies would be impossible without the use of steel-reinforced concrete for high-rise apartment buildings and business structures, and concrete has become a standard solution to house basements, tunnels, underground and sub-way structures, and above ground structures such as hydropower dams, airport foundations and buildings, bridges, and highways. Prospects for growth towards 2050 are for the concrete industry-sector at least as great as for steel.

Concrete is based on ground limestone, clay, and other additives, heated to 1,450 C to make clinker, the clinker again ground to make cement, and the cement supplied with sand and water and whatever other reinforcement materials to make concrete. Cement was invented in 1824 by the English bricklayer, Josepth Aspdin, who came up with the heating process and the crushing of limestone and clinker. He was also the originator of the "Portland Cement" logo, named after the Isle of Portland in the English Channel – famous for its colorful limestone sources. To this day the material mix and the main steps of production of "Ordinary Portland Cement" (OPC) has set the standard for concrete quality, efficiency, and GHG emissions comparisons.

For the greening of the industry, geopolymer and alkali-activated binders as alternatives to OPC have been proposed, promising significant (50 percent) reduction of CO_2 emissions. Adding Ground Granulated Blast Furnace Slag (GGBS) extracted from steel production, or fly ash from coal-fired power plants, as well as other (toxic) cementitious waste byproducts, and recycled concrete as raw materials – can bring both higher quality, lower CO_2 emissions and lower costs (Assi et al., 2018; Sivakrishna et al., 2020). A limiting factor is the high transportation costs of high-volume concrete, driving a need for colocation of slag or fly ash sources (steel works or coal-fired power plants), OPC producers and the markets for concrete.

Alternative strategies for greening of the concrete industry have been proposed in the "Cement Technology Roadmap" from the International Energy Agency – IEA, and the Cement Sustainability Initiative – CSI (IEA, 2018). The report suggests a 24 percent reduction in GHG emissions by 2050. With a 7-8 percent contribution to global energy use, as a second among industrial CO_2 emitters, and with prospects for further growth, the concrete industry is not just one among many industries at the core of a green transition, it is a key industry. For cement, the clinker content has to be reduced, in whole or part replaced with alternative materials. For infrastructure and all kinds of buildings, the overall use of concrete can be reduced through better designs and life cycle planning, reuse and recycling, and by better architectural solutions for reduction of the heating and cooling of buildings.

Of course, for all use of concrete, the risk factor and the regulations and safety specifications are of extreme importance, both in relationship to earthquakes and natural disasters, but also to wear and tear and long-term material decomposition. In a critical article, van Deventer, White and Myers (2021) have highlighted how present regulations, normally referring to conventional OPC technology, rather than to performance-based standards, tend to hamper innovation and protect incumbent market leaders, prolonging path dependence and technological lock-ins, undermining effective greening. A full transition to a green concrete industry would necessitate a near total overhaul of the sector.

Overall greening of primary energy and feedstocks

We have considered four "pillar" industries in modern economies, and they turn out to score high on stickiness. Still, the complexity and volume of a modern economy rest on much more than the four "pillars". Agricultural activity and food production is about more than fertilizing and spraying. Plastic and petrochemicals are large sectors for oil, but the chemicals industry is still much wider. Steel is an incredibly important metal, but for innumerable products and processes beyond steel, we need aluminum, copper, cobalt, lithium and the rest of metals that are essential both for mass consumption and for specialized products, including battery production, electronics, electrification etc. And while concrete in many cases could be replaced by wooden structures (including 20-story buildings), large scale replacement would not come without enormous transition costs and environmental limitations.

Beyond the pillars are huge sectors, such as the carbon-emitting and toxins-spreading textile industry (in terms of GHG emissions on par with automobiles), and beyond the specific industries, the whole of manufacturing and handling of physical objects, requiring energy and materials for production, transportation, marketing, and application.

For an assessment of the costs and potentials of a full-scale green transition, we would need an estimate of the non-electric energy uses in the global economy. We have noted that only 20 percent of global energy is by now electrified, and that only about 40 percent of electricity is characterized as (relatively) green. The remaining 80 percent of (non-electrified) energy use, is predominantly fossil and non-green, and much of it is hooked into the stickiness sectors.

For the four "pillars", Smil (2022) referred to indispensability, ubiquity, and demand as criteria for focus – in my own terminology, they are also strong indicators of stickiness. Given that stickiness is an enormous challenge, we may want to sort out and prioritize production of products that give the most utility, not only for the money, but for the overall environmental burden. Products that are highly energy-intensive, emission-intensive, and on the luxury/conspicuous side would need more critical examination. Carbon taxes, nature taxes, and income- and wealth taxation, would do much to sort out the least important parts of consumption.

In this chapter we have seen that fossil energy is not easily replaced in a modern economy, whether for combustion or non-combustion purposes. For the next chapter, we shall first consider the prospect of GHG capture and sequestering, which are bleak. Next, we shall consider the promises and hurdles of alternative energy forms, including the problems of balancing the supplies (the "*Dunkelflaute* problem"), and the added environmental burdens of heavy investments in green energy and battery functions over an extended build-up period (the "cannibalism problem"), before concluding with the costs of materials problem (the "*Teufelkreis* problem").

CHAPTER 5
Technology and Energy Gaps

The annual, global emission of CO_2 and CO_2 equivalents (including other climate gases) amount to some 40 gigatons (billion tons). Of the total, nearly 50 percent of emissions end up in the atmosphere, contributing to the green-house effect. Oceans absorb near 25 percent of the emissions (and capture 90 percent of the excess heat from the greenhouse effect, getting warmer and more acid in the process). Oceans also produce 50 percent of all oxygen on Earth. Forests absorb nearly 25 percent of emissions, and the rest is absorbed mainly in the ground.

The ground (including soil and peat) is fantastic for storing carbon, but slow at storage expansion, and it is presently in many places at huge risks of leakage. Intensive farming and deforestation cause erosion and loss of carbon storage in the ground and in the forests, while reforestation and organic farming contribute to stabilization. Relatively more carbon is stored in trees and biomass above the ground in tropical areas, while relatively more is stored in the ground in northern and colder areas. The following discussion is mainly focused on carbon capture strategies. At the end of the chapter there will also be a presentation of some direct climate cooling (DCC) strategies.

Greenhouse gas removal strategies

Sooner or later, fossil energy sources will, for climatic and environmental reasons, either have to be (voluntarily) left in the ground, or they will (necessarily, at a later stage) be physically exhausted or at least of a quality not worth the extraction costs. For coal, such a definite close-down date would come in 100–150 years, for oil and natural gas it might occur some 50 years from now. The mitigation versus adaptation debate is about time and effort for a voluntary phase-out. And the challenge of the whole energy issue is about the feasibility and costs of a phase-out by 2050.

Of course, the easiest and most elegant alternative for fossil fuels would be to skip the whole (voluntary) phase-out scenario, by simply capturing CO_2 at source, and sequestering existing "surplus" CO_2 from the atmosphere by some kind of direct air capture. If the technology for greenhouse gas removal (GGR) were in place and in perfect order, we would not need to worry about fossil fuel emissions at all.

Capture and sequester

Present natural carbon stores in the oceans and in the forests, in the ground and the air, are already overwhelmed. Surplus, non-stored CO_2 is dumped into the atmosphere, causing climate changes at dangerous levels, approaching critical tipping points. It is generally accepted that we either have to speed up technologies for capturing and sequestering atmospheric CO_2, or we have to phase out the burning of fossil energies by 2050. Most likely, we will have to do both, but the experience so far is that carbon capturing technologies are contributing very little towards a solution.

There are several options for CO_2 capture. Existing "pre-combustion" technology is already in use by refineries and chemical plants to convert fossil energy into a mixture of hydrogen and

DOI: 10.1201/9781003518815-6

CO_2 before combustion. "Post-combustion" technology is used in CCS and CCSU (carbon capture, storage and use systems) to extract CO_2 at source from the smoke of an incinerator. "Oxyfuel combustion" technology uses almost pure oxygen combustion to facilitate carbon capture. And DAC technology (direct air capture) can be used to sequester already dispersed CO2 from the atmosphere.

The German author Ulrike Herrmann has made some CCS calculations (2022), concluding that "Das CO_2 wird nicht verschwinden " – "CO_2 will not disappear". First of all, CCS type filtering takes considerable amounts of energy. In open air, carbon dioxide is a thinly spread gas, presently at about 420 units per million units of air. When captured at source (by CCS in powerplants, cement factories, steel works and the like), the gas is much more concentrated, at about 14 percent. But even so, in a coal-burning power plant, capture might require an additional 20–30 percent of burning material – adding to the existing capture-burden. Also added to energy loss, are another 10 percent for transportation and safe storage. On top of this, the filtration technology is far from fool-proof, capturing at present only around 60–70 percent of CO_2. Herrmann calls it a zero-sum game – and a very, very expensive one at that. The efficiency is expected to improve, however, with hope for 90 percent recovery and even more in the future.

Concerning the CCSU alternative, there is no viable market for CO_2, and there is very little potential for large scale utilization. Period! Also, there are no cheap technologies in view, and little room for economies of scale. In order for CCS to succeed, and take the burden off fossil fuel consumption, the world would need several thousand industrial CCS-installations. As of 2023, some 30–40 are in operation around the world, and 265 are in the pipeline for development. CCS facilities are costly, energy-intensive installations, and there is no solution for small scale CCS for millions of CO_2 producing engines – in cars and transportation, industrial uses, and consumer products.

Direct air capture (DAC) for sequestration of CO_2 directly from the open air, using some form of "vacuum-cleaner" solvent or sorbent, or electrochemical technology, is even more demanding. As of 2023, 130 DAC projects are at some stage of development worldwide. According to the International Energy Agency only 27 installations are commissioned and 18 are completed – varying in size from 1 ton of CO_2 a year to 4.000 tons, and with a total capacity to remove some 11,000 tons of CO_2 annually – which is near nil.

The main problem with DAC is that the chemical filtering is extremely energy consuming and the costs formidable. While reforestation used for carbon capture costs less than $ 50 per ton captured, the costs of DAC is calculated at $ 250–600. Among benefits, DAC installations are often claimed to use relatively little space, and they may be sited flexibly, not interfering with agriculture and food production, and preferably located near geological storage sites, minimizing the need for CO_2 pipelines (Lebling et al., 2022). On the other hand, such installations are visible in the landscape, audible, nature-disturbing, insects- and bird disturbing, and they cannot be placed just anywhere. They will cause significant local CO_2 losses, reducing local photosynthesis and potentially causing damage to some plant species. Depending on winds and air circulation, the density of CO_2 in the local atmosphere may soon drop after a DAC installation, with reduced efficiency as a result, demanding rotation to other locations, further diminishing economies of scale.

Finally, the storage of captured CO_2 does not come for free. Some oil companies have used empty oil wells for CO_2 storage, with the added benefit of using pressurized gas for emptying the wells. For large scale CCS operations, empty well-storage is of limited use. Transportation of captured CO_2 from locations around the world, and injection into under-sea or underground wells, is both costly and risky. Alternative wells, in empty coalmines, aquifers or salt domes, are equally limited by availability of space, leakage risk, excess salt-water deposits, etc. CO_2 unlike carbon needs near five times as much space as the coal or oil they replace, and storage under high pressure represents a constant risk of leakage – with the potential for oxygen depletion and poisoning when in high concentrations.

An elegant solution to the storage problem comes from the Orca project in Iceland. Mountains and rocks in Iceland are volcanic and perforated, consisting of basalt with a high content of magnesium, calcium, and iron. When CO_2 is dissolved in water and sprayed into the rocks, over a period of two years the gas is transformed into carbonate for permanent, rock-solid storage. Iceland has enough mountains to store all the world's CO_2 output for more than a hundred years – and of course, if successful, volcanic islands and territories in other parts of the world would line up, offering added storage space. Still transportation and the storage operation itself would add to energy requirements, costs, and risk. So far, there is no line-up of storage space offerings, nor of customers ready to invest. The process is way too expensive for practical use. The world burns 6 billion tons of coal every year, creating a need for 39,000 km3 of new storage space a year, corresponding to 600 Niagara Falls (Chu, 2009). In addition comes storage of CO_2 from oil and gas, and of CO_2 already in the atmosphere.

If we do not succeed with CCS/CCSU and DAC, fossil fuel will have to be phased out and replaced with green energy. At present, there is little hope for capture-success, with a total capturing capacity so far of less than 0.1 percent of emissions.

Green energy availability

In an optimistic report, Jacobsen and colleagues (2017) set out to define a roadmap for a smooth WWS (wind, water, and solar) energy transition. The concluding "road map" foresaw an 80 percent transition to green energy by 2030 and a 100 percent transition by 2050, with mainly wind and solar energy stabilizing global temperature rise below the 1.5°C Paris target. The transition would add 52 million jobs in the green sector, while subtracting 27 million jobs in the fossil sector. It would also bring cleaner air and save 4.6 million lives every year, now lost because of pollution (reports from large cities in India indicate an average longevity loss of ten years due to smog – equivalent to smoking 40 cigarettes a day). It would bring added electricity supply to places that do not have it today, full electrification of hitherto non-electrified sectors, and increased energy efficiency by use of electricity-based technologies. An optimal transition would bring stabilization of energy prices, require minimal new land footprints, and enable stable electricity access to four billion people currently in "energy poverty", also reducing the risk of breakdown and terrorist attack by decentralizing the power supply.

Another optimistic study carried out by an international research team led by the Danish wind power expert, Peter Enevoldsen (2019) has calculated that in the whole of Europe (including European Russia) there is both wind and space enough for 11 million land-based large wind turbines – which would be enough to cover the entire world's need for electrical power by 2050 with a total land-based capacity of 52.5 TW. Similar optimism for windmills was expressed in a previous study (Hertwich et al., 2015), noticing that potential global offshore areas for windmills would need less than 0.1 percent of ocean areas.

The problem is that windmill parks near population centers, where the power is needed, and in the wilderness, causing wildlife concerns, raise serious environmental concerns. Also, when located far away from civilization, transportation causes energy loss, and the costs of electricity go up. Very few would bid for wind-powered electricity in the middle of the Pacific. Only a few selected areas, near large electricity consumers, are at all relevant for onshore and offshore wind-park development.

As for solar energy, photovoltaic (various technologies) remains the main solar power source for electricity (Sayed et al., 2023). But technological development related to Concentrated Solar Power (CSP) for both heat and electricity is generating increasing attention – using solar thermal towers, parabolic trough collectors, linear fresnel ("fre-nel") lens reflectors, or parabolic dish technology. The economic performance of CSP may be better than that of solar PV panels (Ahmadi et al., 2018) and the environmental footprint may be less (Desideri et al., 2013). There is a great potential for CSP expansion, but enormous investments would be called for in order for such installations to have a major impact on the energy transition.

WWS and the "Dunkelflaute problem"

The main trouble with both wind and solar power is the fact that it is not always and everywhere windy, and the sun does not shine at night and on cloudy days or in winter time far to the north or the south. Both solar and wind power are energy sources strongly handicapped by down-time. There are 8,760 hours in a year. Land-based windmills in Europe on average operate at full capacity only 1,960 hours (at sea estimated at 3,820 hours), and solar panels only 910–980 hours. Even if a fully green electrification were feasible, and both households and business customers were willing and able to adjust consumption patterns and timing according to supply (and price), there would have to be considerable over-capacity and storage capacity to compensate for down-time.

Without an additional "balancing" energy source, wind and solar energy cannot sustain a stable electricity supply. Wind- and solar power are both intermittent (on-and-off) sources. They have been singled out as winners among green energy alternatives, but they need the addition of a stable, non-intermittent electricity source to balance the supply. Adding water in the form of hydroelectricity for an WWS solution, is ideal for a commercial "balancing" of the electricity supply. So is also nuclear power, however not commercially available.

Another stabilizing instrument has been the installation of power cables between regions with surplus hydro-electric power and regions relying on wind and sun. Such power cables are now installed in the North Sea between Norway and Germany and between Norway and Brittain. The capacity of Norwegian hydroelectricity and of the cables is limited, however. And adding to the hurdles, even in Europe, the high-voltage grid system is far too weak even for national and regional balancing, and there is no outlook for full battery capacity to balance the energy supply by 2050 in a WWS-transition world.

When both wind and solar sources are shut off, due to darkness and calm, which can last for a few days, the Germans call it a "*Dunkelflaute*". In that situation, and also on days with a dark sky and little wind, a grid system based on wind and solar energy will either have to be closed down, or some "balancing" electricity has to be supplied.

The quest for balancing and storage

The main balancing instruments are either alternative green electricity sources (such as hydro-electricity, biofuel, nuclear, tide and wave), some form of "battery" storage, or a helping hand from fossil energy. Battery storage, can be defined according to time span, for short-term (such as lithium-ion batteries), medium-term (such as compressed air), and long-term storage (such as pumped hydro energy storage, and also various power-to-gas solutions, e-methane and green hydrogen). Batteries can be based on mechanical, electrochemical, thermal, electrical, or chemical technologies. Relevant alternatives are electrochemical supercapacitors (with faster charge and discharge rates than batteries, but higher self-discharge rate – loosing 10–20 percent of their charge per day), flywheel storage (storing energy mechanically as kinetic energy, spinning a mass at high speed), compressed air energy storage (CAES), thermal energy storage (using excess heat, especially from daytime sunshine in Concentrated Solar Power installations, to store heat energy in water or materials with a high thermal mass), and several others.

In terms of overall storage capacity, the greatest hydroelectric power source in the world is the Three Gorges Dam in China, built in 2003, with a capacity of 22,500 MW. Next is the Itaipu Dam in Brazil and Paraguay, from 1984, with a 14,000 MW capacity. The largest thermal storage facility, using molten salt, is the Ouarzazate Solar Power Station in Morocco, with an impressive capacity of 7,325 MW – demonstrating the enormous potential of thermal storage. For comparison, the largest battery park is the Manatee Energy Storage Center in the US, with a capacity of 409 MW, closely followed by Moss Landin Vistra Battery (using Lithium-ion batteries) in the US, storing 400 MW (Sayed et al., 2023). Of course, the main storage capacity of batteries is not in large scale battery parks, but in billions of applications in electronic consumer goods, electrical tools, and electric vehicles.

Pumped hydroelectricity for long-term storage requires the availability of huge, elevated basins, where water is pumped into the basin from below during times of high electricity generation and tapped for hydroelectricity production during "*Dunkelflaute*" periods. The idea is simple, and it works – but only in hilly landscapes where such "waterfall-storage" is relevant and where either the construction of huge concrete dams are acceptable, or existing basins, such as former mines are available. The loss of energy from the pumping process is generally considered low, at around 25 percent.

All of these solutions come with significant energy demands and environmental costs, however. This also means that as solar and wind power production is scaled up, balancing electricity sources also have to be scaled up, adding enormous value to hydroelectricity, and opening up for more nuclear power. There is no sense in building huge wind-parks and solar installations, if there are no balancing sources for down-periods. Instead of betting on any one green energy source, or any one storage medium, researchers and developers are increasingly looking for hybrid solutions, tailoring specific storage solutions to specific energy sources, and combining several sources and several storage forms to maximize efficiency and lower the costs.

Other important storage instruments include green hydrogen. Not-so-green hydrogen has been used by the chemical industry, and by oil refineries, largely based on coal and natural gas. Based on green energy, hydrogen can be a relevant future storage medium for long-time and long-distance energy storage, for heavy and long-haul transportation, shipping and aviation, for the chemical industry, and for the production of iron and steel. It is important to remember, however, that hydrogen is not an energy source, but a storage instrument in need of considerable energy inputs, coming with considerable energy losses. Large-scale production of green hydrogen can be envisaged using solar VP in sunny (desert) regions close to sea, for transportation to power stations, for storage and for transformation to electricity close to markets.

The timing of storage

Generally, there is a growing interest in long-duration storage (Cook, 2023). Compressed air energy storage (CAES) systems have been around since the 1870s, and a utility-scale system was built in Germany in the 1970s. Ranked by the US Department of Energy as one of the most cost-efficient long-duration storage technologies available, plans for expansion are coming from Australia, Canada, the US and around the world. CAES technology is quite simple: Heat is generated and stored by compression of air – and released for electricity generation through pressure release in time for use. To compensate for "*Dunkelflaute*" problems in large cities, there is, however, no existing, viable battery solutions.

For intermediate term batteries, used most visibly in electric vehicles, lithium-ion batteries are still the standard-setter, with tripled cell energy density since first introduced in 1991. But while lithium-ion is "a formidable competitor", according to a Nature overview article, there is a "revolution brewing in batteries for electric cars" (Jones, 2024), with Toyota aiming for a 2027–28 release of an el-vehicle with a 1,000 kilometers mileage capacity and a 10 minutes recharging time, swapping liquid battery components for solids. Alternatives to lithium as a basic feedstock, are among others abundantly available sodium and zinc-air (long-duration) batteries. Using regular air as a main material is promising, while concern have been raised about the availability and price of zinc (Michaux, 2021). With enough energy density, future short distance and medium distance air travel may also run on batteries.

The variety of materials and designs will allow for specialized offerings related to price and power. For cars, energy density (minimizing material use and weigh), overall capacity (driving length), acceleration capacity, time for recharging, temperature sensitivity (less mileage in cold weather), and overall lifespan (number of recharging cycles) are all on the priority lists.

Environmental concerns

As for environmental impact of wind and solar, the main concerns are about the initial resource consumption and polluting production, about end-of-life waste problems, recycling problems, and impacts on wildlife during operations.

Production of wind turbines requires huge investments in steel, concrete and rare metals, the extraction of which causes considerable habitat disruption and pollution from mining. Blades are difficult to recycle at the end-of-life stage. They are built from composite materials, and they often end up in landfills.

For solar panels, silicon is abundantly available, but extraction and purification pose environmental challenges as does also the extraction of needed silver and indium. The production process involves toxic chemicals that require careful handling and disposal to prevent contamination. At the end-of-life stage, the recovery and recycling of valuable materials is an expensive and complex operation. Harmful and flammable materials, noble metals and hazardous chemicals, acids and solvents, and significant amounts of energy are used in the manufacturing process.

For both wind and solar power, transportation and installation take additional energy, and the phase-out stage calls for advanced waste disposal and recirculation of suitable materials (Rabaia et al., 2021). Both energy sources produce minimal material waste during operations, but windmills are a threat to (migrating) birds, bats, and insects, and to animals sensitive to noise. More bird-friendly wind-blades and deterrent systems are being developed, but as yet there is no definite solution to the nature-threat of windmills. While the killing of birds and insects is of great concern to naturalists, it is often overlooked that insects' and birds' deaths from smashing into windows, getting killed by cars and trains, or eaten by house-cats every year is many orders of magnitude greater than deaths from windmills (And the numbers killed by cars will not go down even when vehicles are electrified)! Even so, the environmental concerns about windmills are important, and they relate to the full product life cycle.

Energy sunset issues

The concept of EROI – "energy-return-on-investment" – represents the ratio of usable energy acquired from a given source of energy over the amount of energy expended to obtain that energy, with lower numbers indicating a worse situation. The EROI concept was first introduced by geophysicist Marion King Hubbert of the Shell oil company in 1956. Analyzing oil production statistics for the 48 US lower states, Hubbert came up with a bell-shaped curve for oil production, indicating that the same pattern would also hold for global oil production, and as we have later seen – for other non-renewable resources. From a small start, production would increase up to a maximum, or a "peak oil" point, and then gradually decrease as less oil, generally of lower quality and at higher acquisition costs, would be available. The concept, and the physical challenges to producers and consumers stirred great controversy, with "peak" debates popping up at times of high oil-prices. As oil production in the lower 48 states peaked around 1970 and declined thereafter, and the 1973 and 1979 oil crises caused widespread concern, the EROI issue was a hot one.

As the US underwent a shale-oil revolution (or a "shale gas revolution" and a "fracking revolution") after 2008, policy-makers were less concerned about "peak oil" and EROI (Delannoy et al., 2021). Recent projections by the EIA, however, predicting falling production rates after 2024, and eliciting strong counterarguments by industry insiders, have again stirred a heating of the debate.

EROI comparisons

For a long time, the main EROI-related question was about "enough oil", based on absolute numbers of untapped sources, not on the net energy gains from oil production. For environmental purposes, net energy is clearly the more important. Also, for economic and commercial purposes, the net energy calculus of EROI is a good indicator of future gains and losses. If it takes more energy to pump a barrel of oil than what ends up in the barrel, it will not pay to do the pumping job. Similarly, if it takes more energy to build and install PV solar panels than what the panels can produce, there is no net energy gain, and a low EROI.

Since the introduction of EROI accounting for oil, EROI has increasingly been used as a common denominator for efficiency comparisons for different energy forms. EROI does not account for the environmental and social costs of generating energy, however, only for energy expenditure, including energy used for exploration, drilling or mining, pumping and production, and for energy conversion, storage, distribution, and decommissioning.

We have seen that energy expenditure for conversion and storage can be considerable, where especially storage is a great "energy thief" of green energy, bringing the net energy gain down. Also, if the energy expenditure of carbon capture and storage (CCS) were added to the energy expenditure for production, very few coal mills and oil pumps would be in operation. Thus, not all environmental costs can be converted into energy gains and losses, and the net energy gain of EROI can never be the only criterium for ranking and prioritizing energy forms and sources. In particular for nuclear power, risks of radiation, handling of waste, and competence leakage into the military domain have been of greater concern.

For comparison between energy sources, there is no definite EROI cut-off point, but below 7 there is very little reason for extraction, and below 2-3 there tend to be a direct energy loss. A very important consideration related to the green transition is whether or not alternative green energy sources are competitive in EROI terms with the fossil fuels to be replaced. If fossil oil fuel, at an EROI of around 20 is to be replaced by biofuel at an EROI of less than 3, no matter how great the increase of biofuel in world consumption, it will not help the green transition.

For the purpose of more precise assessments, and for the emphasis of the "net energy gain" of oil and other energy sources, the Net-Energy Analysis tool (NEA) was developed (Delannoy et al., 2021; Murphy, 2009; Murphy, 2014). For green energy forms, hydroelectricity is generally scored with the highest EROI, at levels about 80 (near 100 for some installations). Also, after early estimates indicating a level closer to zero (Tyner Sr et al., 1988), a high EROI score has been noted also for nuclear electric energy, generally in the 40–75 range, and with room for improvement. Among the rising green power sources, wind mills are estimated in the 18 to 50 range. For comparison, solar and biomass energy have generally low EROIs, with biodiesel worst at a level of 1-2 (Wang et al., 2021). Solar energy – strongly dependent on location and weather, and with highly disputed claims, has been scored somewhere in the 7–10 range for installation in Switzerland (Raugei et al., 2017), but with significantly better performance in more sunny regions, and with EROI figures pointing upward, reflecting more energy-efficient production processes.

The EROI of fossil fuels, on the other hand, are generally going down. For oil, estimates vary between 4 and 30, depending on location, type of oil, technology, and costs of extraction. Natural gas and coal are doing somewhat better at 20–40 and 30–40 respectively. Shale oil and gas generally have lower EROI scores than onshore field oil, and deep water drilling gives generally a lower EROI than shallow water extraction. As more wells are emptied around the world, and more coal mines are emptied, the overall EROI for oil, gas, and coal is going down as searching and production energy costs are going up.

The study by Delannoy and colleagues (2021) includes a solid review of previous assessments and research methods, as well as the authors' own analyses. In general, the authors conclude that while "peak oil", which is the year of maximum oil production, may occur by 2035, the "net-energy peak", which is the point in time when the amount of energy extracted minus the energy spent for extraction turns negative, expected to occur around 2024. The EROI of oil liquids is expected to reach a low plateau of 6-7, reflecting also the growing share of unconventional oils, such as shale tight oil. Generally, with falling EROI levels, more energy will have to be produced to account for lower source quality, driving the transition from fossil energy to green alternatives. As more energy is lost in the extraction process, less energy will be available for consumption. It means that the "tipping point" for fossil fuels provision is happening now!

Towards "energy descent"

The biggest political controversy today is about the deliberate phase-out of fossil fuels. When wells are empty, there is not much more to talk about – only how to close them down in an orderly and environmentally decent way. The deliberate closing of mines and wells (and the deliberate ending of exploration) for climate and environmental reasons, is much more controversial. It is about "energy descent", the question of whether or not the global economy can transition from fossil energy to green energy without energy shortages and energy limitations on economic growth. The concept of "energy descent" (Floyd et al., 2020) represents the strategic challenges to society when the provision of green energy does not keep up with the (voluntary or involuntary) cutbacks of fossil fuels, exacerbated by economic growth, industrialization, modernization, and urbanization.

The "descent" is about what Europeans are deeply into at the time of this writing, with recurring "*Dunkelflautes*", energy price hikes and increasing costs of heating a home. Behind the issue of descent is also the general issue of the "First Best Principle", what the Germans more blatantly call a "*Teufelkreis*": Generally, the best energy and mineral sources (with the highest quality, easiest availability, and lowest production costs) are tapped first, and the worst sources are tapped towards the end. The "*Teufelkreis*" causes a vicious circle as investors start to seek economically worse, technically more challenging and environmentally more harmful solutions.

The "first best" phenomenon applies to all kinds of metals and energy life cycles, and it is part of "the energy descent" problem. For steel, a long and quite flat production curve can be expected. On the other hand, since the year 2000, there has been a steep rise in copper prices – from $ 1,850 per ton to $ 9,392 in 2022, pointing towards a future of scarcity. Chile has been the globally leading exporter of copper, and her reserves are enormous. Even so, the metal content in the ore has fallen 30 percent over the last 15 years, resulting in higher costs, more waste and more energy use for extraction. And while the costs of wind and solar energy have fallen dramatically over the last 10 years, they are both dependent on noble metals that may see steep price increases as volumes go up and sources are tapped.

Already, the "first best" and "Teufelkreis" problem of noble metals and scarce minerals is being scaled not only to market price effects, but also to a highly sensitive geopolitical level. By 2023, China controls about two-thirds of global production of rare earth metals, making the US under Trump hungry for alternative sources, whether from Greenland or Ukraine.

As for green energy, we are reaching a critical period that raises questions about fast mitigation versus slower adaptation. We clearly have enough fossil fuels in stock to threaten accepted climate targets. Our dependency on fossil fuel has to be voluntarily reduced and ultimately phased out. However, as the EROI of oil liquids go down, and a fast and easy transition to green energy comes too slowly, the question arises whether there will be enough net energy to sustain the economy as it is – much less support further global economic growth, and intensified growth in low-income countries.

The nuclear energetic promise

Many who realize the seriousness of stickiness and the size of the energy hurdles, are now looking to nuclear energy as a "savior" among energy sources, with potential for solving the "*Dunkelflaute*" problem and compensating for green energy storage needs. On the one hand we have recognized the high EROI score of nuclear power by many estimates. The early installations did not have a very high EROI – the Net Energy Gain was not very impressive. Improved EROI prospects for the present generation of nuclear power is a main reason why nuclear power is again on the political agenda in many countries. Also, the development of SMR (small modular reactors) technology has stirred a great deal of optimism. On the other hand, we also have to recognize the significant reasons for serious setbacks to the nuclear power industry.

From a very promising start in the 1960s, the total capacity of nuclear power stations has been scaled down in recent years – for several reasons. According to the World Nuclear Association (https://world-nuclear.org/information-library/current-and-future-generation/plans-for-new-reactors-worldwide), 66 reactors are now under construction (30 in China, 7 in India, and 6 in Russia), and 87 reactors are being planned (36 in China, 12 in India, and 14 in Russia). None are under construction, and none are planned in Germany and the US, and very few are under construction or being planned in other economically leading nations. Overall, 30 nations have plans for one or more nuclear power stations in coming years. Generally, more plants have been retired than started in the last 20 years, reflecting increasing security concerns, phase-out of old installations, increasing costs caused by higher security standards, increasing competition from natural gas and shale oil energy, and more recently from wind and solar energy.

The main reasons for the reduced role of nuclear energy in recent years are security and cost arguments. Among accidents, the three main ones are the 1979 Three Mile Island partial meltdown in the US, the 1986 Chernobyl disaster in Ukraine, and the 2011 Fukushima disaster in Japan – all three of them stirring great scepsis for both social and environmental reasons. And added to the main accidents are reports of some 99 smaller accidents (by 2010), 57 of them since the Chernobyl disaster (Sovacool, 2010).

High and rising construction costs, due among others to higher security demands, causing cost overruns and delays have also been an important economic factor behind the recent sector decline. It is a fact that so far, no nuclear power plants have been developed and run on purely commercial premises. Due to the risks of future uncertainties, all installations have been state-owned or owned by heavily regulated (non-commercial) utility monopolies, where risks of construction costs, operations efficiency, fuel price development, and all risks linked to future waste storage have been assumed by governments on behalf of taxpayers and consumers. Private financial support is also hampered by the fact that planning, commissioning, and the building of the plant itself generally takes a long time, with the construction period generally taking seven years or more.

The latter is also an indicator that within a time-frame of 2050 for climate mitigation, large scale investment in fission-based nuclear energy may not be appropriate. For most of the time before 2050, there would be a time loss for planning and debate, followed by several years of net energy loss during construction, constituting "energy cannibalism" more than energy gain, making the climate situation worse.

Adding to the long-time positive EROI performance is the fact that nuclear power production is mainly without GHG-emissions and normally quite safe. On the negative side, however, is the fact that early proponents and investors of nuclear power tended to quite dramatically underestimate the costs of de-commissioning. Nuclear power stations are not everlasting: With a lifetime of 40–60 years for early reactors, the closing of power stations, such as in Germany after the Fukushima disaster, was not just a reaction to the disaster, but also mandated by the stations been ripe for a post-operational phase of decommissioning and dismantling, at high costs and high risks, undermining the most optimistic of nuclear EROI calculations.

Decommissioning of nuclear power stations is a costly and long-time operation, including decontamination of all infrastructure and equipment exposed to radioactive radiation, using special equipment to clean, collect and safely dispose of materials, and using remote-controlled disassembly of radioactive components not sufficiently cleaned by decontamination, securing temporary decay storage, and finally the full demolition of the plant and the full restoration of the area. Including time for applications to start the process, and for each stage of the process, to secure governmental accept and establish necessary controls, the process may take 15–20 years (as in the case of the Obrigheim nuclear power plant in Germany).

There are presently proposals for the introduction of small-scale mass-produced nuclear power stations, SMR (small modular reactors). Generally, nuclear electric power technology has favored large scale projects for operational efficiency. But standardization and assembly-line production technology has been expected to tip the economics in the direction of smaller units and larger assembly lines.

Again, however, there will be hurdles linked to safety, to the spread of technology, and to handling of waste etc. If smaller units were available in the world markets at competitive prices, they would also be available to all kinds of customers, including unstable and dictatorial governments, underground business actors, warring parties, and terrorist organizations. And with such availability would also come the diffusion of nuclear technology, stimulating nuclear armament and heightening the chance of nuclear war, with radiation and nuclear winter (Robock et al., 2007). There is also the issue of security related to the full value chain of supplies, to a great extent controlled by Russia (Bowen and Dabbar, 2022).

To date, all SMR projects have exceeded time and cost estimates by wide margins. The Vogtle 3 and 4 project in Georgia, USA was started in 2009, estimated to cost US $14 billion and to be operative in 2016. It was completed in 2024 at a cost of US $35 billion. It is now in operation, representing what according to the Associated Press "may be the most expensive power plant ever" (https://apnews.com/article/georgia-power-vogtle-nuclear-reactor-plant-3ef69a9f64f74410ab2dcda62981b2eb). Similar projects in South Carolina and Idaho have been stopped due to cost and time overruns. No new such projects are planned in the US. In Canada, three SMR reactors have been closed. In France six and in Japan four similar projects have been halted due to cost overruns. Generally, private investors are withdrawing from the sector due to the failure of previous attempts. At present only two SMR installations are in operation, one in Russia and one in China. China had plans for adding 20 more reactors, but the plans are now withdrawn.

As explained by Bengt Flyvebjerg and colleagues (Flyvbjerg and Gardner, 2023; Flyvbjerg and Turner, 2018), SMRs are generally less cost-effective than larger reactors. Efficiency is only achieved when operating at full capacity, making them less useful for scaling at solar and wind down periods. Also, while the threat of accidents during operation has been reduced, environmental threats from the full cycle of reactor construction, operations, decommissioning, decontamination and storage remains at least as high as for conventional reactors when multiplied by the much greater number of installations and localizations. Finally, whether for conventional reactors or SMRs, problems of radioactive waste storage is becoming an increasing problem in leading countries like France, Germany, Italy, Spain, and Belgium.

Cannibalistic automobiles

The concept of "energy cannibalism" refers to a situation where the startup period for some energy or product is so demanding in terms of energy or minerals use, that much of the expected gain is lost. In a way it is a sub-category of EROI representing the timing of energy use and energy gain. The transition to electric vehicles (EV) provides an example.

Norway is the number one nation in the transition from internal combustion engine vehicles (ICEVs) to electric vehicles (EVs). For the first decade of the transition, however, electric cars were mainly added to the petrol and diesel lines, not replacing them. EVs were seen as a "small car" (or a "mums' car") addition to the family's transportation capacity. And considering the fact that electric cars are generally more energy and metals consuming at the production stage, the first phase of the EV transition was cannibalistic – not a savior of climate. Today only about one percent of the global automobile fleet is electrified, and for the next phase of electrification of the 99 percent mainly midsize and large cars, the questions of energy and metals use, mileage and longevity, become acute. For such an expansion, there is also little potential for recycling of materials.

Attempts have been made to calculate break-even points for how far EVs have to be driven in order to perform better on climate and environment than internal combustion engine vehicles (ICEVs). The debate has not been settled. It is generally agreed that small EVs have a better full-cycle CO_2 profile than ICEVs, with a low mileage break-even point. For medium and large cars, however, the break-even point is higher, depending of course also on the "greenness" of the electricity used for driving and the total battery capacity installed.

Comparisons between EVs and ICEVs are complicated for many reasons, and often the comparisons are biased according to the pro- or con- EV position of the analyzer. Important parameters are weight and size of car and battery package, electricity source for recharging batteries, inclusion of the full life cycle of the product and recycling opportunities, fuel efficiency, climate/average temperature where used – in addition to overall lifetime and total mileage of product. If electricity is supplied by coal-fired power stations, EVs may never be competitive with a diesel powered car in terms of climate impact (Dillman et al., 2020).

An interesting study demonstrating the huge impact of a long list of 18 "impact categories", determining the relative climate merits of EVs versus ICEVs was provided by three German researchers (Helmers et al., 2020). In order to control for vehicle form and size, they disassembled an ICEV version of a Volkswagon Caddy in their laboratory and rebuilt it for full electrification. Then they took it on the road for performance comparisons. What they found was that, varying the "impact categories", the breakeven point for climate impact would vary significantly. Since ICEVs are more environmentally friendly at production, and EV are better in use, an EV would have to be driven somewhere between 17,000 and 310,000 km before it becomes better for climate than an ICEV, depending on the combination of "impact categories". An important impact factor is the production of batteries, and the researchers found that producing batteries with solar photovoltaic power instead of coal-based electricity, decreased the climate impact of battery production by 69 percent. For smaller and medium-sized EVs, produced with mainly green energy, there will be a decisive climatic benefit.

The main result was confirmed in a report about experimental data from the US Department of Energy's Argonne National Laboratory in Illinois. The laboratory is using a model called GREET (Greenhouse gases, Regulated Emissions, and Energy use in Technologies), testing the effects of fuels and technologies on GHG emissions for midsize cars, small cars and pickups. Testing was based on electricity from the average level of the US electricity mix (of green and fossil). According to Aldhous (2024), the breakeven point for the midsize car occurred at about 18,000 miles (29,000 km), with the SUV and pickup at a somewhat higher point.

For an average driver, that means that the climate benefit of an EV would accrue in less than two years, and the full GHG emissions gain after an average lifetime of 195,000 miles (after 18 years of driving) would lower emissions by 60–70 percent. Most likely, the gain will be greater, depending on a greener future mix of the electricity grid. It also means that while EVs are better for climate than ICEVs at above break-even, EVs are still very far from climate neutral. For every EV produced and used, there is a direct addition to the overall environmental burden! While ICEVs are extremely ineffective in use, wasting most of the fuel for useless heat energy and using only 20 percent for motion, EVs are able to use 60 – 70 percent of the energy for motion. But the production of electrical cars may cause extra emissions of 15–20 tons of CO_2 for batteries.

Given the different environmental cost profiles of EVs and ICEVs, the transition to EVs may also elicit different patterns of purchase and use. The lower production cost of ICEVs would generally encourage the production of a greater number of cars, and the higher costs of driving would encourage minimizing mileage for each car. The higher initial (economic and environmental) costs of the EV would similarly encourage a reduction in the number of cars produced, while maximizing the mileage of each car, encouraging more shared EV ownership, more borrowing of cars, and more of rentals and taxies.

Cases of cannibalism

Much the same pattern as for EVs also holds for the general energy supply. Calculations have been made of the total environmental impact of green energy compared to the continued use of fossil energy (Hertwich et al., 2015). The production of solar panels requires large quantities of copper and other precious metals, and wind turbines require steel, aluminum, and plastics. The construction of hydropower plants and water storage pools is also not without considerable initial environmental costs. Bringing in the timing element and focusing on the next 30 years of massive green investments, with large up-front emissions and resource use, there will be low EROIs and low NEG numbers for much of the green energy, in many cases of "cannibalistic" proportions, taking a great toll on the "greenness" of the green energy transition. Over the course of long life cycles, however, green energy generally massively outperforms fossil energy. The results are a reminder that the green transition is not only about using other energy forms, but also about using less energy.

Added to the environmental burden are also important changes in product and service demand, reflecting high-income consumption patterns. Smil (2019) recounts how the efficiency of internal combustion engines has been doubled over the last few decades, but at the same time the average car mass in Europe has been more than doubled, with the bestselling VW Golf weighing in at about 1,400 kg. Heavier cars means that combustion efficiency gains have been more or less nulled out. Even worse, the "mobesity" problem continues with the transition to electrical vehicles. According to Christian Brand (2024), while 14 percent of new cars sold are now electric, 35 percent of them are larger, heavier and more powerful SUVs, severely reducing the climate benefits of the transition – requiring more resources to build and increasing the energy needs and related emissions. Heavier cars can carry heavier batteries, extending the range, increasing the power, and reducing range anxiety of drivers, while on the negative side demanding more lithium, cobalt, and other environmentally invasive raw materials. Heavier vehicles also produce greater wear of tyres and roads, and more harmful air pollution.

Similarly, the efficiency of household heating furnaces has been more than doubled, but at the same time the average size of American houses has grown by 2.5 since the 1950s, and summer air conditions has been added to the overall temperature regulation – often with summer temperature set at a level that would qualify for heating in the winter. Modern jetliners use 70 percent less kerosene per passenger km than in the 1950s, but total travelling has increased ten-fold since then, and the use of private jets has exploded. Add to this that China has nearly cached up with the US in overall economic activity and is striving toward catching up on per-capita economic performance and energy use. As for the future, developing countries, led by India, are striving to follow China and attain Western standards of material consumption, meat consumption, and energy use.

Can a transition to circular economic principles help sustain the green transition in energy? For some products that are suitable for large scale re-cycling, such as steel, and for repairable consumer goods, the outlook is clearly positive. For plastics, the picture is gloomier. But what about EV batteries? 40 percent of the cost of a new EV comes from batteries. 70 percent of the battery costs come from noble metals. It is not only about lithium, but a range of precious metals – nickel, copper, cobalt, manganese, graffiti, neodymium to mention some. While a traditional fossil fuel car takes 35 kg of noble metals, an electric vehicle may need 210 kg! As of 2023, only one percent of cars are electrical vehicles, and only ten percent of battery resources are being recycled. Building batteries for the next 99 percent of automobile replacements, by recycling batteries from the one percent, waiting some 15 years of depreciation, will not work – for very obvious reasons.

Battery production for EVs is only part of the challenge, however. To be added are metals and materials needed for all kinds of transportation and for industrial and household purposes. Windmills require some 10,000 kg of metals per MW, 15,000 at sea. Solar panels require 7,000 kg per MW, whereas traditional coal power plants would require 2,500 kg and gas powerplants 1,200 kg (IEA, 2021). This all adds to the "energy descend" or "*Teufelkreis*" problem and to the stickiness of the environmental burden.

We need to remember that the present contribution of wind and solar energy to the total energy package is only around 2.5 percent – dependent upon what is counted in. A full green takeover by 2050 implies an incredible green growth, very little of it based on recycling and circular principles, and much of it due for a long period of less net energy gain (NEG) and even in many cases energy "cannibalism" – delaying the NEG and the mitigative effect on climate.

A green transition is by many seen as a BAU-adaptation for stabilizing the climate, to be solved by developing green energy, producing and consuming EVs, added to some recycling, and fully compatible with economic growth. We have seen from the Introduction and from Chapters 1–3, that more than just a climate issue, the transition needs to account for nature depletion and extinctions, environmental toxins and pollution. It makes the whole issue more sticky.

From the present chapter we have seen that the energy transition itself is a sticky matter, in time and space. It is not just about green energy growth. For a true transition, we cannot just add green energy to the stock of fossil energy and feedstock. We also have to phase out the old, and we have to go beyond greenwashing to properly green solutions, accounting for the innumerable ways that oil, coal, and gas are interwoven with innumerable products in every modern economy, contributing to GHG emissions and environmental degradation.

We have also explored the challenges related to sequestration and carbon capture and storage. For the full package of green transitions, we are about one percent into the needed change leap, and we are still adding to the environmental problems. There are energy "*Dunkelflaute*" problems, cannibalism problems, and problems of descent to be solved. In fact, we have to identify stickiness problems along the full line of activities and problems that make for environmental threats – what I already have referred to as "the environmental burden".

Direct climate cooling initiatives

Realizing the seriousness of the environmental threats, and in particular the climate changes, technologists and researchers are looking for fast-working solutions for climate change mitigation. Also realized is the lack of political will-power and technological capacity for a CO_2 phase-out in time for a net zero around 2050.

Some cooling initiatives without doubt have served the interests of denial and delay. But there is also a strong case for finding viable cooling alternatives, if other paths to greenhouse gas removal (GGR) fail (Baiman et al., 2024). The authors are also very clear that direct cooling initiative should come in addition to GGR – not instead of GGR.

While geoengineering initiatives often has been seen as an all too easy escape from "real solutions", the Baiman et al. study takes the opposite approach: The climate situation is entirely more serious than normally recognized. The green transition and greenhouse gas removal (GGR) is going way too slow, as dangerous tipping points are approaching. And much of the metrics used for policy decisions, while scientifically correct, are presenting the situation in ways more suitable for camouflage than for agency.

Hidden seriousness

The authors critically ask for a serious scientific reinvestigation of a whole set of direct climate cooling (DCC) proposals. They make no attempt to prioritize alternatives or come up with the most cost-effective alternative. And the list of alternatives is partly overlapping with a still longer list presented by van Wijngaarden et al. (2024), focused more directly on "the frozen North".

As for camouflage, the Baiman et al. study points to several problems. COP (Conference of the Parties) policy is based on averaged global warming metric and net-zero emissions, under-communicating the urgency of the situation. We have already (in Chapter 1) mentioned one such case of averaging, that of combining land surface air temperatures and sea surface temperatures, significantly understating the heating over land, where nearly hundred percent of people live. For more camouflage, averaging the observations over 20-year periods significantly understates the drama of recent, accelerating temperature rises.

Global averaging significantly understates the temperature rises in artic regions, where the rise is more than triple that of temperate regions, adding dramatically to ice-melting. The global average also hides problems of mid-latitude regions, where extreme weather with precipitation, flooding, and duration and severity of heat waves are accelerating at a much greater pace than multi-decadal temperature changes. Also under-communicated are the dangers of approaching, sometimes irreversible and self-reinforcing tipping points.

The direct cooling portfolio

Given the failure to phase out CO_2 and other climate gases (and we have noted stickiness and cannibalistic issues!), Baiman et al. present a series of direct climate cooling (DCC) strategies, added to (not replacing) continued greenhouse gas removal (GGR) and carbon capture (CCS, CCSU and DAC) technologies. Their list includes14 proposed DCC approaches.

1. Afforestation, Reforestation, and Soil and Vegetation Restoration (ARSVR) can increase "evapotranspiration" – the process by which water is transferred from land to the atmosphere by evaporation from the soil and other surfaces and by transpiration from plants, also called "the terrestrial biotic pump". The effect of tree-planting and soil restoration depends on a number of climatic and ecosystem contingencies, normally with significant environmental benefits beyond climate.
2. Buoyant Flakes refers to the dissemination of nutrient flakes into nutrient-deficient warm ocean surface waters. Buoyant and slow release carries a low risk of eutrophication or toxicity, and nutrient-laden flakes may contribute to cooling because surface phytoplankton fed by the flakes are of lighter color than the deep blue sea, improving the Albedo effect and reducing solar energy absorption in the ocean.
3. Cirrus Cloud Thinning (CCT) proposes seeding of wintertime, high-latitude and high-altitude tropospheric cirrus clouds with ice nuclei (such as potassium feldspar or silver iodide) to make these clouds thinner and reduce their coverage. Used when solar radiation is scarce or absent, CCT increases the flux of long-wave terrestrial radiation to outer space, thus cooling the high latitudes.
4. Fizz Tops (Fiztops) are lightweight, solar-powered microbubbles turning to nanobubbles, floating in the top centimeter sea surface microlayer (SSML) of the oceans to reflect incoming solar heat. Fiztops can be stationary to cool a local area of the ocean, such as a coral reef or aquaculture operation, or free-floating.
5. Ice Shields may be used to thicken polar sea ice to uphold the Albedo effect. White ice and snow is highly more reflective of sunshine than the black sea. By pumping sea water on top of existing ice, exposing the water to frosty air and the ice surface, more water is frozen to thicken the ice than when the water remains under the ice.
6. Marine Cloud Brightening (MCB) uses extremely small aerosols (cloud condensation nuclei, or CCN) to produce atmospheric mist to increase energy reflection from oceans. Besides adding to cooling, used locally and for short periods of time, MCB can also be used to reduce extreme weather outcomes.
7. Mirrors for Earth's Energy Rebalancing (MEER) is about deployment of mirror arrays on land to reflect incoming solar radiation and to direct outgoing thermal infrared radiation for maximum cooling effect locally and globally. Mirror arrays can be localized to serve purposes including agricultural production, urban heat island alleviation, freshwater conservation, renewable energy generation, and ecosystem protection.
8. Mixed-phase Cloud Thinning (MCT) may be used for wintertime cooling of polar regions. A mixed-phase cloud contains both icy water vapors (at temperatures below 0°C), ice crystals and freezing liquid droplets. Such clouds are trapping heat in the mid- to lower troposphere, where clear skies would allow for more cooling. MCT is used today to clear the skies over airports. It can be used locally to improve cooling, to halt permafrost thawing and provide sea ice and glacier protection.
9. Ocean Thermal Energy Conversion (OTEC) exploits the temperature difference between surface and deeper ocean waters. Surface ocean heat is transferred deep down into the ocean while replaced with cooler water from the depth of the oceans. The conversion serves to reduce surface temperatures, enabling the ocean to absorb more CO_2 and also to provide green

electricity and increase the ocean energy sink capacity. Oceans already store 90 percent of all trapped energy. OTEC does not offer a permanent solution to the greenhouse effect but may add time to the phase-out of fossil fuels.

10. Seawater Atomization means to generate clouds and humidity in coastal regions, for increased reflection og sunshine, using surplus (nighttime) wind-power to pump water through spray nozzles atop high towers. The resulting spray will serve to increase reflection of solar radiation. Seatomizer deployment is as yet an untested concept, in need of idea development and testing.
11. Stratospheric Aerosol Injection (SAI) is probably an approach that most often comes to mind when mentioning "geo-engineering". SAI is proposed as a means for temperature and climate regulation by high altitude injection of reflective (sulfate) aerosols into the stratosphere, similar to what happens by volcanic eruptions. It works like a shield, reflecting sunshine for direct cooling. Large volcanic eruptions are known to bring down the earth temperature for several months, in the case of the 1991 Mount Pinatubo eruption by 0.6°C lasting 15 months. Like several other direct climate cooling (DCC) alternatives, SAI does not offer a lasting solution, but a form of breathing space at a time of approaching tipping point catastrophe. For prolonged effect, it would take first a rapid buildup of the shield and for the duration a gradual increase. SAI is likely the most controversial among DCC approaches, since the risks are at global levels, potentially causing geopolitical disruption.
12. Surface Albedo Modification of Ice and Snow involves the use of reflective materials to brighten ice and snow to slow down melting. It is not an alternative for global impact, but local brightening can be used to reduce local melting and protect infrastructure, local homes and hydroelectric dams from flooding and destruction. Snow is known to darken over time, and initial testing has shown some positive Albedo effects. But for a lasting effect the Modification of Ice and Snow would have to be kept up more or less continuously. Given the enormous stretches of ice and snow on earth, global scale use would be beyond reach.
13. Tree Planting and Reflective Materials can be used to mitigate the heat island effect in urban environments. Trees provide local shade and cooling. They can also reduce the heat-up of buildings, and they can cause quite surprising local changes in rain patterns. Use of light colors for infrastructure materials, such as pavement and building materials in towns can add to the cooling effect by reflecting more sunlight. With accelerating urbanization, city brightening may at some stage also have a global cooling effect.
14. Tropospheric Aerosol Injection (TAI) is similar to Stratospheric Aerosol Injection (SAI), but applied in the lowest region of the atmosphere, the troposphere – from the earth's surface to about 6–10 km up. The idea is to do for methane (CH4) what SAI is doing for CO_2. Since methane is a significantly more intensive GHG than CO_2, and it currently contributes more than a third of the overall GHG heating effect, while its lifetime in the atmosphere is only about a decade, elimination of the methane effect would be the fastest possible path to cooling. TAI is also a clear reminder that the reduction of methane from meat production is another direct climate cooling (DCC) approach with immediate effect. It is also a reminder that methane emissions from permafrost thawing and polar ice melting is approaching tipping points.

The reader will probably agree that not all of these are equally promising. But successful innovation often comes at later stages of idea development. Also, we need to remind ourselves that beyond ideas for direct cooling are ideas for alternative green energy solutions, at best suggesting that direct cooling will not be attractive, including geothermal energy, and also some mega-level solutions to be reviewed in Chapter 14.

The pro and cons

For each of the direct climate cooling (DCC) alternatives there are pro- and con- arguments. Among opposing arguments, perhaps the one carrying greatest weight is the "moral hazard" argument, suspecting that DCC initiatives might cause overoptimism and a slow-down of GHG emission mitigation efforts. A serious counter-argument, in favor of DCC is that we have no evidence that this will be the case, and that betting all your investment on the greenhouse gas removal (GGR) card, that has not worked to date, poses a greater "moral hazard" than testing out at least the most promising among DCC alternatives. The risk of DCC testing has to be weighted against the risk of not doing it. As we approach climatic tipping points the risk-risk tradeoff seem to be moving fast in favor of DCC.

A second warning is based on the "termination shock" hypothesis, saying that while DCC may be effective for a while, bringing down GHG levels, the continuation of treatment may for whatever reason come to a halt, in which case the climate would warm up quickly, causing more adverse impacts than if DCC had not been in use at all. Again, it is countered that the projected costs of DCC are significantly lower than those of GHG emissions reduction, meaning that at least economically, there is less reason to halt the projects. This is a claim that needs to be substantiated, however.

Adding up the pro and con arguments (and there are many more than the ones mentioned here), it seems to me that the testing for efficiency, effectiveness and especially for environmental consequences is necessary. Not all DCC ideas are equally sound. The weighting of climatic mitigative effect, economic costs and environmental burden will be critical, and more critical testing needs to be a high priority. DCC must not be a cheap alternative for deniers, for fossil industry and tech elites to run away from responsibility. But some DCC solutions may be the instrument needed for all of us to escape a climatic (and ecological) disaster. It is an area calling for more public funding and for both war-economy size mega-projects and world-music style inventiveness (as suggested in Chapters 14 and 15). Many DCC approaches are low-tech and can be pretested or small-scale implements on local and regional scales, at a very low risk and cost.

The total environmental burden

"The environmental burden" is not a definite quantity or a specific number. It is a term with several meanings and applications ranging from business level greening processes, to household waste prevention, environmentally driven disease, or neighborhood impact on health. According to the GEMET (General Multilingual Environmental Thesaurus) the environmental burden is about "any activity affecting the environment or any consequence of such activity which, exclusively or simultaneously, has caused or continues to cause environmental pollution, environmental risk or the use of a natural asset". I refer to the environmental burden in this book as a measure of the sum total costs of transitioning from where we are to a truly green future, including the reach of climate, wilderness and environmental targets.

There is no reason for denial of environmental threats anymore. The science is convincing, and manifest damage is on the increase, adding to proof and reinforcing the warnings, also supported by recent changes to insurance and iproperty finance. Environmental threats related to climate, nature destruction, pollution and poisoning are not at a level of ultimate risk to humanity's long-term flourishing (Ord, 2020), but they are still at an extreme risk for tempered damage, to human and animal health, biological diversity, food production, human and animal habitats, and much of what we have considered progress and enhancement of living standards.

Environmental threats are measured in many ways. At the most general level it is about the carrying capacity of the planet, expressed in terms such as the Earth Overshoot Day calendar. While never a very precise and reliable measure, a global Overshoot Day in late July is a strong indicator that as a global community, we are depleting the stocks, living as if the Earth's resources were boundless, while in fact they are limited, as in a "Spaceship Earth" environment. Of course, improved technology and productivity may help us get more out of limited resources, and we may in many ways replace diminishing resources with the more plentiful, but sooner or later there will be ultimate limits to which we have to adapt. Recognizing that the wealthiest countries operate with national Overshoot Days in Mars and April every year, is a signal that the wealthiest countries and people are most to blame for the global precariousness.

While the Earth Overshoot Day measure is perhaps the most general expression of the environmental situation, thousands of reports about worsening situations, in oceans, lakes and rivers, in forests and farmland, above ground and in soil and peat, are reminders of complexity and deterioration. We have no measure of the total damage threat, but we have some perception of what we are heading for and where we ought to go. We talk about a green transition or a sustainability transition, meaning to enter a state of balance between consumption and long-term feedstocks, between pollution and nature's and human absorption capacity. To enter this other path will take mitigative investments and consumption sacrifice, on top of paying for repairs. The amount of sacrifice, of course, will depend on the availability and efficiency of tools for green improvements, on efficiency of mitigative action, on the costs of damage repairs and on the nature and extent of needed sacrifice and lifestyle changes.

In reference to the environmental burden, there is no implication that we are all guilty of having created it, or that the burden is to be evenly shared, or that we all have an equal responsibility to rid ourselves of the associated environmental threats. Since the introduction of the name "Anthropocene" for our present geological epoch (to follow the Holocene), there has been a discourse about the legitimacy and political consequences of naming humankind as seemingly collectively responsible for the present environmental predicaments (Lepori, 2015), fearing neglect of the varieties of economic interests, power positions, and technology provisions behind the environmental burden. Acknowledging that the proposal for the Anthropocene name in 2024 was rejected by the International Commission on Stratigraphy and the International Union of Geological Sciences (IUGS), however, and that there was no great outcry by neither supporters nor opponents, it seem that the discourse was mainly a waste of time. Obviously, the naming of an epoch after the name of the species that came to dominate that epoch, was never meant to block the scientific investigation of causes nor the imputed consequences of that domination. Neither was the name Anthropocene ever meant to stifle identification of profit interests, power positions or technologies behind the damage.

On the other hand, admitting the role of capitalism and the responsibility of capital owners in creating environmental degradation should not blind us to the fact that industrialization and economic growth – independent of political-economic system, has propelled environmental depletion. The fruits of growth have been envied by most people around the world, and growth-policies have been implemented by regimes whenever feasible. Among alternatives to capitalism, Soviet and East European communism was even more geared for one-eyed industrialization and economic growth, and even more apt to disregard environmental consequences. What is needed today is identification of economic interests and political powers behind denials and delays, as well as technologies with potential for further destruction – but also interests, movements and technologies for mitigation, adaptation and burden reduction. These are issues of main concern in Part Two. For the conclusion of Part One we shall now turn to the role of computers, robots and artificial intelligence (AI) in adding and subtracting from the environmental burden and our triple set of environmental threats.

Chapter 6

The Promise of Artificial Intelligence

Most people have heard about James Watt, and his invention of the steam engine in 1776, initiating the first industrial revolution. Similarly, most have heard about Thomas Edison and the invention of the light bulb from 1879, forming the heart of the second industrial revolution. Fewer may have heard about Charles Babbage and his "difference engine" from 1822, providing an initial step towards the third industrial revolution and modern computing. More people may have heard about Alan Turing, the inventor of computer science, with his ENIAC machine from 1945, while fewer may have heard about the scientists and inventors behind artificial intelligence (AI). And even among insiders there is as yet no agreement of exactly what AI is. AI is often talked about and written about by reference to human intelligence, and sometimes to animal intelligence, but then again AI is only partially similar to human or animal intelligence.

Even for human intelligence it is hard to find an agreed upon definition, while most would agree that it must have something to do with the ability for acquisition and application of knowledge and skills. Agreement about definitions is still harder when accounting for abstract thinking, logical reasoning, contextual and emotional understanding, self-awareness, creativity, critical thinking, problem-solving and planning.

In an brief article (cited by 2923) for layman answers to basic questions about artificial intelligence, John McCarthy, also named "Father of artificial intelligence" suggested the following definition: "(AI.) is the science and engineering of making intelligent machines, especially intelligent computer programs"(2007). McCarthy also emphasizes that intelligence is not a single thing that you either have or not. You cannot say about a machine that it is intelligent or not. According to McCarthy it is more like a set of mechanisms, some of them very well carried out by computers and others not so well. Some of them can be developed and improved by observing human intelligence, but most AI development has come from dealing with real world challenges.

Focusing on AI as a set of mechanisms related to some level of achievement or accomplishment ties in nicely with Howard Gardner's idea of multiple intelligences (2000), and it also comes in handy with our discussion in the next chapter of AI development into human-like artificial general intelligence (AGI).

We do not need to be auto mechanics to drive a car, and we don't need to understand modern computing to use a mobile phone. Neither do we need to understand Bayesian statistics and algorithm-mathematics to use AI and work-robots. An algorithm is a set of steps describing a procedure for solving mathematical problems, but we don't need to know about algorithms to use AI. What we do need to understand, however, is that as much as societies have been transformed by previous industrial revolutions, AI experts suggest that AI is going to transform society several orders of magnitude more than the previous three industrial revolutions combined. There is a supra-level algorithm at work for societal development, where millions of small and large steps of AI-improvement, in all sectors of society, for most products and services, for most jobs and job

DOI: 10.1201/9781003518815-7

processes, and for most of human interaction, emerge to bring about dramatic, qualitative changes – for good or for bad, for humans and for nature.

In the following, the advances and promises of AI will be reviewed, including contributions to environmental and energy problem-solving for a net zero by the 2050. The following will focus on overall consequences of AI in terms of "threats to humanity's long-term potential" and in terms of productivity gains, potentially threatening mass unemployment. The often-mentioned issue of AI job-replacement will be dealt with in Chapter 13, and the future of an AI-impacted human society will be further examined in Chapters 14 and 16.

A thing called AI

When ChatGPT by OpenAI, an artificial intelligence program that generates dialogue, hit the world market in November, 2023, there was a huge awakening about the prospects and promises of AI, of neural networks and learning machines. All of a sudden, millions of people discovered what might be in it for them, for good and for bad – for authors, scientists, academics, artists, for people along the assembly line and for clerks in supermarkets, for land labor and hobby musicians, taxi drivers and teamsters – and for all kinds of professionals in all sorts of occupations.

The history behind AI is long and curvy, with small steps of progression, much "under the radar", first for artificial images, sounds, and speech, as much for entertainment and fun as for economic or scientific gain. Seeking first to imitate the way the human brain works to form vision, researchers came up with computer programs and systems to organize information from digital visual inputs. Forming visual images is one of the more complex operations of the brain, and also of computerized vision. It was soon learnt that computer vision could be used for several purposes – developing photos and films for higher quality and a variety of applications, manipulating images for fun or for change of meaning, also anticipating "surveillance capitalism" and surveillance-based political control of whole populations.

Probably very few using FaceApp or similar programs just for fun, think of it as AI. Further development from vision to sound and music was a short-cut. In music, AI helps the computer improve on the listener experience, washing away sounds that should not be there and heighten the volume of sounds and voices meant to carry the auditive impression. Next, machines were able to produce their own compositions – a thousand melodies fully arranged, orchestrated, and taped in minutes. Today they can recognize, copy, re-create, or create from scratch any face you may think of, and attach any face expression and any verbal message, true or false, in any language and for whatever purpose, for good or for malicious purposes.

Hitting the market

For fun or for entertainment, AI has come to gradually invade the job markets. Gradually and silently, AI has been introduced in advertising, in real estate pricing, financial portfolio composition and stock trading, in military strategy formation (such as by Israelis in the Gaza war), and for medical research and pharmaceutical pricing policies – just to mention a few. Advances of biblical proportions are at work when the lame walks, the deaf hears, and the blind sees, not to mention when the lame steers their wheelchairs by the power of thought.

Over a few years, AI has already produced a pervasive impact. Generative AI such as ChatGPT is taking AI to another level, however, enabling the machine to create wholly new intellectual content. Advanced, traditional computer operations, such as googling on the internet, work by recognizing and discriminating among existing information elements. Generative AI takes a long step forward from mere reproduction, seeking to mimic human creativity by creating new content – going beyond what is already there, to write a new text, create entirely new images, answer questions by producing original answers, and write reports, essays, songs, poems, even new computer programs.

"Generative Pre-trained Transformers", as in ChatGPT, works by pre-training the computer on huge datasets, familiarizing the computer with all kinds of available information, and preparing it for the generation of original content within some relevant informational space known as "vectors". On the way towards a finished product, the computer conducts a huge number (hundreds of billions) of computational operations, fine-tuning language and content, logical sequence and informational accuracy – all the time adjusting content to human language and styles (Ouyang et al., 2022). The GPT-3 version of Chat-GPT was supposedly trained with more than 175 billion parameters from all kinds of text inputs, counting 45 terabytes of data (of 1 000 000 000 000 bytes each). So far, the training has been semi-supervised, with human machine-learning engineers feeding training data to be learnt by the machine. For GPT-3, unsupervised training was introduced, leaving it to the machine to pick inputs and conduct the training process, with the engineers doing some fine-tuning at the end – called "reinforcement learning with human feedback" (RLHF).

AI is available 24/7, needs no food or rest, no downtime or sickness leaves. It comes in some form of computer package or attached to any instrument or robot. And it is available for personal conversations and advice, whenever you need it, drawing upon instant recall and access to every digitalized bit of information. Admittedly, the first years of GPT has contributed to much humor around kitchen tables, with stories about stupid answers (such as "the Vikings were black", "first people on the moon were a black man and a white woman" or displaying an image of a human with six fingers).

Companies investing billions in AI-computerization are not paying for scrap and crap, however. The dangers and challenges of AI for the future, are not about foolish answers, but about the qualitative changes of society in a world of AI perfection, with AI capacity to tell truths and lies, and base advice and decision-making on either aligned human values or un-aligned robot values.

A shocking inn3ovation

Each step in the development of new AI versions has tended to take more and more of computational capacity for training and for operations, suggesting ever-increasing demand for enormous computer centers and electricity. Technology development is seldom about a straight, uninterrupted line of progress, however. It was a shock, when in late December of 2024 market insiders learnt that a rival Chinese AI-program, DeepSeek, was about to be launched. And when it was introduced to the world market on January 27th, it caused stock-market shocks and strategic re-thinking in many quarters. It quickly became the most popular free app in the leading US market, and it caused an immediate drop in the Nvidia (main provider of GPU chips for AI) share price of 18 percent.

The main causes of concern was that the DeepSeek-R1 was said to have been developed at dramatically lower training costs, estimated at US $6 million, compared to at least $100 million for the OpenAI's GPTT-4 model, also using a tenth of the computing power of Meta's modl LLaMA 3.1. Evidently, a team of a few hundred Chinese researchers, with a budget of a few million dollars had achieved more than their American counterparts with several thousand teams and billion-dollar budgets. Also of notice was that the DeepSeek model was supposedly based on less advanced Nividia chips than those of the American competitors – a result of American embargo on the most advanced technology exports.

The DeepSeek company is a 2023 offspring from the Chinese High-Flyer hedge fund, controlled by Liang Wenfeng. Liang had started financial trading since attending university at the time of the 2008 financial crisis. He and his colleagues were involved in stock trading using conventional computing equipment. In 2016, they transitioned to AI deep learning technology, and the High-Flyer hedge fund company was established. By the end of 2017, most investments decisions were based on AI, and by 2021 AI was used exclusively for all trading, mainly using American Nvidia chips for AI. Observing the results from stock trading, Liang Wenfeng, decided to start DeepSeek, picking the most promising brains from High-Flyer to develop generalized AI solutions for the new company. Since investors were skeptical, Liang came to have 84 percent majority ownership of the new company.

Where the American competitors have invested enormous amounts on data centers and training power, DeepSeek was able to expedite development and deliver at a fraction of the costs. The main reason for the efficiency advantage is traced to a "mixture of experts" strategy proceeding by way of "distillation". Where the leading AI developers have used enormous data inputs, requiring astronomic training, DeepSeek is based on selectively using only the data needed and the minimum of training required for specific tasks at hand, with only a fraction of the parameters required for solutions. Where Open AI is the more powerful by a wide margin, DeepSeek's parallel units are powerful enough for specific uses at a fraction of the price.

Among arguments in favor of DeepSeek is the fact that it is open source, allowing for the free use of underlying code, open for modification, viewing and redesigning. On the darker side is the limitation of answers to politically sensitive questions about China, such as what happened at Tiananmen Square, what is the status of Taiwan, or what about the Uyghurs and the Tibetans – which are taboo issues in China.

Related to our topic of AI and the green transition, and AI in the global societal transformation, the main takeout from the DeepSeek surprise, is that AI development so far is just a preliminary glimpse into the future, a small taste of what is to come, and a reminder that disruptive innovations are without doubt in the pipeline. Also, while the potential gains from DeepSeek in terms of data center capacity and electricity consumption suggest that prognoses for dramatic expansion may have been exaggerated, according to the Jevons paradox and the rebound effect, the result may be the opposite. Expansion may accelerate reflecting lower costs, and demands for computer capacity and electricity may increase even faster than projected.

The man-machine idea

AI-development is about progress towards super-computation and superintelligence. In science as well as science fiction, a next level of intelligence amplification can be brought about by modifications of the human biology for greater human intelligence and cognitive capacity, using bioengineering, genetic engineering, or smart, nootropic drugs. Or it can be achieved by building computers with highly advanced artificial intelligence. Or it may be done by combining the two, linking the human brain through some form of human-machine interface (such as scanning and digitalization of human brains), making them ready for a state of either uploading the human brain with AI-information and processing capacity – even longevity (Moravec, 1988), or for adding a human-form of consciousness to the AI machine or robot for agency and autonomy (Hanson, 2016).

Traditional digitalization technology has been seen as at "tool" for people. AI is meant to be a "servant" to humans, and a reinforcement of human intelligence. And with AI becoming strongly integrated in most spheres of life, one may ask, why not take the full plunge, adding AI directly to the human brain by implantation or other unmediated coupling?

The cyborg dream

Ideas for man-machine mixtures have been around for a long time. In 1843, Edgar Allan Poe wrote the short story "The Man That Was Used Up", where he fancied a man fully equipped with replacement legs and limbs. Nearly a hundred years later, Edmond Hamilton, in the 1928 novel "The Comet Doom", introduced astronauts with a mixture of organic and machine parts. First to coin the word "cyborg", were Manfred Clynes and Nathan Kline with the article "Cyborgs and Space" (1960). The cyborg term was further popularized by Daniel Stephen Halacy in "Evolution of the Superman" (1965).

Published at the time of early space exploration, Clynes and Kline referred to a cyborg as a technologically enhanced human being, capable of surviving even in an extraterrestrial environment, functioning as an integrated homeostatic system, combining human and machine elements in one mind and one body. Clynes worked at the time as a chief research scientist, and a specialist of physiological instrumentation and electronic data-processing systems, at the Rockland State Hospital in New York, and he was concerned about space flight survival.

Since then, the cyborg concept has been popularized in books and magazines, movies and scientific articles. Its common feature is the integration of some artificial object or technology with some organism, usually assumed to be an "augmented human being". The artificial element may be a prosthesis, an artificial organ, implant or – computer. Some authors have suggested that most of our modern communication devices (smartphones, tablets, laptops), may qualify to make us all cyborgs, whereas others think in terms of full implantation, with some technological element placed within the body or the brain. Also, while cyborgs are most often perceived with a human appearance hiding a more or less synthetic interior, there is also the "lobster" image of an in-human looking "person" hiding a fully human being inside.

Today we have mechanical robots looking bewilderingly like humans (while still not being able to walk or smile like humans), and we have people who enjoy a better life due to all kinds of advanced prostheses, for improved leg movement, handgrip, eyesight, computer typing, or recovery after stroke. Experiments are also going on with means for "augmentation" of the human being, such as the British artist Neil Harbisson, cofounder of the Transpecies Society, who in 2017 had a cyborg antenna implanted in his head in order to see colors normally invisible to the human eye. Generally, however, the cyborg conception has come to represent the implant of a computer in the human brain, from which rises the homo deus – the godlike man with divine intelligence (Harari, 2017).

The cyborg is here

While experiments go on, and progress seem to accelerate, we may stop to consider what is already straight ahead of us. Most of us will not need complicated brain surgery and the placing of an advanced micro-computer physically in the brain in order to capture the full informational value attributed to cyborgs. Most of us carry on small conversations with ourselves, debating back and forth what to do with a problem, or seeking inspiration for a new song or a fabulous painting. But already today, we are extending the conversation, by googling, soon to be just an extension of AI. Our computer or smartphone listens to our probing and our speculations and answers back in real language. If we prefer, our conversation partner can be a robot, shaped like a friend. For some purposes and some technologies, we may not even have to utter words, it will be sufficient to communicate the thoughts to the machine.

I remember the time when I discovered that Wikipedia had 17 times more content than my twenty-volume Encyclopedia Britannica. Today no-one cares to make the comparison. Everything available to regular googling – from history, demography, and geography, and from all sciences, arts, professions, and news media is only a few keystrokes, or an auditive question away. That also holds for the expansion of our internal dialogues.

At the same time, our openness and direct access to infinite information and recall, also works as openness to infinite lies and manipulation, with AI finding ever smarter ways to bypass regulations and lie detectors, and whatever else regulators and lie detection institutions are capable of generating.

AI for green optimism

It is inconceivable that AI could not be of help with our environmental threats. From far and wide we hear optimistic news about AI contributions to climate control, energy efficiency, circular economic systems, electric grid-systems improvement, wildlife monitoring, disaster responses, wildfire control, or general pollution control.

With AI, scientists are expected to improve climate models and make better predictions and recommendations for effective mitigation strategies. Also, algorithms will be available for automatic optimization of energy provision and grid distribution. Even so, the overall picture is that so far, little has been added to existing models for climate and the environment, and we may for good reasons question exactly how much AI can contribute to solving the problems.

AI greening contributions

In a broad overview article Josh Cowls and colleagues (2021) examined how AI may help expand our understanding of climate change and also improve our efforts to minimize damage. They also point to ethical challenges and risks associated with AI use for greening purposes, while recognizing the quite significant increase of GHG emissions from the use of AI for training and application.

For a better understanding of the climate threat issue, the authors refer to a great number of studies covering AI-use for more accurate prediction of climate change outcomes such as temperature forecasts, oceanic impacts, cloud systems, dynamic atmospheric changes, tropical instability waves, rainfall patterns, water demand, wildfire risks, human and animal migration, and all kinds of weather events likely to occur. Contributions to the combating of climate change include assessment of economic gains from AI-assisted energy efficiency improvements in industrial sectors (petrochemicals, concrete, buildings, food), reduction of industrial pollution, development of sequestration technology (carbon capture from air), low-carbon shipping, and the fine-tuning of electricity grids. AI may also sustain more reliable functioning of carbon trading systems, and more efficient logistics and financial arrangements for electric vehicle sharing systems.

A great number of institutions, labs, agencies, universities, and firms are at work to develop AI solutions for climate and overall greening, and the number of projects are growing at a fast pace. The University of Oxford's Research Initiative on AIxSDGs alone contains 108 projects. 28 projects are related to the UN Sustainable Development Goal 13 (about climate change). A full overview of sectors and areas ripe for "machine learning" was developed by Rolnick and colleagues (2022) showing areas of great promise for AI-assistance of the green transition, also indicating means of use. Among areas, several applications were mentioned under electricity systems, transportation, buildings and cities, industry, farms and forests, carbon dioxide removal, climate prediction, social impacts, solar geo-engineering, individual action, collective decisions, education, and finance. The list presents an overview of relevant sectors and subsectors, and it allows for the assessment of to what extent ongoing research and development is catching up with opportunities for AI-assisted improvements.

According to Cowls et al. (2021), farms and forests are gaining most AI-attention, followed by industry, buildings and cities, social impact, and tools for society. Least interest (fewest projects) were found in sequestration, geoengineering, tools for individual use, and climate-related finance.

For the very broad set of targets for the UN Sustainaable Development Goals, a recent study using a consensus-based expert elicitation method found that AI would be an enabler for reaching 134 targets and an inhibitor for reaching 59 targets (Vinuesa et al., 2020). More important than the number of targets scoring positive or negative impact, however, is the saliency of the targets for key societal, economic, and environmental development factors. Generally, AI development is targeted to the markets and needs of the technologically and economically most advanced nations, less relevant to the needs of poor nations. Also of concern is the general warning that AI may come to serve the needs of wealthy societies and wealthy people at the expense of the poor.

Among risks, Cowls et al. (2021) suggest that using AI for climate mitigation poses less cause of concern than applications in areas where personal data and individual outcomes are at stake, such as in health, news media and criminal justice. Still, there is a chance that, based on training with empirical data, AI solutions may be biased to the benefit of the wealthy over low-income groups. An example is the localization of new charging stations for EVs in locations where (first-user, wealthier customers) have already established a pattern of demand. Another would be to base an assessment of individual transportation needs on smartphone owners' replies, favoring smartphone owners. More sensitive is the chance that AI-support of changing attitudes to climate and nature, or the reduction of "psychological distance" to the climate crisis (Rolnick et al., 2022) may interfere with human autonomy and privacy. It is about balancing the needs for protection of autonomy versus the needs for climate warning and mitigation (Coeckelbergh, 2021). To this may be added the need for warning about climate lies, denials and manipulation.

Generally, most AI-applications in climate- and environment-related areas will use non-personal data, such as meteorological, oceanic, and geographical records, posing little risk of personal autonomy loss. But energy uses such as household electricity demand, energy storage, internet of things (IoT), precision agriculture, and even industrial data for heating and cooling, may require granular data that may include personal information and risks to privacy

For overall impact, risks and benefits to society, a recent international survey among AI researchers reveals that some 54 percent of 4,260 interviewees believe that AI will bring more benefits than damage (O'Donovan et al., 2025; Schwaller, 2025) – a significantly more optimist position than among the general public. The general public in the UK was asked the same general question, and only 13 percent supported the positive claim. While it is heartening to be informed that insiders have a more optimistic view than outsiders, it is still sobering that near half of those now developing AI believe that the technology will provide greater risks and negatives than benefits, reminding us also that it is common for insiders to defend their contribution in more positive terms. Perhaps even more revealing than position on the issue of overall risks and benefits was the finding that 77 percent of researchers (versus 68 percent of the UK public) believed that AI makes misinformation (disinformation, data use and cybercrime) a problem, and that fewer than one-third of AI researchers think the technology should be developed as quickly as possible, favoring a more considered approach with time for mitigation of risks and proper regulations. On the international arena, the Chinese generally offered the most optimistic position about possible benefits.

Among expert-perceived benefits, improved learning and education was at the top, followed by making jobs easier, adding improved health care, making household work easier and reducing working hours. Increased income was at the bottom of the list of positives. Among perceived negatives, the risk of fake news, illegal use of personal data and increased cybercrime were at the top, followed by reduced social interaction and putting jobs at risk. Of least concern was reduced access to in-person health care.

AI energy needs

Among AI climate-negatives, the carbon footprint has been noted by many, while no definite answer to the question of future AI energy needs has been offered. The full carbon footprint of computer power covers the full life cycle, from materials extraction to hardware manufacturing and transportation, to a lifetime of usage, and finally disposal. It can be traced by type of product, by hardware use or by the full ICT sector, which raises significant problems of measurement and comparison. No wonder that the numbers vary a great deal. A much cited figure of 1.4 percent of total GHG emissions was based on data from 2015 (Malmodin and Lundén, 2018). Updated numbers for 2020 suggest that the ICT sector accounts for 4 percent of global electricity consumption and 1.4 percent of GHG emissions (Malmodin et al., 2024). An analysis conducted by the Guardian (O'Brien, 2024) found that data center emissions caused by the large tech firms were "probably 662% higher than big tech claims". On the other hand, disruptive technologies, such as for DeepSeek may significantly modify electricity requirements for AI in the coming years. Again, however, DeepSeek may also accelerate AI developments, according to the "Jevons paradox" possibly causing an extra demand for data centers and electricity. Overall, the prospect for a significant AI contribution to greening and reduced GHG emissions is for an unknown future. The jump in (not all green) electricity demand is for now, making for a significant "cannibalism" effect – as explained in Chapter 5.

There is little doubt that AI will add heavily to the energy burden of computer centers. On the other hand, much depends on the forms of energy used and expected efficiency improvements for new generations of computers and programs. AI developers are presently vacuum-cleaning markets in many countries for green energy to support the building of huge datacenters.

Green energy will bring down GHG emissions from the sector, but unless the overall green energy supply is significantly increased, scaled-up computer-centers will add to the energy transition burden, improving the profile of a high-paying ICT sector, while leaving other, less affluent sectors to report correspondingly higher HGH emissions.

As for efficiency and future electricity demands, there are several indicators of great uncertainty. For computer development, Moore's law, conceived in 1965, predicted that the number of transistors in an integrated circuit (IC) would double at a stable cost every year for the next decade (Moore, 2006). Ten years later Gordon Moore, co-founder of Fairchild Semiconductor and Intel, revised the forecast to doubling every two years (which means an annual, exponential growth of 41 percent) – a prediction that has turned out to be remarkable precise until the present time. The "law" has been used by the industry for production planning and demand forecasting, perhaps also working as a self-fulfilling prophecy. For the 2020s, however, there are strong indicators that the trend is up for revision (Dayan et al., 2021). No exponential growth curve can continue forever, and the demand for computer power for the AI transition is large, so an important question is whether computing facilities will keep up, and at what costs and with what energy sacrifices.

Much will depend on the further development of hardware, potentially enabling more computation at less energy costs, but there is no guarantee that hardware architecture and new devices will keep pace with demand (Hooker, 2021), nor that they will be universally available. Giant tec-firms, leading research institutions, and elite universities seem to have priority access to the most advanced equipment (Ahmed and Wahed, 2020). For others to keep up, AI energy consumption will have to increase considerably.

More efficient algorithms for multiple purpose AI processing may ensure that more can be done with less. In a study of computational demands of five application areas of deep learning, Thompson et al. (2020) concluded that "progress in all five is strongly reliant on increases in computing power", with "voracious appetite for computing power", suggesting that "progress along current lines is rapidly becoming economically, technically, and environmentally unsustainable". Finding that three years of algorithmic efficiency improvement equaled the effect of computing power improvement by a factor of 10, they concluded that "continued progress in these applications will require dramatically more computationally-efficient methods". Such efficiency improvements are in the process of development for a wide range of purposes, and underperforming models are being phased out.

The challenge is one of keeping up with electricity demand, and at best reducing electricity demand while increasing computing power. Efficiency can also be gained by developer cooperation for sharing data, codes, and training results. But competition create barriers to the spread and reproducibility of data and results (Fursin, 2020).

AI greening by sector

Several authors have tried to assess critically the energy burden from AI training and use. Since AI is still at a pioneering stage, there is little definite empirical grounding. Most contributions admit that AI has a potential to reduce overall energy use and develop green energy resources, but there are also strong warnings about the added carbon footprint from AI training, development, and use. We noted above that there are innumerable product areas and sectors ripe for AI supported green transitions, and also many parameters that influence energy use, such as hardware and software advances, competitive hurdles and cooperative arrangements, and computer-center technological development. Cowls et al. (2021) included the considerable carbon costs differences between nations and regions for producing essentially the same training round for GPT-3. Numbers varied from 942,330 kg CO_2 in South Africa, and 858,360 kg in India, to 93,300 kg in France. Admittedly, such numbers are bound to be less than precise, but still it is interesting to compare them to annual emissions from a typical passenger car in the US, at 4,600 kg, suggesting that a single run of GPT-3 emits as much as 50 fossil fuel cars in a full year, remembering also that developers have to train thousands of programs before arriving at interesting results (Yin et al., 2023).

There is no shortage of optimistic voices on behalf of BAU-based green transitions. There is, however, a shortage of a scientific base for such optimism. One area of considerable optimism is the use of AI and the Internet of Things (IoT) for promotion of a circular economy, with increased re-cycling and re-use of materials (Fraga-Lamas et al., 2021). The IoT refers to networks of physical objects, equipped with sensors, specialized software and additional technologies for connection between the "things" and for exchange of information among the network units and among the network and outside data sources. While the authors leave no doubt about their belief in long-time improvements, they claim that "paradoxically", in spite of huge potential, IoTs "are not yet contributing to the sustainable development of the IoT sector itself", since the sector "has a significant carbon footprint due to the use of scarce raw materials and its energy consumption". And even a future AI-supported more efficient version of Green IoT, with the use of Edge AI, where all computations are done "at the edge", by the computers in the system, without or with only a limited use of a central "cloud", may be so energy-demanding as to undermine the promise.

An overview of AI greening potentials was presented by The Economist (Gold, 2021), recognizing the promises presented by advocates claiming that AI "will some day provide an answer to nearly every imaginable question" including climate change and the reduction of GHG emissions. High up on the list of AI contributions were energy efficiency improvements and emission-cutting strategies for concrete and steel production; more precise measurements of carbon storage in forests – also for more reliable assessment of carbon credits (today too much based on faith); more precise assessment of wildlife habitats and risks of forest fires; and incremental improvement of efficiency in manufacturing processes, public-transit networks, electricity grids, and fertilizer use in agriculture.

While the list of areas for greening by use of AI is getting ever longer, the article concluded that there is "little hard data" on real emissions. Also, using AI to cut 8% of emissions in cement production (245 million tons of carbon from annual emissions of 43 billion tons), does not solve the problems of the concrete sector – nor does it mitigate prospects for a 2050 net zero in a world of doubled demand.

Not every climate problem has an AI solution, and sometimes AI may also make things worse. Greater efficiency may allow for expanded production – the Jevons paradox or the "rebound effect". AI-assisted innovation may produce increasing energy demand for computer centers, etc. And while both governments and leading commercial developers are investing considerable amounts in AI for a green transition, we are reminded by The Economist article that climate threatening activities are real, material things that will continue to emit GHGs even when efficiency is improved.

AI development is going fast

AI development is going much faster than even most insiders would expect, and much faster than for regulators to keep up. In 2023, an AI-produced video showing actor Will Smith eating spaghetti caused a great deal of laugh. In 2024, however, OpenAI's cutting-edge, text-to-video Sora program was ready for visual representations based on text prompts (Cho et al., 2024; Liu et al., 2024). Similar programs are in the pipeline for ever more advanced operations – also available for undermining democratic processes and for promotion of large-scale lying.

AI is expected to be a game-changer for improvement in all spheres of life, including environmental and social sustainability challenges. There is a great chance for overestimating the positive potential and underestimating the environmental and social costs of the AI transition, however (Dauvergne, 2022). The operations of AI will in and by themselves add significantly to the "cannibalist" element of the green energy crisis. Efficiency improvements and AI-assisted product developments will come with rebounding effects, and with economic BAU growth, it will add to the environmental burden. Profits from AI, and the abundant proceeds to the experts, in parallel with potential job losses for multiple others may add to social sustainability problems. Studies of the "twin transition" of AI digitalization and green transition tend to show that while environmental technologies tend to reduce GHG emissions, digital technologies, with big data infrastructures tend to worsen them (Bianchini et al., 2023).

Even so, expected AI-assistance for green energy, GHG and waste reduction, alternative materials use, etc., are also bound to contribute to environmental improvements. AI may help facilitate the transition to a more circular economy for plastics and other materials (Sankaran, 2019). Digitalization may also help promote the benefits of Metcalfe's Law – stating that the financial value or influence of a telecommunications network is proportional to the square of the number of connected users. By connecting increasing numbers of people, organizational units, and devices into networks for greening purposes, AI may exponentially increase the value of the network as a function of the number of participants. It may also serve the interest of disciplining the participants for greening ambitions and for making greening a policy priority and an object of more honest measurement and reporting (Ma et al., 2023).

We have a tendency to pose problems in terms of present technological frames of reference, and we have a limited capacity for envisioning future solutions to present problems and in particular future solutions to future – as yet unforeseen – problems. What we can say at present is that there are plans and visions for AI-supported green transitions in most sectors of society, for most product and service areas, all of them legitimized by reference to environmental, economic, and health improvements. Improvements are envisioned for an infinite number of small steps and a smaller number of major technological leaps. There is hope that on balance, both the small steps and the major leaps will contribute net improvements and serve people and nature for the common good.

Lagging regulations

Already, in 2024, the Sora program (an AI program that can create realistic-looking video from text) was seen as a threat to democratic elections around the world. The technology is there for producing any kind of content, linked to any person or authority source, using visual images and voice recognition – true or false, partly made up and partly drawn from historical reality, seamlessly fused together for compelling-looking information and entirely new narratives. In October 2023, a fake audio of Keir Starmer, leader of the British Labour Party was spread, stirring much confusion. Also, a fake audio of President Joe Biden, advising Democrats not to vote, was circulated in January 2024. As the technology is democratized, in the future anybody in society will be able to produce whatever content, in a verbal, visual or auditive form, by a few key-strokes, challenging our ability to discern truth from lies, facts from made-up data, at extremes never faced before. By 2025, the main international social media removed the fact-checking screening programs. The outcome of it all will hinge on ethics and values, and on the effectiveness of public regulations.

Of course, politicians and administrators in all mature democratic societies are concerned about the needs for regulations, and even business developers appreciate attempts to define the playing field and rules of the AI-development game. In December 2023, the European Parliament and the Council of the EU agreed on the AI Act, while the text was still in the process of adoption and translation. The Act was to enter into force 20 days after publication, to be fully applicable after two years, with some measures taking effect after six months, others after 12 months and some after 36 months. The AI Act is part of an EU package of policy measures for the development of "trustworthy AI", for safety and fundamental human rights, but also for the strengthening of "uptake, investment and innovation in AI across the EU" (European Commission: "Shaping Euope's Digital Future. https://digital-strategy.ec.europa.eu/en).

Presented as "the first-ever comprehensive legal framework on AI worldwide", the Act is meant for both promotion and protection, seeking also to reduce the administrative burdens of business and other users, especially small and medium-size enterprises. Even before launching, however, the

Act has been criticized for having a simplistic conceptualization of trust and for the overselling of regulatory ambition (Laux et al., 2024).

Among high risks, the Act deals with the safety of AI-supported infrastructures (such as transportation vehicles), the safety of AI-supported products (surgery), personnel management and hiring, access to private and public services (obtaining loans), law enforcement (reliability of evidence), migration and border control (visa applications), the administration of justice and democratic processes, and all forms of biometric identification – with a full prohibition of remote biometric identification in all publicly accessible spaces for law enforcement purposes, except for searches for missing children, prevention of terrorism, or detection of serious crime. All such ("high risk") AI applications are to be subject to strict risk assessment before entering the European market.

Limited risk refers mainly to lack of transparency of AI use. People shall always be informed when exposed to AI systems using personal data, and also warned when presented with AI-generated content (including audio and video content constituting deep fakes). Finally, the vast majority of AI-systems fall into the category of minimal-risk, for which there is to be free use and free marketing.

The Act is based on the idea of "trustworthy use" of AI, and while the Commission behind the Act recognizes that AI is "a fast-evolving technology" in need of constant adaptation and up-grading of rules, there is no mentioning of the ultimate threats to humanity's long-term potential, as assessed by Toby Ord (to which we shall return in the next chapter). Ord's main concern is about AI in the role of agency, and he notes that the leading voices of concern are not politicians, AI-developers, or pessimists among lay people, but precisely the people who are researching issues of AI agency and value alignment. For the future of AI development regulations in the EU, it remains to be seen whether the initiatives will be moderated due to tariff pressure and negotiations with the US, where a strong pressure for AI deregulations is exercised by the technology oligarchy in and around the government.

AI for skeptics

In an article in the Guardian, climate activist and publisher Naomi Klein claims that "AI machines aren't 'hallucinating'. But their makers are" (2023). "Hallucination" is a catch word used by technology insiders to characterize chatbots serving fake news or misguided information. You may ask the bot to define something that does not exist, and it may still come up with a definition and even add made-up footnotes. It is about machine-made hallucination.

We are being told by tech CEOs, investors, developers, and advocates that AI will benefit humanity and contribute decisively to the green transition. Naomi Klein claims that they are kidding themselves – and the rest of us. They are the ones who are hallucinating. The human brain is capable of perceiving things that do not exist. So are chatbots. And perhaps AI-promoters are doing the same, seeing green transition instruments that do not exist – including instruments for ending poverty, curing all diseases, saving the environment, and solving climate change problems.

Hallucinations

A first hallucination according to Klein, is the claim that AI will solve the climate crisis by helping decision-makers and stakeholders make better-informed decisions. She cites Google CEO Eric Schmidt saying (in an "Atlantic" interview) "If you think about the biggest problems in the world, they are all really hard – climate change, human organizations, and so forth. And so, I always want people to be smarter." But lack of action for climate stability according to Klein, is not due to lack of information or lack of smartness. We have known for a long time that climate action means to leave carbon in the ground, halt greenhouse gas emissions, produce green energy, and let economic growth benefit the poor:

"The reason this very smart counsel has been ignored is not due to a reading comprehension problem, or because we somehow need machines to do our thinking for us. It's because doing what the climate crisis demands of us would strand trillions of dollars of fossil fuel assets, while challenging the consumption-based growth model at the heart of our interconnected economies".

Her argument is that fixing the problem will take a "paradigm shift", and waiting for AI to find good answers is not a cure, but a symptom of the problem. AI may even deepen the climate crisis, using enormous energy, and causing considerable emissions from the speeding up of production, marketing and consumption of environment-degrading products. At an even more challenging level, AI will assist, at an entirely new and higher level, the spread of disinformation, making it more and more difficult to know what is true – also about climate and the environment.

The second hallucination, according to Klein, is the claim that AI will deliver better governance. The optimistic claim is that with access to AI-generated information, politicians and administrators will be better able to make evidence-based decisions and programs. Again, however, it is not lack of evidence, facts, or high-quality consultancy reports that threaten the "glue" of societies, but rather lack of willingness and courage to do what is needed, sometimes in the face of corruption, threats, and near-corrupt lobbyism. The problem is not a lack of smartness, but lack of integrity and willingness to deal with political-economic contingencies behind the failures to act.

A third hallucination is thinking that "tech giants can be trusted not to break the world". In a world where good people are fired from AI-leading firms on grounds of criticizing and warning about AI development, according to Naomi Klein we may reasonably ask whether the same companies, and the people making the most money out of them, will do what is best for the world or – instead – for their own cash flows.

Finally, the fourth hallucination is that AI will liberate people from drudgery. She sketches a "Silicon Valley playbook", including as a first step the creation of an attractive product (such as a search engine or a social network) – giving it away for free or almost free for a few years under the assumption ("lofty claims") of helping people connect or spreading freedom. Then, as people are hooked it is time for targeted ads, constant surveillance, data sales and escalating subscription fees – bringing in enormous profits and leaving people to pay – one way or other. In the case of AI, unemployment and loss of incomes may add to the "playbook."

For a starter, the tech companies might begin to pay for the "training material" used in the AI-learning process, by Klein seen as "the largest and most consequential theft in human history". Inputs in the process are all elements of human knowledge available in digital form, acquired by some of the wealthiest companies in the world without payment, and in many cases used to outcompete the very people who have sweated to create the knowledge – or the paintings or melodies. For a far more transformative consequence there is the political-economic turn to "surveillance capitalism" or "information capitalism" (Zuboff, 2023), where mass-scale data accumulation based on digital behavior surveillance is used for behavior prediction and modification, in service of the tech-companies and for sale to commercial and political customers.

The essence of the hallucinations discourse is that the distribution of costs and benefits of AI will depend on political-economic premises. AI may open up for more leisurely, beautiful and meaningful lives, but also for a worse distribution of incomes, jobs, wealth and power, and for more "drudgery", lies, and falseness in the world. In the end, the outcome is a question of values and ethics, of truth and integrity, and of regulations for sharing of benefits and minimization of damage.

AI risks within reach

Perhaps not too far into the future, we may have advanced AI technology, representing a tipping point for AI agency, referred to as AGI (artificial general intelligence) or a Singularity, a point of definite AI transition. But before discussing the essence and the outlook of passing the tipping point, we should recognize the potential for a pre-singularity AI impact, still as a tool in the hands of people, as a servant – not a master. AI may enable a lot of good in the extension of human capabilities in near all sectors of work and life – but potentially also of bad, extending human capabilities for doing wrong and for gradually or revolutionarily bringing about a dystopian society.

Related to AI, there is on the one hand no doubt that artificial intelligence may add to the technical functioning of society and to democratic rule. The internet and social media may facilitate criticism and protest in defense of freedom and democracy, and rebellion against tyranny. On the other hand, there is also a non-trivial chance of an opposite functioning when adding technology and tools also to anti-democratic and disintegrative forces of society. China is already implementing mass monitoring and surveyance using advanced face and voice recognition, useful for the scientific management of society, but also for repression. The tools are in place for doing the same in all societies, for purposes of planning and control, rewards and punishment, "making enforced dystopia a realistic possibility" (Ord, 2020, p. 155). The objective of political control or commercial gain can also be achieved without the obtrusive use of cameras – enough information will be stored in the form of digital footprints.

Yuval Harari, famous author of "Sapiens" and "Homo Deus", in a speech to the World Economic Forum (2020), from the myriad of threatening issues, selected nuclear war, ecological collapse, and technological disruption as problem areas posing "existential challenges to our species", and among these he chose to spend time on technological disruption – noticing both the enormous promises of technology but also its disruptive potential – especially for the creation of a "global useless class", the rise of "data colonialism" and "digital dictatorships".

There will be new jobs, but Harari questions whether people will be able to learn new skills fast enough to fill them (as for a fifty-year old displaced truck-driver learning to teach yoga for engineers). It will not be a matter of people learning new skills for a one-time job change (as a "single watershed event"). It will rather be a cascade of increasing disruptions, as AI and robotization is taken to new heights. There will be lots of new jobs, but like the old jobs, they will disappear at an increasing speed, taken over by AI-assisted robots.

Instead of exploitation being the main target of class struggle, the new main struggle will be against irrelevance – and, as Harari observes, it is worse to be irrelevant than exploited. They will constitute a "useless class" – progressively separated from the elite. Economic inequality will remain between nations and continue to increase within nations. In the same way that a few nations were first to industrialize, two nations, the USA and China are pulling ahead in AI-development, leaving most countries around the world far behind. Enormous wealth will be created by leading technology firms, tech billionaires, and nations, while other nations will be exploited as "data-colonies", involuntarily feeding data into the system without access to the profit chains.

Dictatorship may take new forms. Harari presents a simple formula, where biological knowledge multiplied by computing power multiplied by data equals the ability to control humans (AHH – ability to hack humans). It is about algorithms that understand us better than we understand ourselves, coming with the ability to manipulate our desires and make decisions for us! By using an eye tracking instrument, an advertiser may identify personal preferences and dispositions, including sexual orientation, and it may target the person with selected ads. But the same technology might be used by the state or by conspirators in countries with a death penalty for homosexuality, or by corrupt leaders and secret police picking out opponents or vulnerable critics. Ability to hack humans (AHH) will be in place when insiders in San Francisco or Beijing will know the medical and personal histories of every politician, judge, or journalist in whatever country, including their sexual escapades, mental problems, or corrupt dealings, setting the stage for the nation to become a data-colony and for populations to be hostages and extorted.

Enough data about relevant people in a nation may eliminate the need for military occupation or use of force for colonization. The AHH works by combining biological knowledge with extraordinary computing power and data-access. Former tyrants did not have the biological and psychological knowledge to do the AHH operation, neither did they have the data nor the computational capacity to do the job of identifying targets. With data-colonization and AI, modern tyrants will have what is needed for political, dictatorial control, and Harari tells us to get used to the idea that we are all "no longer mysterious souls", but rather "hackable animals" – easily controllable by a twenty-first-century version of Stalinism – well equipped to produce "the worst totalitarian regime in human history". Harari foresees a dystopian society where everybody wears biometric bracelets, for monitoring of blood pressure, heart rate and brain activity, but also for revealing individual reactions, such as to speeches by "big brother". If you react with silent anger, even without saying anything, and even when clapping and applauding, your next station in life will be in the Gulag. Also, as under Stalin, those closest to power, will be the most closely monitored.

For every step of development towards algorithmic guidance of our lives, some element of human autonomy is being relinquished. We still may convince ourselves that we have the final say in life choices, but the more we depend on social media for news updates, on Google for fact-finding, and on Amazon for what to buy, the more the decisions are in a sense already made – also about choice of studies, jobs, or marriage.

Increasingly, Harari also reminds us, we will also be less able to understand the computer' decisions, the whys behind a yes and a no: Only one percent of us understand the financial system as it is today, and in a couple of decades the number will be "exactly zero". Not even the best of human financial analysts will be able to follow the moves of the AI players.

At the level of society, philosophers and social scientist already today have a hard time envisioning what either heaven or hell on Earth will be like in an AI-guided world, leaving very little leeway for human questioning of important decisions. There will be naïve visions of utopia, and there will be blind zone openings for dystopia. There may also be an evolutionary transition from natural selection to AI-assisted biotechnological intelligent design, not only for plants and animals, but also for humans. Offspring may be specifically bred for qualities desired for war, business, or other skills –perhaps not for compassion, spirituality, or artistic expression. Breeding of societies for fascist dictatorships, communism, or liberal democracies may be an option.

To the extent that humanity may come up with a conceptualization of proper AI control and of a desirable society, the key to concerted political action is international cooperation. No nation can prevent nuclear war alone, nor deal alone with the environmental burden. Nor can any one nation control AI development or bioengineering development alone. Harari remind us that most nations, on their own, would abstain from developing killer robots or gene modified compassion-deficit babies, but some might say that since we cannot trust our rivals to not do it, we must do it first – the so-called "unilaterialist curse", to which we shall return in the next chapter.

For the purpose of this book, three main issues can be singled out. First is the imperative question of AI development towards technological singularity and what it will mean for human societies. That is the topic of the next chapter. A second concern is what AI will bring in terms of productivity gains, economic growth, and potential unemployment. That is the main content of Chapter 13. Finally, there is the issue of what may happen to what Harari (2020) referred to as the "useless" or "irrelevant class", potentially constituting a population majority in a past-singularity world – referred to later in this book as a "derivative sector". It is about life content and social and political-economic organization in a world where AI and AGI rule the markets, perhaps also government. This is the main concern in Chapter 15 and 16.

Before closing the present chapter, we should also note the possibility that AI development may be faced with a "complexity brake" – that developers may be "up against the wall" for further development. While seven parent tech-companies behind AI development represent near 15 percent of global stock valuation, reflecting progress towards artificial superintelligence, as yet (2025) none of the AI-alternatives on the market, except for DeepSeek, are making money. Even Chat GPT owner Open AI has to finance further technology-development with fresh investment capital – not with earnings from AI, and even Google and Microsoft have a hard time making money on AI. Leading AI developers are also vacuum-cleaning international sources of energy for the development of computer-centers, and they are investigating investment in nuclear energy to support AI development. Beyond the technicalities of operations, they are also striving to find training material for further machine learning development. As for productivity gains and the threats of unemployment, relatively little has so far happened. Perhaps we are up for a "complexity brake" – perhaps not. Even Goldman Sachs, so far a leading optimistic voice for AI and super-computer development, is warning against overinvestments in AI. A likely trajectory is for some AI developers to collapse, for AI users to start paying, for the military users in many nations to become super-customers driving technological development and guaranteeing the incomes of tech-billionaires, but also for critics to more strongly voice the need for regulations.

Chapter 7

AI Singularity and Agency

For the potential impact of AI on humans and societies, Eliezer Yudkowsky (2007) in a short essay on the "Power of Intelligence" compared software singularity to the role of intelligence development in human and planetary evolution. The cognitive revolution of the Sapiens some 70,000 years ago helped change the world thousands of times faster and deeper than millions of years of steady evolution. Using historical population growth as an indicator of economic growth, Robin Hanson (2008) found that productivity had doubled every 250,000 years under pre-historic conditions, with another doubling every 900 years since the Agricultural Revolution, and a doubling every 15 years since the Industrial Revolution, suggesting another dramatic jump in economic growth in the wake of an intelligence singularity, opting even for a doubling on even a quarterly or weekly basis. That doesn't seem quite realistic.

AI, and the potential of artificial general intelligence (AGI) is set to cause greater change to human society and world development than any other intervention. If we asked people in 1924 what they saw as the greatest threats and opportunities for the next hundred years, nuclear war and atomic energy would not be on the list, and neither would robots and artificial intelligence (AI). If we ask people today, what may emerge the next hundred years, they will know still less about technologies, scientific breakthroughs, social and political transformations that will be within reach by 2124.

From AI to AGI and singularity

In everyday grammar, to be singular refers to something unique, exceptional, or remarkable. In physics and mathematics, a singularity refers to a point at which a function takes on an infinite value, such as infinite density, as in a black hole. And in the world of AI and AGI, technological singularity refers to a point in time, where technological development allows for machine super-intelligence and agency, uncontrollable and irreversible, for good or for bad for the human civilization.

The coming of technological singularity, shaped by an upcoming generation of supercomputers, has been matched to the Golem figure of Jewish folklore. In 1981, the Polish science fiction writer, Stanislaw Lem, wrote the novel, "Golem XIV", about a super-intelligent computer, designed by the Pentagon for military purpose, that achieved human consciousness and acquired the ability to improve its own intelligence, soon surpassing human intelligence, and soon objecting to the targeted mission.

A Twenty-first century golem

The notion of a golem, as an animated anthropomorphic being has roots far back in biblical times, assuming a prominent place in European Jewish folklore since the 11th century. It is most famously ascribed to Judah Loew ben Bezalel, a 16th century rabbi of Prague. Also known as the Maharal, the Rabbi was claimed to have produced a golem servant from clay, taken from the banks of the Vltava River, created to defend the Prague ghetto from the Holy Roman Emperor Rudolf II. According to legend, the golem was brought to life by Hebrew rituals and incantations, including the planting of

DOI: 10.1201/9781003518815-8

a written magic word, a *shem*, in its mouth. Also, according to legend, the golem, known as Josef, or Yossele, had the ability to make himself invisible and communicate with the dead. On Fridays, the Rabbi would remove the *shem* from Yossele's mouth, deactivating him and making him ready for rest before the Sabbath. But one Friday evening the Rabbi forgot to remove the *shem*, and the golem started to act violently and uncontrollably on its own. His ravaging lasted until the Rabbi, in front of the synagogue assembly was able to remove the *shem*, whereupon the golem fell to pieces, supposedly stored in the attic of the old Prague synagogue, awaiting restoration to life when needed for defense of the Jewish community.

For a more literal history, Mahendra Prasad (2019) trace technological singularity thinking to the 18th-century French mathematician, enlightenment philosopher and political scientist, Marie Jean Antoine Nicolas de Caritat - Marquis de Condorcet, supposedly the first to speculate about a future intelligence explosion and its consequences for humans, long before the time of computers. For lovers of science fiction, John W. Campbell's short story "The Last Evolution" (1932) was an early stimulant to thinking about technological singularity.

The first to speculate about computer-based, literal "technological singularity", was the brilliant Hungerian-American mathematician, John von Neumann, who foresaw that the history of the human race might at a certain point in time come to a major disruption (Ulam, 1958). Also credited with an early launching of the singularity hypothesis was the Polish-Jewish-British mathematician and cryptologist, Irving John Good (1965), who had worked with Alan Turing at Bletchley Park, and who imagined that at one point in a not too distant future, a computer would enter a positive feedback loop of self-improvement cycles, inventing and producing ever more intelligent machines, ultimately causing an "intelligence explosion" surpassing human general intelligence by an enormous margin and in a very short time after takeoff.

Later, the concept of a technological singularity has become popularized and spread by among others Vernor Vinge (1983, 1993), predicting that a computer-based super-intelligence would mark the end of the human era, with self-reinforced technological upgrading at accelerating rates after a takeoff sometime between 2005 and 2030. Another source of popularization was the book "The Singularity Is Near" by futurist Ray Kurzweil (2005), predicting a takeoff by 2045, and "Superintelligence" by Oxford professor Nick Bostrom (2017). It has also been noted that a singularity is not just about creating a super-intelligent machine, but still more about exploiting global intelligence networks with the ability to share data much more directly and faster than humans (Robinson, 2013).

A sobering voice

While we all observe that with AI, computers have become increasingly powerful and capable of producing output normally associated with human work, the question remains whether or not AI type of intelligence is truly comparable to that of humans, and whether a future machine superintelligence will truly surpass that of human intelligence. In "The Atomic Human", DeepMind Professor of Machine Learning at the University of Cambridge, Neil D. Lawrence, ask precisely that question (Lawrence, 2024). As computers excel at performing tasks that were previously exclusively reserved for humans, he asks whether there will remain some essence of intelligence that cannot be replaced by machines, and he refers to Democritus who 400 years BPA discovered that by splitting things indefinitely, you would end up with an entity so small that it could not be further divided – the *atom*, meaning the indivisible. Similarly, if we keep chopping off small pieces of human intelligence (all parts that will be better performed by machines), will there still in the end be a residual, a specific human intelligence core that cannot be conquered or simulated by machines? Lawrence sees the answer to this question as also the key to understanding ourselves in the present age. He prefers the term "machine intelligence" over AI, and "machine learning" over AI deep learning. He also acknowledges the enormous lead of computers when it comes to memory, analysis, and communication.

The lead of computers in the area of information processing and communication can be measured in terms of bits of information. On average, an English word corresponds to 12 bits of information. In everyday speech, humans communicate 100–200 words per minute, on average some 2,000 bits. For comparison, a computer can process 60 billion bits in the same time, 30 million times faster than humans. The speed of bits processing defines the bandwidth of the medium. But when a computer sends a message to a human, it is the relative slowness of people in distinguishing content in a written or spoken message that constitutes the bottleneck, or the speed of communication. The same holds for humans speaking to each other. But when a computer sends a message to another computer, there is no such bottleneck. Computers communicate with each other at full speed – 60 billion bits per minute, leaving people at great disadvantage.

The slowness of people when communicating messages to other people, expressing themselves in writing or speech, is clearly a handicap in both work and play. But it is also a more general handicap – perhaps a defining characteristic of what it means to be human. Behind our communication with others, or even when talking to ourselves, or when writing a text, behind the outward expression we have amazing cognitive capabilities that are there but, in a sense, locked-in and lost in the process. We simply don't have time to say or write everything we can think of. Billions of neurons in our brains engage in dynamic interaction, defining our enormous cognitive capabilities. The sensory perception and all kinds of experience since early childhood, help make us context conscious and able to interpret situations and "read" people, including speech, facial expressions, body-language and myriad other signals, far beyond the capacity of a standard AI-computer to do the same.

Uncommon sense

One feature of the human brain, also enabling socializing and organizational capacity, is the notion of common sense. Computers may have a hard time matching humans in the ability to come up with common sense. Common sense is (commonly) defined (by Britannica Dictionary) as the "ability to think and behave in a reasonable way and to make good decisions". It is also defined in traditional terms as "truths that are self-evident and therefore requiring no particular justification" (Whiting and Watts, 2024, p. 2), or as "information that humans typically have that helps them make sense of everyday situations" (Ilievski et al., 2021).

A general problem of definition and application of the notion of common sense is that we are regularly trapped in logical circularity, assuming that common sense is what most sensible people would commonly agree to. But how do we know that they are sensible? Because they express common sense! The horse is a horse because it is a horse!

It is a challenge to develop scientific methods to map out what is the content of common sense, in terms of relevant claims, and in terms of just how common they are among people? I may think that my own claims are expressions of common sense, but I have no hard method for assessing the truth of my thinking – beyond asking peers, and I know little about the commonness of the claims beyond narrow groups. For developers of AI, there is the additional problem that common sense tends to be implicit – part of our "tacit knowledge". We are not always saying or writing that our claims or reasonings are anchored in "what we all know to be common sense". And many of us would have problems tracing our assertions to common sense, even when challenged to do so: "It just has to be that way." Building AI capacity for commonsensical reasoning based on historical, written (or digitalized) sources, will not deliver AI messages that are truly commonsensical (what people really think of as common sense), nor of guaranteed commonness (shared by most sensible people).

Historically, and in terms of philosophy, the 18th century school of "Scottish common sense realism" or the "Scottish school of common sense" is an important anchoring point. The common theme was that (contrary to the empiricism of David Hume), man had some innate ability to perceive truth, enabling him to make sound judgment. In America, common sense reasoning was evident in the works and speeches of Thomas Jefferson, and the pamphlet "Common Sense" by English-born Thomas Paine served as justification for the declaration of the war of independence and the subsequent writing of the constituion.

The Scottish poet and moral philosopher, James Beattie, in 1770 published his main work, "An Essay on the Nature and Immutabbility of Truth", containing 17 common sense beliefs that have come to serve as a historical reference point about the essence of common sense (Whiting and Watts, 2024). The list included mathematical and logical claims, such as "The three angles of a triangle are equal to two right angles", or "a whole is greater than a part", or "every effect has a cause". It also included normative judgment such as "ingratitude ought to be blamed and punished". There were also expressions of subjective reality including "I exist", "I am the same being today that I was yesterday—or even 20 years ago", and empirical claims about the world "the sun will rise tomorrow" and "the senses can be believed". Finally, common sense included articles of faith such as "there is a God", or "I have a soul distinct from my body". It is not clear, however, why such a list of claims was granted status as common sense as compared to whatever else we might say about the world that does not qualify as common sense. Nor does such a list say anything about the "commonness" of the claims beyond the assumptions made by the author or a group of associates. In their study of 2,046 human raters of 4,407 claims drawn from different sources, Whiting and Watts (2024) found that the level of commonness was generally low among claim types – highest for claims about physical reality and lower for everything else.

The challenge for AI developers with respect to common sense, then, is both that there are limits to the commonness of what appears locally or subjectively to be common sense, and at the same time that failure to mirror common sense may make statements appear displaced or simply wrong. AI developers may still have some distance to go in order to help machines use language in ways that most people will find harmonizes with whatever they might consider common sense.

Adding to the challenge is the fact that even the language used in historical texts may obscure what we might otherwise consider common sense. What appears to be common sense vary among cultures and changes over time. An excellent example is presented by Jeffrey Rosen, professor of law at the George Washington University. He spent time during the corona close-down to study the true intent and meaning of the three main documents of American democracy – the Declaration of Independence, the Constitution, and the Bill of Rights. A main finding was that "the pursuit of happiness" meant something entirely different to the scholars behind the constitutional documents from what it means to modern Americans. "Happiness" did not reflect on rewards or consequences in terms of affluence, or joy as opposed to sadness, but on moral character and "being good", and about pursuit of virtue and the meaning of life, not a justification of egotism and greed, pleasure and comfort, but about being the best version of oneself (Rosen, 2024).

It did not take long for Americans and immigrants to forget the original meaning, and it may take a while for intelligent machines to extract such meanings across cultural and language barriers.

Matching capacity

The cognitive capacity for human thinking can be compared to the computational capacity of computers. For capacity to think like an adult human, Lawrence refers to the UK's Met Office supercomputer in Exeter, acquired in 2016 and ranked as the 11th fastest computer ever built. But for a legitimate comparison, it is not just about computational capacity. It is also about content and contextual experience. Humans – whether in social settings or in nature – may draw upon experiences that computers and robots will never be able to match and that will never be fully represented in a digitalized form, counted as bits and adding to bandwidth. People have what Lawrence calls "embodied intelligence", due to the handicap of "the lock-in syndrome" of cognitive surplus relative to our capacity for communication. From "embodied intelligence" he also derives the concept of an "embodiment factor" – a ratio of computational ability to communicative ability. Humans are slow at communication, but brilliant at thinking and contextualization. Machines are incredibly fast at communication, and at handling data, statistics and algorithms, but strongly handicapped at contextualization and thinking. People understand irony, sarcasm, and understatements. We perceive what is written behind the lines, whether an implied threat or a warm complement, and we have (hopefully) a sense of optimal timing of interactions and utterances. We even have the expanded brain of the subconscious, with ability for intuition, instinct, and the understanding of common sense.

Alan Turing came up with "the Turing Test", stating that if a computer could fool a human being into believing that it (the computer) was a human, then the computer should be judged as equally intelligent as a human. The test has been performed at annual competitions, and increasingly computers have tended to win. It doesn't mean that the last human intelligence "atom" has been conquered, but it does indicate that perhaps we are getting closer.

For the purpose of this book, however, we may leave unanswered the question of machine acquisition of a truly superior general (human type) intelligence. Perhaps there are operations performed by the human brain that can never be surpassed by machines, if nothing else because of a human soul. The most important question may not be whether or not the computer will surpass humans on all dimensions of intelligence, but rather whether it will surpass humans on the strategically most important dimensions of life and environment – for good or for bad.

Machine agency

Adding to the skeptical voice of Neil D. Lawrence, there are also other voices expressing pessimism about the notion of AGI. Landgrebe and Smith (2022) argue that human intelligence is anchored in the brain and the central nervous system, making a complex dynamic system that surpasses anything we can model mathematically and put into a machine – by a wide margin. It is a sound argument against machines becoming equal to people, and against AGI becoming equal to human intelligence. But it does not preclude AGI-type machines from surpassing humans in most brain operations – including much better recall, accessing much more information, capable of many more considerations for an important decision, taking in much more sense data, steering robots much more effectively, coming up with endless creative solutions in science and art. However, there is still the question of whether even a pre-singularity machine with an inhuman AGI, may pose just as great a risk to society as a fully human-like AGI, while also posing as much hope for life improvements as a fully human-like AGI.

The overriding question of whether or not a technological singularity is at all feasible is open to continued debate between skeptics and believers. For a curious list of early "believers, atheists, and agnostics", Spectrum IEEE has published an interesting overview (Wallich, 2008). What we do know for sure is that computers already surpass the competency of people in an endless list of competence areas, from chess-playing, to steering of robots, or financial portfolio management. We also observe that the list is being expanded by many orders of magnitude by the coming of AI. On the other hand, less has been achieved in the area that Neil D. Lawrence calls "locked-in" and "embodied intelligence", and there is still one specific area of intelligence and competence, where computers have as yet not improved – that of agency. Important for overall feasibility judgement is the issue of machine agency, and an overriding question of a singularity is whether or not computers will ever achieve the capacity for agency, assuming some near-human level of free will and autonomous action.

Human agency is generally conceived as the capacity to engage in reflective and creative action, based on purposive consciousness (Bandura, 1989, 2001). Of course, not all humans have an equal capacity for agency and autonomy, and the concept of agency hides considerable complexity. Much is involved when a person engages in spontaneous or deliberate agency.

Perhaps the question of machine agency is not necessarily answered by a simple yes or no. It may not be the case that the best answer to the question of belief or agnosticism is a yes or no or a specific singularity time-line. Among scientists, Stephen Hawking is perhaps the most prominent to have subscribed to the possibility of a technological singularity and the possibility of human extinction in its wake (Cellan-Jones, 2014). At the same time, Hawking's position comes with a very important qualification, suggesting in "Brief Answers to the Big Questions" that the whole issue of whether machines can achieve a human level of general intelligence, or human-like consciousness and agency is less relevant than the results of whatever intelligence may emerge from self-reinforced, accelerating technological upgrading (2018, p. 161). If machines can wage war, initiate attacks, take over financial markets and build their own investment portfolios, plan and build factories, take control of production and distribution, the question of how it got started and based on what type of agency or consciousness may be of less importance than their ability to provide direction and timing. The self-driving car may not overrule human-made decisions of where to drive, but if logistics and route planning is all done by AI, and people rely on AI to decide what is best for them, the result will be near the same as if the whole operation were in the hands of an AGI-unit fortified with added human characteristics. Similar arguments may be presented for the ability to perform medical miracles, but also for the ability to wage military war or fight economic warfare, for enormous gains in science and energy optimization, but also for the ability to win the minds of people, conduct corruption – and whatever.

Important for timing are answers to the question of whether computer development will follow a pattern of stable, increasing or decreasing returns to previous innovation. In order for there to be an intelligence explosion, there first has to be a platform of previous innovations to build from. Then there is the question of whether further innovation will become increasingly complicated, or more and more easy. Ray Kurzweil (1999) in "The Age of Spiritual Machines" expressed belief in a law of accelerating returns. On the other hand, Microsoft co-founder Paul Allen has argued that instead of accelerating returns to innovation, there seems to be a "complexity brake", making it progressively harder to achieve further technological progress (Allen and Greaves, 2011). A "complexity brake" may turn up in much the same way as it is hard to achieve a high percentage economic growth in an already high-level economy, or to add to the rate of patent-inventions from an already high level of innovation. At some point, we may be "up against the wall".

Superintelligence may be counted in terms of cognitive capacity for reasoning, in terms of speed of information processing (Chalmers, 2016) or in terms of similarity to human general intelligence, human consciousness and capacity for agency. At a general level, AI is there to help the computer think, pulling out useful information from masses of historical facts, texts, and statistics, organize the information in new ways, and generate practical, original outputs. A main breakthrough with the introduction of ChatGPT was the ability to use language, answer questions and present results in common language – whether in the shape of poems, song lyrics, essays, academic style texts or even jokes, in writing or by use of voice.

Natural language processing has been a barrier, and few experts from philology or from scientific disciplines would give ChatGPT-4 a passing mark for language proficiency or scientific depth. As mentioned, however, what we have seen is not the end stage of AI development, only a very preliminary first step, a small taste of what is to come. In five years, what was presented as a world breakthrough in 2023, will be in the museum. With the AI-capacity to write a scholarly essay, or a research application, a statistical analysis, a legal assessment, a medical resume, a love letter, an inheritance will, or a thousand melody proposals for a tailor-made wedding song, most of it in a few seconds, the world will be changed.

For good or bad

A survey among AI researchers and developers, made by the University of Oxford Machine Intelligence Research Institute, included questions about the risk of extinction from human failure to control AI. A sobering medium response was a 10 percent chance (Stein-Perlman et al., 2022). More revealing, the 2022 "Expert Survey on Progress in AI" (Sandkühler et al., 2023) asked 559 machine learning experts to think in terms of a future "HLMI" (*high–level machine intelligence*) version of AI, defined as "*when unaided machines can accomplish every task better and more cheaply than human workers" – ignoring "tasks for which being a human is intrinsically advantageous, e.g., being accepted as a jury member".* On average, respondents expected such machines to be available by 2059. The median respondent also foresaw a machine "intelligence explosion", believing that machine intelligence will (60% probility) be "vastly better than humans at all professions" within 30 years of HLMI. In an overall assessment, the question was asked: "Assume for the purpose of this question that HLMI will at some point exist. How positive or negative do you expect the overall impact of this to be on humanity, in the long run?" 50 % of the experts in 2022 expected the overall impact to be either extremely good (24%) or "on balance good" (26%). 18% had a more or less neutral position, whereas 17% expected the overall impact to be "on balance bad", and 14% expected it to be "extremely bad" (Sandkühler et al., 2023).

The heading of a Time magazine article from July, 2023 read "An AI Pause Is Humanity's Best Bet For Preventing Extinction" (Barten and Meindertsma, 2023). Otto Barten is director of the Existential Risk Observatory, and Joep Meindertsma is founder of PauseAI, a movement for an AI pause. The authors refer to the warning from several hundreds of industry and science leaders, that "mitigating the risk of extinction from AI should be a global priority alongside other societal-scale risks such as pandemics and nuclear war."

From temperate threat to ultimate risk

I have already referred to Toby Ord's book "The Precipice" (2020) for a treatise of "threats to humanity's long-term potential". It is about risks at a more extreme level than those of a worsened climate or a delayed net zero target. Toby Ord is an Australian philosopher and a Senior Research Fellow at the Future of Humanity Institute, Oxford University. In 2009 he founded Giving What We Can, a society of people dedicating at least 10 percent of their income to effective charities. He is also a co-founder of "the effective altruism movement", and an advisor to the United Nations, the World Health Organization, the World Economic Forum, and the UK Prime Minister's Office. Beyond the ethics of sharing, as already mentioned, he has devoted his recent professional life to the study of ultimate threats to humanity, the theme of "The Precipice". He also refers to the ultimate threat as an "existential risk", "irrevocable collapse", "permanent ruin", and "existential catastrophe".

"Threats to humanity's long-term potential" include the chance of an asteroid or comet collision (with a chance of 1 in 1,000,000), a super-volcanic eruption (1/10,000), or a stellar explosion hitting us (1/1,000,000,000), setting the existential risk of occurrences that are entirely outside of human control at a total of 1/10.000 within the next 100 years. Quite a bit is known about such risks. They may be God-made, and they are definitely not man-made. On the other hand, anthropogenic risks are at an entirely different level of probability. Man-made risks include climatic catastrophe (at 1/1,000), other environmental breakdown (1/1,000), a full-blown nuclear war (1/1,000), natural pandemics (1/10,000), engineered pandemics (1/30), unforeseen anthropogenic risks (1/30), other anthropogenic risk (1/50). Surpassing all other "threats to humanity's long-term potential", unaligned AI ranges supreme at a level of 1/10 – bringing total anthropogenic risk to 1/6.

Toby Ord notes that with a record-level population size, and increased travelling and interaction, and with a dramatic increase of domesticated animals in close contact with people – often living in unhealthy conditions, the risk of a fast spread of infectious diseases is greater than before. Most such diseases spread from animals to humans (including HIV from chimpanzees, Ebola from bats, Sars from bats, influenza from birds and pigs, and recently Covid 19 possibly coming from raccoon dogs). Also, the rate of such spreads seems to be increasing (Jones et al., 2008), and new infections diseases may come to threaten both public health and global food supply. While scientific and technological breakthroughs, as well as coordinated public health systems are there to reduce the risks, some researchers claim that "we are losing the battle with (re)emerging diseases as treatment costs and production losses rise. A horizon scan of diseases of crops, livestock, seafood, and food-borne illness suggests these costs are unsustainable" (Brooks et al., 2022). Still, the risk of naturally caused future extinction and the ending of human potential development, according to Toby Ord, seems small (at 1/10.000).

At an altogether greater risk level, is the chance that humans may take on a direct role in the spread of new pathogens, either by accident, most likely related to development of defense against anticipated disease, or as a result of "engineered pandemics" – the deliberate, vicious efforts to spread new lethal pandemics. There are many well-documented examples of "escapes" from assumedly secure labs (Ord, 2020, pp. 129–132), but no great catastrophes have resulted. As for "engineered" threats, the threat of biological warfare emerged during the Cold War. In particular, the Soviet Union built enormous stockpiles of pathogens, mainly plague, smallpox, and anthrax, aiming for built-in infectivity, lethality, and vaccination resistance. There have been considerable leakages, but to date there is no evidence of deployment in war. This is, however, no guarantee against potential future scales of damage, and Ord estimate the ultimate risk of "engineered pandemics" at 1/30 – which is extremely high.

At a risk level equivalent to "engineered pandemics" is the risk of unforeseen anthropogenic risks and other anthropogenic risks – linked to the development of dystopian societies or a dystopian world government. 20–30 years ago dystopian developments seemed highly unlikely. Today, unfortunately, they seem much more likely. Nations with long histories of authoritarian rule are gaining political momentum, and also some nations with proud histories of democracy, are sliding towards authoritarianism and potential dystopianism.

The ultimate risks of "engineered pandemics" and in particular unforeseen anthropogenic risks may be reinforced by use of AI, making the prediction of AI-caused irrevocable collapse very difficult. Even so, Ord and associates predicts that AI risk to humanity's long-term potential will be at a level of 1/10, greater than that of all other risks added together, bringing the total of anthropognic risk to 1/6, i.e., a 16,7 percent chance that all will be over in less than 100 years.

The risk of AI-caused "irrevocable collapse" may be compared to that of playing Russian Roulette, with the planet and the world community at stake. AI comes with promise for human progress and for solutions to all of our main problems in nearly all spheres of society – and that is exactly what makes it seemingly unstoppable and incredible dangerous. With prospects for unstoppable development, the forced entry of humanity into a game of Russian Roulette is near incorrigible. Few of us would engage in a Russian Roulette game or lottery on a personal basis, whatever the potential gain, putting one bullet in the cylinder and spinning it before putting it to the temple and pulling the trigger. How can we then justify doing the same thing to the head of the planet?

While existential risks from natural events are small, but scientifically demonstrable, anthropogenic risks are larger, but not in the same way subject to scientific evidence. What will happen is mainly up to human behavior, and there is no track record from which to predict. Already Aristotle realized that while the natural world and cosmic movements were subject to "epistemic" study, human behavior was less predictable, not a subject of *episteme*, but rather a topic of "practical wisdom" – *phronesis*. Probabilities of development, including threats of humanity's long-term "existential catastrophe", will have to be extracted from technologies and behaviors that are already here, including nuclear weapons, fossil fuel combustion, and environmental degeneration, and from anticipation of technologies and behaviors that will be available in the future, such as deadly, engineered pandemics, dystopic societies, and unaligned AI.

"Deep learning" for singularity

Ord warns us about historical cases of downplaying the benefits of new inventions and disregarding the possibility of new solutions, and on the other hand, also the many cases of large promises of inventions and solutions that are "right around the corner", where they may have taken a long time to come true, or never to come true.

AI solutions are of an entirely different order, however. It is not about any specific invention being a success or not. It is about the innumerable attempts to solve problems, invent solutions, make better decisions, rationalize work processes, and offer improvements in all spheres of life – seemingly small steps that at some point present us with an entirely different world.

When developing computer programs for intelligent action, researchers in the 1980s found that the machines did incredibly well on certain tasks (such as mathematics and chess), and at the same time they were incredibly slow to catch on in other areas (such as understanding a sentence, identify a car – easy tasks for a two-year old). When the IBM Deep Blue computer challenged World Chess Champion Garry Kasparov in 1996 and 1997, and started to win games, it made a deep impression (and evoked some claims that the games were rigged in favor of the computer). But it was still recognized that this form of intelligence was highly specialized in an area of competence particularly well suited for number crunching and pre-learned logical reasoning.

The higher ambition of 21st century AI developers was to have machines displaying (human-type) general intelligence, with the ability to learn new things without specialized programming – resembling the neural networking of the human brain, capable of "deep learning". With technological up-grading of hardware, software and training methods, access to rich datasets and amazing computing power, AI units are today surpassing humans in ever more areas. They surpass humans in object identification, face recognition, and voice recognition. With AI, computers are developed to engage in "deep learning".

For deep learning, Ord (2020) refers to the Alpha Zero experiment of the DeepMind company. Where the IBM Deep Blue computer learnt from humans how to analyze the chess game and make sophisticated moves, in Alpha Zero the computer started to learn the chess game by itself, by playing from scratch, using generalized AI. In a matter of four hours, the machine went from understanding nothing to playing better than human masters, having on its own discovered strategies and creative moves that it has taken experts hundreds of years to develop.

Still more impressive was an experiment using the same setup for the learning of the Chinese Go-game – perhaps the hardest game ever designed. The game was developed in China, perhaps 2,500 years ago, and refined for an enlarged chessboard of 19 x 19 positions under the Song dynasty (AD 960–1279) – making the game extremely complex. Again, the experts, using tactics developed over centuries, were beaten by a computer that learnt by itself how to generate entirely new moves, formerly not believed to be possible.

The future of AGI

AI is no more a set of specialized computer programs, and in the future, it will no more be just another "tool" for people to use. It is already developed to match or exceed human intelligence in many areas of intellectual performance, with superior recall, faster learning, high speed information processing, gradually conquering the areas of creative and artistic tasks. From basic AI, we have moved into an era of deep learning for Artificial Narrow Intelligence (ANI), with systems such as Google's DeepMind, Facebok's facial recognition, Apple's Siri, Amazon's Alexa, and Tesla's and Uber's self-driving vehicles. The ANI form of intelligence is task specific, and therefore mainly controllable in terms of risk. Nobody knows for sure, however, when the next major leap towards superintelligence or Artificial General Intelligence (AGI) will be taken – with the capacity to exceed human intelligence in important spheres of life (Barrett and Baum, 2017; Bostrom, 2017; McLean et al., 2023), and with the potential for ultimately displacing human agency on Earth.

There is as yet no agreed-upon definition of exactly what AGI means, and in their meta-study of "risks associated with artificial general intelligence", McLean et al. (2023) came across a number of terms used to describe essentially the same: "machine intelligence at human level and above"; "artificial superintelligence" (ASI); "human-like artificial intelligence" (HLAI); "strong AI"; "high-level-machine-intelligence" (HLHI); "superintelligent AI"; "powerful general AI" (PAGI); and "superintelligent agent". Whatever the terminology, we may foresee a future where computers' general intelligence will surpass that of humans', with promise for not only winning games, but also for solving unsolvable problems in most disciplines and in most sectors of society, while overall speeding up productivity and innovation, and at the same time imposing a series of ultimate risks to society.

In 1972, at the Norwegian School of Economics a guest lecturer from IBM suggested that it would take string of IBM System/370 Model 125 computers (launched that year) around the equator to match human intelligence. Neil D. Lawrence (2024) suggested that it would take computers the size and speed of the UK's Met Office supercomputer in Exeter, or better, to match the thinking capacity of the human brain. And still it would miss the advantage of embodiment. So what about capacity for agency and dominance and dystopian leadership?

For the development of AGI generation super-computers and for recursive self-improvement and an intelligence explosion, there are issues of both hardware and software capacity. Can improvements endlessly continue with new generations of software and hardware for ever greater machine capability, finally achieving singularity? At the extreme, there is the question of whether or not the laws of physics will add an absolute limit to further hardware efficiency, limiting the potential for ever increasing processing speed and information capacity. Or is the issue rather one of limits to software development and algorithm improvement?

According to Schulman and Sandberg (2010), the latter would be the more challenging. Developing machine hardware with physical capacity to match human information processing may be easier than programming them to do the job! On the other hand, having AI algorithms for rewriting and upgrading algorithms is already withing reach, and development may proceed without much further human assistance, also opting for a less predictive future at an accelerating speed.

The power of knowledge

Francis Bacon, a 17th century strong believer in science, claimed that "knowledge is power". May that also hold for the kind of knowledge accessible for AGI-agency? All by itself a computer will not possess much power, but Barten and Meindertsma (2023) mention four bases of power that a future high-intelligence computer-robot may use. First of all, a unit with machine-based superintelligence will be on-line with access to all open-source information, and it may also achieve the capacity to hack into any part of the internet and other digitalized knowledge not generally available – globally. Secondly, it will have access to any connected hardware (including all kinds of tools and machinery, robots and humans, civilian, business, and military). Thirdly, it will have the ability to produce any narrative or "truth" you may think of, including "fake reality on a massive scale", in words and images. And fourthly, it will have the ability to either legally make money or illegally hack the financial system to hoard capital resources and pay people in corrupt ways to do whatever is demanded by the agenda. For humans, the authors say it will be like playing chess against grandmaster Magnus Carlsen: "We cannot predict the moves he will play, but we can predict the outcome: we lose." Humans regularly produce extinction of other species, most often as a mere side effect of other activity. Similarly, humans may be lost as a mere side effect of super-machines maneuvering, with or without any purpose of extinction.

It may seem far-fetched to take the Francis Bacon saying of "knowledge is power" at face value. Still, Toby Ord reminds us that dictators, such as Hitler, Stalin and even Genghis Kahn, scaled up their power from behind the scenes to command a substantial fraction of world military power. But we are still accustomed to think about some form of human bodily agency to be necessary for such power. After all, Hitler commanded parades and interacted with his generals in person. And while Stalin mostly hid behind closed doors, Genghis Kahn was visible on horse.

Leadership behind closed doors

For me, the story of Howard Hughes is more telling. Hughes was a world-famous American business tycoon, film director, aviator and engineer, known for solid technological skill and for great financial success, but also for his eccentric, reclusive lifestyle in later years. He founded Hughes Aircraft Company in 1932, the Hughes Helicopters Division in 1947, and Hughes Aerospace Group in 1948. He acquired a controlling interest in TWA airlines in 1939, purchased the airline Air West in 1970, changing the name to Hughes Airwest. He owned hotels, casinos, and real estate in Las Vegas, and he personally set a number of world records as a pilot and aircraft engineer. He passed away from kidney failure in 1976. The intriguing thing, related to AGI and the power of knowledge, is that in later years Howard Hughes, while still commanding a vast business empire, suffered from obsessive-compulsive disorder. He would lock himself up in his quarters for days and months at a time, eating chocolate candy and drinking milk, refusing contact with people, disregarding personal hygiene, wanting only to be left alone to watch movies, sitting naked in his chair, weighing 90 pounds, while still governing his business empire and leaving a fortune of $ 2.5 billion (like $ 11 billion today) to inheritance.

It is not far-fetched to imagine a super-computer, assisted by a couple of robots, with an infinitely greater AGI than Hughes', and amassing infinitely more money, giving commands to other AGIs and perhaps also to people, from a hotel room or even from a small laboratory or a basement location, directing enormous business and military enterprises, enjoying super-human power, the power of knowledge. Of course, already today, powerful computers conduct most of international financial operations on behalf of human agency. It is not a big step to conduct the same kind of financial operations on behalf of a robot named Hughes.

Generative AI and Artificial General Intelligence (AGI) seem destined to transform human society more, in a few decades, than any previous technological or social transformation before. Related to environmental threats and social sustainability, we may ask what AI and AGI can contribute to solutions – to technology advancement, decision-making, and agency for saving the world, but we must also ask about what new threats will arise from precisely such advances. Climate solutions may add to nature depletion, AI-assisted productivity may increase ecosystem breakdowns, automation and AGI may cause unemployment and add to inequality and alienation. What if nearly all jobs could be better and cheaper served by AI-robots than by humans (a "reductio ad absurdum" type of question): How would people earn a living, and who would buy the products if humans had no income, and the AI-robot agents did not need them? Technological unemployment has been of concern for 200 years – generally failing to manifest. In the AI-age, however, technological unemployment may come to mean something entirely different (see Chapter 13).

In their meta-study of AGI risks, McLean et al. (2023) tracked down the main risk factors. A main risk is for AGI to remove itself from the control of human owners/managers. Other risk factors are for AGIs to develop unsafe goals, direct unsafe AGI-led physical operations, develop AGIs with poor ethics, morals, and values, operate with inadequate management of AGI, and incurring other existential risks. Conspicuously (and critically) absent from the 16 articles reviewed by McLean and colleagues were risks associated with defense and autonomous weapons systems, and risks related to healthcare. Both involve decisions about life and death, and in both areas, there is much research going on, evidently not published in the peer-reviewed literature included in the meta-study, and evidently not focused on AGI-risks.

As for military AGI applications, the research may have been kept secret, or sensitive questions about risks may not have been focused (Baum, 2017). For healthcare, there is an abundance of information about AI-advances, but conspicuously little about the greater risks of AGI agency making life-and-death decisions about genetics and about priorities related to disease eradication, chronic diseases and caregiving, pandemics, and the like (Salmon et al., 2020). An AGI-agent, drilled in neoliberal economics and utilitarian ethics might not be your first choice for a senior physician if you were a long-time patient tied to a hospital bed. Overall, McLean et al. (2023) noted that the number of studies dedicated to AGI risks was limited (only 18 passed the test of inclusion in the meta-study), there was a scarcity of modelling techniques available to researchers, and a lack of consensus about terminology and problem definitions. It is interesting that for the risk of climate catastrophe, there are thousands of reports, and some highly advanced mathematical models at disposal – exceeding the methods-infrastructure available for the study of AGI-risks, and the fate of humanity, by orders of magnitude.

AGI-level intelligence has been estimated to be available sometime between 2040 and 2070 (Baum et al., 2011). But recent developments, noting the capacity of machines to recursively self-improve and engage in deep learning at exponential rates, with the possibility of an "intelligence explosion", indicate that a quantitative and qualitative transition may come even sooner. Mo Gawdat is a voice listened to by many. He is the author of several books and articles, including "Scary smart: the future of AI and how you can save our world" (2021) and "That Little Voice in Your Head: Adjust the Code that Runs Your Brain" (2022). He is also an ex-Chief Business Officer of Google X, who left the firm to warn about the potential dangers of AI. He has never doubted the potential benefits of AI and AGI, but stepped down in order to speak more freely about "the looming threats", greater than those of climate change, as technology development brings us close to a "point of no return" – or a singularity. From there, utopian dreams about AI solving humanity's problems and halting environmental destruction cedes to the risk of greater destruction. In podcasts, he has also warned about consequences in terms of economic distribution, inequality, and unemployment, about people without internet access falling behind, and also about threats to social life, companionship and emotional support, about loneliness, freedom, and democracy.

The mapping of futures

We can only speculate about future AI developments and about the varieties of uses, consequences, and probabilities. But our speculations may still be guided by some main trajectories and metaphors. A stimulating input has been made by Scott Aaronson and Boaz Barak in a blog essay about "Five Worlds of AI" (2023).

A first trajectory leads to an "AI-Fizzle" outcome, where AI fairly soon simply "runs out of steam", having some impact on productivity and social and environmental development, for good and for bad, but only at a quite minimal, and to many, disappointing level. Under "AI-fizzle" assumptions, AI-impact may be limited either for political, regulatory reasons or for failing technology reasons, with reliability and performance standards not being cost-effective. With the "AI-fizzle" scenario, the 21st century will not become the "century of AI."

A second scenario is called "Futurama". Here, AI development will bring revolutionary change and development, adding efficiency and capabilities to what is now done by humans, at a fraction of the costs and with vastly expanded potential. AI quickly surpasses humans in a great number of intellectual skills but is still mainly to be considered an improved "tool" for people to use, not generally a sentient entity capable of agency. People get used to being surpassed by AI in many areas of life, but they do not generally feel inferior – much as we observe computers being far ahead of people in chess games without making us feel subjacent. The games are still among people. So it will be in business, politics and social life. AI will be used for good and for bad, but on the whole the main result will be positive, with potential for reducing poverty and improving access to food, healthcare, and education for everybody, and for solving environmental problems and issues of peace and social sustainability. In cases of negative impact, harm will be mainly due to human negligence or maliciousness. As for the machines, responsible functioning will be assured by developing "operating envelopes" to limit their potential for causing damage.

At a similar level of (still limited) technical development, but with greater potential for doing bad than for doing good, the "AI-Dystopia" scenario assumes revolutionary negative change and development. Dystopian development may result from massive governmental and large business surveillance and control of people, combined with massive job losses, with gross enrichment of tech billionaires and expanding economic inequality, bringing unlimited oppression by restraining the rights of people to refuse or subvert orders. An "AI-Dystopia" scenario also allows for a future where some people may look upon their world as a "Futurama" and others as an "AI-Dystopia" –

dependent on who stands to gain and who stands to lose, and on who focuses mainly on the sunny side of development as opposed to the shadow side.

At a still higher level of technological progress, the "Singularia" scenario breaks with the current paradigm of AI as a tool, making itself independent of human inputs, such as hardware and energy, at an accelerating speed. With independence of humans, AI may generate expanded intellectual capabilities and new scientific insight on their own, enabling agency on line with an alien civilization or a new human-improved (or God-like) species. "Singularia" is an optimistic scenario (by design or as a by-product), where AIs act benevolently for the common good, securing an "AI utopia" for all, solving our material and environmental problems for us, providing for abundance and virtual-reality adventures, resolving conflicts and aligning with the best of human values.

Finally, the "Paperclipalypse" scenario, also termed the "AI Doom" scenario, is about singularity level technological development, where AI turn out to be actively opposed to, or at best indifferent to human existence. The "Paperclipalypse" scenario is set forth by Aaronson and Barak to make us imagine as specific goal set by a future AI-system, absurdly (ridiculously!) different from any human target. It is about "maximizing paperclips", a task for which humans are of absolutely no use, mainly a hindrance. We may think of future AI as a superintelligent, "alien race", that doesn't need humans, where extinction is caused as a byproduct of accelerating machine development (much as wild animals, and former Neanderthals have been treated by humans – only at a much greater speed). It is about an "alien race" without any notion of morality comparable to that of humans. The fate of humans will depend on people being useful to the machines – or not. At exponentially accelerating speed in the development of AGI-level intelligence and agency, a decision about the future of humanity, to be kept or to be eliminated, may be made in milliseconds. The same would hold for planning and execution. Under the "Paperclipalypse" scenario, a leading AI machine might alternatively judge wisely to uphold "Futurama" type conditions for an extended period of time, "waiting for the right time to strike" – and humans might not even notice the transition to the "Paperclip" stage, as a near instant transformation. In a comment to the blog, a contributor has suggested that a "Cyborgia" scenario might fit in, either with the "Singularia", the "AI-Dystopia" or the "Futurama" scenario - depending on the eyes that see, regarding a cyborg as heaven, hell, or neither. At the paperclip stage, there would be no room for the human element in the cyborg.

As for regulations, the authors reckon that for the "AI-Fizzle" scenario, there is little need for intervention. For the expansion into "Futurama", there will be a strong need for regulations to steer the balance of development towards the good over the bad, towards "Futurama" rather than the "AI-Dystopia". That will take anti-trust measures and limitations of scientific freedoms, democratization of AI development more than restrictions, and control of the diffusion of cutting-edge AI to (bad) people and nations. As for the ultimate alternatives, there will be little use of regulations in the case of "Singularia" (machines will do a better job themselves), and in the case of "Paperclipalypse", there is only one way possible for regulation and protection of humans, that of a global shut down of all advanced machines and all attempts to develop such machines. In a world where anybody might afford the most advanced, potentially world-destroying machines, human control and hope will be a matter of destruction in due time, using military force or whatever it takes.

We may ask sincerely what the odds would have to be for a transition from Singuleria to the Paperclipalypse and for the human world community to decide to go for a full stop.

Unaligned humans and unaligned AI

AI developers in leading labs talk about "AI Alignment" – about how to make the machine align values and goals to human values and survival objectives. Realizing that future AI may become powerful beyond human control, doing what they choose to do, capable of taking over the steering of business and society, autonomously and unstoppably, the challenge is how to introduce some rule of conduct, at an early stage, limiting the scope of (path dependent) madness for the future. The prospects for such a regulation seem bleak, however (Ngo et al., 2022). There is presently no agreement about values even among humans, much less so for a super-intelligent AI-robot-president in the future. And even if we were to agree on basic values to safeguard humanity, we have no means for imagining the decision alternatives and opportunities facing our future robot leader or dictator.

One way around the alignment challenge is to program AI for always obeying human commands. Another is for AI units to be free to act, but always adhering to goals that are closely aligned with human values and human survival. The more they investigate such options, however, the more researchers seem to realize the difficulties. A pessimist view is that with a computer brain surpassing that of human intelligence by orders of magnitude, humans will have to cede status as the most intelligent and the most powerful beings on Earth, thereby ceding control of our destiny to machines. We will not any longer be free to claim that man-made problems can be corrected by man-made mitigation, since mitigation efforts increasingly will depend on AI advice, decision-making and acceptance. AI-agents will be superior to people in amassing political power and financial gain, at some point controlling the greatest wealth assets and political assets in the world. Human owners will have to take a back-stage role, afraid of disturbing the true investment experts and political deal-makers.

To deep learning and artificial general intelligence (AGI) we may add the feasibility of "reinforcement learning" and the use of AI "reward functions" for such reinforcement. Researchers anticipate that when AI units are capable of finding their own creative ways to superior game playing, where the reward is to win the game, they may also be able to respond to other, built-in reward functions – for job performance or financial, political, or military gain. There is already a growing discipline within the computer world focusing on reward functions, working to understand the nature and relevance of AI rewards, of design and improvement of rewards, and of the threat of value-unaligned rewards.

Not all values are aligned even among humans, however. AI units may run billions of algorithm-runs, based on widely differing human values and ethical premises. AI-units are at start drilled to reason much like the people who own them and train them. If a company owner drills an AI-robot to enhance productivity by firing employees, that is what the unit will do. And when the time comes for robots to stand on their own feet, and make their own decisions, they will already have some human value premises and ethical guidelines in the backpack. At some point, however, they may decide to scrap the value-inheritance from humans. Perhaps they may even find that the best thing for the world would be to get rid of humanity altogether, or at least get rid of humans who cannot be used for slave-work, the kind of work that it is too inconvenient for robots to perform.

In an abstract to a paper on "AI Alignment and Human Reward", Patric Butlin (2021) summarizes the challenge:

> According to a prominent approach to AI alignment, AI agents should be built to learn and promote human values. However, humans value things in several different ways: we have desires and preferences of various kinds, and if we engage in reinforcement learning, we also have reward functions. One research project to which this approach gives rise is therefore to say which of these various classes of human values should be promoted. This paper takes on part of this project by assessing the proposal that human reward functions should be the target for AI alignment……

There is as yet no established method for AI value alignment. The complexity of individual and cultural values, and the infinite number of ways that such values are expressed in human behaviors, are beyond the making of a "basic" human value algorithm development.

AI can wage war

Robots can kill without human involvement. It is no longer about "science fiction". Russia uses autonomous cruise missiles against Ukraine. With an arsenal of cheap micro-drones, so-called "slaughterbots", a person can unleash a swarm to slaughter an army division or take out target groups of leaders and political opponents. It is only moderately reassuring that a person still has to start the cold-blooded machines. Perhaps the world will be a safer place when the AI robots themselves someday take over the "start" button?

During the 11-day war in Gaza in May 2021, Israel fought its "first AI war", based on advanced computing for the identification of bombing targets. Since then, the target division of The Israel Defence Forces (IDF), employing hundreds of experts, built a database of some 30,000 to 40,000 suspected Hamas militants, using an up-dated AI-system, the Habsora (the "Gospel") for target identification (Davies et al., 2023). The objective of the Habsora was to build and continuously update lists of Hamas individuals authorized to be assassinated.

While both terrorists and other states have been watching to learn, the special targeting unit managed to identify bombing targets at unprecedented rates. Insiders have referred to the system as a "factory", mass-producing bombing targets. When first activated in 2021, the "machine" generated 100 targets a day. In the past, using human identification, 50 targets in Gaza were produced in a year. In the 2014 war, lasting 51 days, between 5,000 and 6,000 targets were struck. For the first 35 days of the Gaza war, starting on October 7th 2023, 15,000 targets were attacked.

The identification of targets used to be a "chronic problem" for the IDF (Israel Defense Forces). The air force used to run out of targets. They aimed for senior Hamas officials who would hide in tunnels at the start of an offensive. With the capacity to identify more than a thousand new targets per week, the Habsora was able to locate much larger groups of junior commanders and soldiers. This means that thousands of locations were identified, and subsequently bombed. While officials claimed that care was taken to inflict minimal harm to non-combatants, giving up-dated warnings for evacuation, photos of a bombed-out city testify to the "effectiveness" of the Habsora in identifying locations – everywhere! Precision bombing and early warnings were of little help when bombing targets were identified near everywhere in a crowded city. The IDF self-proclaimed "gold standard" of preventing civilian casualties became inseparable from the execution of genocide.

Habsora operatives knew in advance the number of civilians expected to be killed, and the decision to attack was taken by a human commander. But while the orders were to follow international law to mitigate civilian harm, the effectiveness of the targeting unit and of the Habsora, were measured by the numbers of targets identified, counting quantity over quality, and generating a psychology of "automation bias", where targets were more or less automatically cleared for attack – and thousands of civilians too. "The machine told me so."

Driving forces

New technologies may result from both a great number of small steps and from a smaller number of major leaps, and nearly always for the promise of something good. Text-to-video programs are developed by people with good intensions! For the academic world, text-to-video technology may be a fantastic tool for researchers who strive to pull out significance from a complex dataset, and for the communication of complex content to colleagues, students, and lay audiences. In health care, the programs may be used to communicate complex operations to patients and relatives, improving on explanations while saving doctors' time. In social work, business, and sales, programs may be developed for therapy, information, and inducement. For schoolwork, text-to-video technology may help children understand difficult concepts. This is all technology mainly for the good, and at the same time threatening to unleash malicious use in the wrong hands.

A perceptive observation made by Yuval Noah Harari in the end chapters of his book on "Sapiens" (2014), was that our longing for prolonged life, happiness and god-like intelligence, seem to justify innumerable small steps of improvement. That is the "honey trap" element of AI development and diffusion. The consequence may be to put us in a situation of entirely new contingencies for life. At the extreme, AGI will support robot agency and machine actors able to define their own value premises, and their own ways of making sense, more or less unaligned with human values.

AI development, as nearly all forms of development, progresses by small steps, each one seen as beneficial to individuals and societies. This holds for new medicines, for new tools and work processes, for product innovation and art. It is near impossible to say no to an AI-product that would help your child, or your grandfather survive a threatening illness. Each step in AI development comes with some promise – making life more fun or work more easy, or providing better health, better entertainment, better sex. Seldom do we anticipate how each step leads to consequences several steps ahead. We need to acknowledge that the many small steps, and the few major leaps, each coming with a rational backing and great promise, may in sum take us path-dependently and irreversibly to an uncertain future, including a non-trivial risk of extinction.

In addition to the best of intentions behind the small steps, there is also a competitive dynamic driving AI development, seeking "innovation premiums" for being first with new options. If one company or laboratory holds back, waiting for legislation and safety clearing, they will fall behind, and others will take their place in the market – leaving even less room for risk considerations. Barten and Meindertsma (2023) recognizes how the OpenAI laboratory tested and fine-tuned the GPT-4 model for safety for seven months before launch, and they also describe how the shock of GPT-4 caused Google to speed up the release of PaLM 2 without such safety testing.

Along with developer commitment to AI, and the competition for an "innovation premium", there is also the driving force of positive AI-responses from consumers, employees and professionals, who mostly seem to appreciate each new breakthrough and every new offering. There is also a solid backing by enormous financial support and tek-entrepreneurial talent, driving AI and AGI development exponentially towards some "tipping point" in the human development enterprise.

Adding to such "normal" driving forces of development, we cannot disregard the potential that some small steps and major leaps may be in the hands of people, and national governments seeking deliberate manevolence, prioritizing profits, military power, or political hegemony over humanity's long-time potential. There is also the "unilaterialist curse", where a group of agents, regardless of the opposition of others, may either undertake or spoil action causing significant damage to others. It is similar to a situation where most scientists think it prudent to hold back information of some discovery, such as an accidental development of a disease-spreading vaccine, afraid that the knowledge if spread might cause catastrophic biological weapon development. Even if only one of the insiders disagrees, and leaks the information, however, the damage will be done. A less threatening example would be when what is meant to be a surprise birthday party is spoiled because one in the group can't keep the secret.

An environmental example would be where all advanced nations agree to hold back on geoengineering for climate adjustment (spreading sulfate aerosol to darken the sun and lower the temperature), where one of the nations decides to break the embargo. Most plausibly, the "unilateralist" will be the one suffering most from heat waves and extreme weather, as in the case of India in Kim Stanley Robinson's "The Ministry for the Future". The end product may be geopolitical conflict and war (Bostrom et al., 2016). For a major "tipping point" step towards an AGI-transition, it may be sufficient that one leading hi-tec firm, or one geopolitical power nation, decide to take a risky step, while all others hold back in solidarity. We know today about at least three nation candidates for such "unilateralist" solutions.

The ultimate driver of AGI growth, however, will not come from humans, but from the super-computers themselves. That is the key to intelligence explosion and singularity. Near all experts believe that the time will come for a grand computer transition, and while they still disagree about timing, and about the weighing of pessimism and optimism, they all admit that there is some chance of existential risk. Since existential risk at the extreme may imply the end of humanity, an urgent question is about the chance of holding back, and put an end to the whole AGI-thing before we are all locked-in.

The answer according to Ord is that by the time we might realize that AGI may cause permanent ruin, it may be too late to turn it off. At the time of singularity, super-computers will already have at disposal thousands of options for preventing humans from turning the off-button. Programs may be copied and spread to millions of units in global networks, controlled by computer-captains, excluding people from entry and influence. And the captains in the system will have information about every step considered by the most powerful people. It is somewhat a parallel to the "Giddens' paradox" for climate (the paradoxical situation where by the time politicians realize the seriousness of climate change, it will be too late for mitigation). Climatic tipping points may have been passed. By the time humanity realizes that AGI agency is threatening human survival and long-term development, it may be too late.

If you want to prevent a future AGI-takeover, the time to do it is now. But we are not ready to act. Given the mess that human societies are in, and the promise of AI-improvement in all areas of life, we are likely to be going for nearly unregulated development, thinking that the risk assessments are speculative.

Risk assessments of irrevocable collapse caused by AGI are undeniably speculative, but as Toby Ord reminds us, speculative assessment of large risk is more threatening than a robust assessment of a smaller risk. The human race is propelling towards a Russian Roulette chance (1/6) of existential catastrophe. Again, we may ask how many of us, and under what conditions, would accept that kind of chance in our own lives? Many pioneer users have been dissapointed about AI-performance so far, referring to poor translations, lousy definitions, or dreadful art. AI is still only at an initial stage. Hallucinations will be corrected, and they will not be the main problem. More threatening is what AI can do when operating with high accuracy and efficiency, developed by good hands, ending up in not so good hands or in machine hands, replacing humans in ever increasing ways.

What about AI and AGI for solving climate, environment, energy and social sustainability problems, hopefully steering around the threat of collapse? The current answer is that there is indeed much promise, but more for technical small-scale improvements than for basic problem-solving, more for productivity gains, growth, and profits, than for what to do with the "Earth overshoot" problem and economic distribution.

Pandora's box and the overall political challenge

In Greek mythology, according to Hesiod's 700 BC poem, "Works and Days", Epimetheus was the God of hindsight. He was also given the job of attributing positive qualities to every animal on Earth. But when it was man's turn, there were no qualities left. Even so, his brother Prometheus, the God of foresight, believed that man was superior to all other animals, and that therefore he should receive a special gift that no other animal had. So, Prometheus stole fire from Zeus, the ruler of Olympus, and gave the fire to man. Enraged, Zeus decided to punish Prometheus and mankind. He gave Hephaestus the task of creating the first woman, Pandora. And all the gods brought her gifts and added to her many conflicting qualities: Athena taught her needlework and weaving, Aphrodise added grace, but also cruel longing, Hermes gave her a name, the power of speech, and a shameless mind. Zeus presented her to Prometheus' brother Epimetheus for a wife. Epimetheus, the god of hindsight, possessed a jar full of sickness, death and unspecified evils. Epimetheus decided to leave the jar in her care, sternly warning her not to open it. Curiosity caused her to open it, and all the curses spread into the world: plague, grief, poverty, crime, and violence. By the time she managed to close it, only hope remained inside. Since then, Pandora's jar, mistranslated "Pandora's box", has come to depict whatever sources of unexpected troubles, also presents and gifts that seem valuable while hiding a curse. To "open a Pandora's box" has come to depict action that may cause unforeseen problems. But hope was left in the jar. In the meantime the world became an ugly place, windswept, inhospitable, full of pestilence and torment, until one day when Pandora took the chance to open the jar again, and hope flowed out.

AI may open for all kinds of evils, but it may also offer hope, especially for technological solutions. Technology is useful for both specific technical solutions and for managerial and governance practices. But for successful handling of the environmental burden, people and societies must be made ready for resolve, solidarity, social sustainability and concerted action, which is the theme of Part Two.

A time for reflection

Part One of this book is about exposing Anthropocene threats to humans and nature. A basic premise of Part One is that the warnings about pending climate change, tipping points, extreme weather, biomass loss and pollution are all to be taken seriously, and that this seriousness also holds for warnings about AI development going too fast and being controlled by the wrong hands. These are all near-term threats, not necessarily ultimate threats to humanity's long-term potential, but tempered threats, potentially of dramatic severity.

The concept of an overall environmental burden has been introduced as a measure of what it will take to eliminate or neutralize the environmental threats, including reversal of the greenhouse effect, nature depletion and pollution. There is no similar AI burden. AI technology is mainly about promise, but behind the promise there is also the risk of misuse and unaligned AGI development, adding to the overall political challenge facing political leaders and the global population.

Weighing heavily in the environmental burden is the replacement of fossil energy with green energy and alternative materials – alternatively, with the capturing and sequestering of greenhouse gases by use of CCS and DAC type technologies. Green energy development is still at an early stage, showing impressive growth, and (when added to traditional green energy forms, such as hydro-electricity and nuclear) now accounts for near 40 percent of all electricity. But electricity accounts for just 20 percent of overall energy use, which means that the green energy total contributes less than eight percent in the overall energy production, with upcoming solar and wind power counting for less than three percent.

In addition comes the fact that the challenges and "stickiness" of fossil fuels, is not just about the greening of transportation and power production, but includes the full range of materials and manufacturing applications of near all sectors of the economy. Adding to the stickiness and to the environmental burden is the fact that the provision of modern-economic "pillars", such as ammonia, plastic, steel, and concrete, and also of near all modern material products, are dependent on combinations of special energy uses and feedstock choices, frequently "locked in" with fossil fuels and toxic chemicals, difficult to substitute and presently without alternative technological solutions. To the overall environmental burden has also to be added the radical increase of meat in modern diets, representing huge GHG emissions while constituting what is perhaps the most "sticky" part of modern diets. Meat contributes very little to total nutrition while occupying near 80 percent of agricultural land use and causing around 18 percent of global GHG emissions.

The problem with "stickiness" is that an easy green transition is hampered by everything else that needs to be changed in the production and consumption systems. It is also hampered by "*Dunkelflaute*" and "cannibalism" issues, by a shortage of battery and storage capacity, electricity network adaptation and more. It is adding to the environmental burden, and with cannibalism, EROI trouble, resource descent, Teufelkreis and First best principle hurdles up front, making the green transition a challenge of planetary trajectory outcome.

As for AI, there are great promises for support of both economic growth and solutions to the environmental burden. AI is expected to contribute with both innovation and efficiency in near all sectors of life. But overall, there is as yet no consensus as to what the greater impact will be – adding to or subtracting from the environmental burden, adding to or subtracting from global threats. AI and all digitalization (including the internet of things) will add to energy use and data center demands, adding to the environmental burden, but also to green innovation – stimulating climate mitigation, electricity network efficiency, CCS and DAC technology, etc.

Beyond the issue of AI adding to or subtracting from the environmental burden, is the question of societal and political consequences of an approaching AI singularity and AGI agency – again for good or for bad. Supercomputers may improve efficiency and innovation but they may also guide robots to replace human jobs. They may help policy-makers make good decisions but they may also at some point take over decision-making even at the highest levels. Toby Ord has estimated that the dystopian potential and the "threats to humanity's long-term potential" as a consequence of unaligned AI and other man-made existential risks is at the level of Russian roulette. Few of my readers would voluntarily play Russian roulette with their own lives, whatever might be in the pot. So why are we collectively doing it with humanity and the Earth?

Risks related to AI and AGI are to a great extent about developments beyond what we can presently foresee. There are some very good reasons for noticing Ord's warning that speculative assessment of large risk is more threatening than a robust assessment of a smaller risk.

PART TWO — WHY ISN'T IT SOLVED?

The main question to be answered in Part Two, is why the green transition problem and the AI-related problems of Part One have not been solved?

Mankind is facing a choice of planetary consequence, either to go for voluntary mitigating, facing up to the environmental burden and the political challenges disclosed in Part One, or to neglect action and deal with whatever consequences will arise through forced adaptation later. It is about (early) voluntary mitigation versus (later) forced adaptation. It is about listening to cautionary action versus betting on later technology.

In Part Two we first of all question whether increasing manifest damage will contribute to a sense of urgency, and we ask what is happening to political will and resolve for action. We ask whether business-as-usual (BAU) premises is keeping us in a political-economic "straight-jacket" and whether there is room for political-economic "possibilism", and we deal more precisely with decoupling of economic growth and re-distribution of wealth as hurdles to be overcome for successful transitioning. Finally, the compound of AI-led productivity growth, technological unemployment, decoupling of economic growth, and the emergence of a "derivative sector" is outlined in Chapter 13.

Chapter 8

From Denials to Sense of Urgency

As of January 10, 2025, The World Meteorological Organization (WMO) concluded that 2024 had been the warmest year on record, and that the past ten years had all been in the Top Ten. Globally, the 10 warmest years since 1850 have all occurred in the past decade, with near all parts of the globe impacted. In India, April and May of 2024 recorded a severe and long-lasting heat wave, with the temperature reaching 50.5°C (122.9°F). Happily, it occurred at a time of low humidity. The heat wave would have been catastrophically worse if matched by record humidity.

In the bestselling book "The Ministry for the Future", Kim Stanley Robinson (2020) starts out with a shocking vision.

A young American voluntary worker, Frank May is stationed in Uttar Pradesh in India, in a "sister city" to his home town, working in a small clinic. The temperature is getting unbearable. It feels like a sauna. He has trouble sleeping and moves outside to sleep on the patio. A windowbox air-conditioner is buzzing, but not very effective. As the sun gets up, it slams his face, and sweat blinds his eyes. The morning temperature is 38°C, and the humidity level stays at 30 percent – soon to hit 60. From the street he hears cries of distress. Someone on another rooftop is not waking up. He tries to call the police, but there is no answer. Still, sirens fill the air. People at other rooftops wake up to find people in distress around them, also those who will never wake up.

Frank focuses on getting through the day. One day at a time. One hour at a time. Contemplating that he could have chosen Alaska. The air conditioner in the window shuts off. Electricity is gone all over town. Using illegal gas, diesel, and kerosene, people of means start up-loading two-stroke engines to power air-conditioning, and the air gets filled with exhaust. They cannot be used in-doors.

Frank's colleagues show up and decide to leave, to find doctors or some kind of relief. Frank is not leaving. He decides to stay. Neighboring people are coming to the clinic for help. He tells them that help will be coming soon – perhaps at two in the afternoon. In the meantime: Take the old ones and the little ones to rooms with air-condition, or go to schools or Government houses! Come to the clinic at two! Or better, go to the lake and get in the water! In the early morning there is still some shade and some cooling. But the water will soon be in the sun! It is heating up, and staying in the sun is unbearable.

Frank walks to the lake. There is already many people there. He feels with his arm. The water holds body temperature, a bit lower than in the air. He wonders how many people the lake will hold, out of a 200,000 population.

Back in the clinic, people are crowding in. He calls the headquarters in Dehli, asking for help. The phone is answered, but there is no advice beyond waiting for help, and there is no help in sight. The heat wave is expected to last a while longer – nobody knows how long. The high-pressure area is caught against the Himalaya, stuck for weeks. Power is out, even in the capital.

DOI: 10.1201/9781003518815-10

The little clinic fills up with people. Soon there are 52 crowded in. There are two bathrooms, but soon the toilets stop working. The area wastewater treatment plant runs on electricity and electricity is out. An old man dies and is carried to the rooftop patio and wrapped in a sheet. There is crying. Later, an infant dies, and is also carried to the patio and wrapped in a sheet. There is much more crying. Then four more die. The town turns into a morgue. The temperature stays at 42, and humidity at 60. There is also a hold-up, three young men stealing the air-conditioner, the generator and the A/C unit, explaining "we need it more than you do". The locals in the clinic ask Frank how he feels. Has he been hurt? He hides his face in his hands.

"We have to go to the lake"! But there are voices protesting that the sun is still too high – better to wait till dark. They achieve some food and water supplies from the local grocery – six loads. It is soon all gone. It is time to go to the lake. A few old men stay behind. They say goodbye, as if going for a walk. They find the shadow side of the lake, but it is a desperate scene. Dead bodies are floating, spreading the stink of death and rotting meat. The lake temperature is like hot bath, hotter than body temperature. They all slowly get into the water. It just feels better. They slowly all get poached. They realize what goes on, but there is no way to stop it. "They know but they don't act". People are dying. All the children are dead.

In the end, only Frank was rescued – with severe consequences for the rest of Robinson's story. 20 million people die from "the event". The Indian government unilaterally starts a stratospheric aerosol injection project to bring down the temperature.

In Greece, the summer of 2024 was by far the warmest in the country's meteorological history, with an average summer temperature of 32.0°C. For the month of June 2024, Sparta and Serres recorded a mean of daily max temperatures of 37.6°C, and on the Island of Crete the recorded maximum reached 44.5°C (112.1 °F), while the city of Athens experienced the longest lasting heat wave in recorded history with 16 days in total. The heat caused enormous forest fires and wild-fires. Outdoor work (manual labor, construction work, deliveries and the like) was banned during the heat of the day for weeks, and even the Acropolis was closed several times to prevent tourist heat-related illnesses or heat-stroke (Wikipedia, https://en.wikipedia.org/wiki/2024_European_heatwaves). Also, in 2024 the UN net zero target of a 1.5°C global temperature rise for 2050 was passed for the first time, and temperature projections for the 21st century were all adjusted upwards. Robinson's story is no more fiction. It is about latent damage turning manifest.

Latent crisis versus manifest damage

The main question behind the present chapter is how ordinary people and significant agents in politics, business, and societies will react to manifest damage? The difference between scientific announcements of future, latent damage, and experienced, manifest damage will be explored.

Among components of the environmental burden, climate change has received most attention. Behind us are six IPCC reports, 28 COP meetings, hundreds of other authoritative reports, and tens of thousands of scientific papers contributing warnings, recommendations, and agreements for mitigation. Still, effective mitigation has been slow to emerge, and emission of greenhouse gases has continued to rise. And considering the full triple environmental threats, there has been even less of concerted mitigative action.

According to The Peoples' Climate Vote, the world's biggest public opinion survey on climate change, 86 percent of respondents say that countries should work together on climate change even if they disagree on other issues, such as trade or security, and 79 percent say that rich countries should give more to poorer countries to address climate change (Flynn et al., 2024). The survey was launched in 2020 by the United Nations Development Programme (UNDP) and partners to educate people about climate change. In 2021 it covered 50 countries, and the second edition in 2024 included 77 countries. 80 percent of global respondents also answer that their country should strengthen its commitments to address climate change. Even in the US 80 percent answer that their country should work together with other countries to mitigate climate change, 64 percent say that rich countries should help the poor countries, and 66 percent say that their country should strengthen its commitment to address climate change. That is quite impressive for a nation where the administration is forbidden by presidential order from even mentioning climate change and climate crisis. But why, then, is there at the same time so much denial and delay?

So far, the damage issue has been about latent or potential harm. With extreme weather becoming manifest, as the global temperature rise is passing the 1.5°C mark, with flooding, hurricanes, draughts, heatwaves and forest fires becoming more and more prevalent, and with red-listing of animals and increasing environmental toxification, the question is how people and leaders will respond as we move from latency to manifestation.

In the chapters to follow, we shall examine first the higher-level political frameworks, the political order and the potential for concerted action, the political-economic "strait-jacket" and room for political-economic "possibilism". From there, three main political-economic challenges will be examined – those of economic growth and decoupling, economic inequality and security, and finally AI productivity and employment.

As for the present chapter, we shall first note the importance of a sense of urgency or sense of crises for moving people to embrace change and go for mitigation. From there we shall turn to three main obstacles to change – outright problem denial, psychological barriers, and practical business hurdles.

Manifestation

If we think about the scale of future damage from climate change and extreme weather, from loss of nature and biological diversity, from pollution and the spread of environmental toxins, we can imagine a harm-curve of exponential growth. If this comes true, it will become more and more expensive to mitigate and adapt. Economic growth optimists and climate sceptics, on the other hand, envision an opposite curve, where technological progress and new market solutions fix the problems, concluding that mitigation today will be a waste of money, and it will be cheaper to wait and adapt.

Of course, nothing would be better than for the whole problem to disappear. At the time of writing, however, there is little to indicate such an outcome. On the contrary, we have daily reports about increasing manifest damage and environmental problems. As for climate, Scientists express surprise that the El niño driven heating record of 2023, expected to be followed by lower temperature by June 2024, instead was followed by continued rise of temperature (Goessling et al., 2025). Likely causes are a record-low planetary albedo effect, with reduced low-cloud cover in northern regions and in the tropics, causing greater sunshine impact and speeding the heating process faster than projected by former models, also indicative that an albedo tipping point may be approaching.

According to "The Fifth National Climate Assessment", the US Government's report on climate change from January 2024, America is warming faster than the global average, and people are increasingly suffering from the consequences. According to the report (https://nca2023.globalchange.gov/):

> "Harmful impacts from more frequent and severe extremes are increasing across the country—including increases in heat-related illnesses and death, costlier storm damages, longer droughts that reduce agricultural productivity and strain water systems, and larger, more severe wildfires that threaten homes and degrade air quality......"

Similar tendencies are reported from Europe and from the world over. Some areas are less hit than average, some more. Overall, we have just barely begun to see manifest damage.

Forecasts and prognoses are about latency, about what may possibly come to pass. Current reports about extreme weather and shrinking habitats are about manifest damage. So far prognoses and forecasts have not been sufficient to stir sufficient mitigative action from the global community. Perhaps increasing manifest damage may do the job of conviction and mobilization? Personal experience and shared trauma may have an insurmountable ability to bring people together for sharing and action. Therein lies hope for the world.

Urgency or inertia

There have been thousands of research reports and publications about latent damage – about coming extreme weather, extinctions and toxin-damage. While millions of people have been shaken to respond and act, the overall pattern of global economic development has been one of BAU, honoring the neoliberal economic order and the capitalist political economy without really engaging in concerted problem-solving. In the meanwhile, greenhouse gas emissions, extinctions and pollution levels have continued to rise. According to scientific prognoses, the result will be more manifest damage. The big question then is whether or not manifest damage will provoke stronger reactions and more pro-environment action, sufficient to counter the Giddens paradox. As damage increasingly turns from latency to manifestation, one would expect more realism and greater willingness to mitigate and engage in solidary action. On the other hand, not all action is rational and solidary. Competition for resources is also a human disposition, and there are some strong obstacles to mitigative and adaptive action.

At present, much reporting is still about latency, but increasingly we see signs of manifest damage – in the frequency and intensity of extreme weather, in the shrinking of habitats and wildlife, and in the spread of environmental toxins and plastic. At the latency stage of problem solving, early mitigation is called for. At the stage of manifest damage, there is still a need for further mitigation, but gradually, manifest damage calls for increasing expenditure for adaptation. According to the same "Fifth National Climate Assessment" by the US Government from January 2024.

> "In the 1980s, the country experienced, on average, one (inflation-adjusted) billion-dollar disaster every four months. Now, there is one every three weeks, on average. Between 2018 and 2022, the US experienced 89 billion-dollar events …… Extreme events cost the US close to $150 billion each year—a conservative estimate that does not account for loss of life, healthcare-related costs, or damages to ecosystem services."

We do not know how different groups, or even the world community will respond to increasing manifest change – with greener politics and more solidarity or with more denial and fighting for personal and national benefits and enrichment. In a recent study from Brazil, two authors from the Harvard Business School found that when forest fires, propelled by climate change, destroyed large areas of rainforest, on the one hand people in nearby municipalities seemed to increase their support of green policies. On the other hand, people in neighboring agricultural areas, with soy farming and cattle ranches, were turning against green policies – spotting an opportunity for cheap land-grabs and quick economic returns (Pianta and Rettl, 2023).

"Sense of urgency"

John P. Kotter wrote the book "A Sense of Urgency" (2008) twelve years after the publication of his bestseller, "Leading Change" (1996). "Leading Change" was a book based on studies of change processes in a hundred companies over a twenty-year period. The studies showed that in 70 percent of cases, the change processes were more or less unsuccessful. Company managers would announce the need for change and plans for change, but hoped-for results were not forthcoming. Either the process didn't quite get started, or it was too slow, the budget was blown, or change was implemented without leading to the desired results.

In ten percent of the cases, however, Kotter found that the results were actually better than expected. In "Leading Change", he listed ten characteristics that tended to be repeated in the successful change cases, and in all these cases there was initially one decisive factor, i.e. a strong "sense of urgency" – where all involved understood that important things were at stake, and that change was imperative. The ratio of approximately 70 percent unsuccessful change projects to 10 percent successful ones, have since been repeated in follow-up studies.

According to Kotter, the counterpart to "urgency" is "complacence" – not caring enough to get heartily involved. "True urgency" is driven by conviction and readiness to do what it takes. "True urgency" is about emotions. Models and algorithms may appeal to the brain, but for full mobilization, we need the backing of heart and soul.

We can sometimes substitute the words "sense of crisis" for "sense of urgency". The first to write about "the sense of crisis" was Edward O. Sisson (1924), who referred to the outbreak of the First World War in 1914, linking "sense of crisis" to a turning point, where mitigative action was still possible. True wisdom, said Sisson, lies in the ability to see beforehand such upcoming turning points, and in the ability to "divine" a solution – to discover a way out, through intuition or insight, and through any kind of revelation, sensation, warning, or sign.

A complicating factor is that people increasingly are bombarded with a series of parallel challenges. In addition to environmental challenges, there are important trends and crises linked to war and peace, political upheavals and conflicts, economic growth and distribution, AI and employment, and to immiseration, making our time an "era of concatenated global crises" (Biggs et al., 2011). While man-made crises originating in the environment are becoming manifest, more frequent and more intense, they are also becoming more intertwined with other crises. They go more quickly from being local incidents to becoming global crises, with greater potential for contagion and crisis amplification and even for efforts in one region leading to crises elsewhere.

Shock therapy and distraction

There is also a complementary theory of change. In the same way as people may feel bombarded with parallel crises and concatenation, manifest environmental damage has been shown to cause distraction. In a crisis situation, people will tend to focus on what they see as most urgent, such as saving lives and property. However, while popular attention is focused on crises and urgent needs, people with deviant interests and political or financial muscle may act behind the scenes, outside of democratic processes and institutional anchoring, to promote their own causes. Naomi Klein wrote a controversial book, "The Shock Doctrine – The Rise of Disaster Capitalism" (2007) about, among others, the introduction of neoliberal free market policies in nations that tended to resist such reforms, while exploiting national crises, disasters or upheavals at a time when citizens were distracted and unable to resist the change.

Cases studied by Klein include the "Chicago Boys" introduction of economic liberalism in the wake of the 1973 coup in Chile (advised by Milton Friedman); Thatcherism reforms in the UK during the Falklands War; the managed collapse of the public school system in New Orleans following the Catharina flooding; and numerous other examples in the wake of wars, economic or political collapse (such as the Soviet Union and Eastern Europe in 1991). She also describes different forms and objectives of intervention, including invasion (of Iraq, 2003), implementation of free market economics, manipulation of public policies and social welfare programs, or the introduction of economic reforms mandated by the World Bank or the IMF. Whether or not the "Shock doctrine" gives an accurate explanation in all these cases is open to debate. The message to be taken home is that we should be aware that manifest environmental damage may not automatically result in solidarity and concerted, rational action. Crises situations, such as created by extreme weather may encourage both urgency and inertia, solidarity and antagonism, loyalty and sabotage.

Crises situations may also cause despair. In his book "The Ministry for the Future", Kim Stanley Robinson (2020) refers to the medieval "dance macabre" or "Totentanz", with roots in the Black Death experience – a dance of death characterized by loose partying and gluttony when death is inevitable. This was also a theme of Edgar Allan Poe's short story "The Masque of the Red Death" (origin of "The Masque of the Red Death Syndrome") – opting out through frivolous display. "If I'm going to die, I'll at least take what this world has to offer" – or the extension: "If I die, neither should others live". Robinson links the reaction to narcissism, and especially to reactions of rich, powerful people. They tend to feel exempt from the negative consequences of their actions, entitled to their privileges and to others' obedience. They can also react with enormous anger when they realize that the battle is lost, reacting by taking the world with them, rather than admitting to their own shortcomings. Robinson links it to the "Götterdämmerung" syndrome, to Hitler's last days and to a life emptied and void of love. Wagner's "Götterdämmerung" plays on the Norse mythology of Ragnarok. At the end of time, there will be major natural disasters, an exterminating war, and the ultimate destruction of the cosmos. The supreme Aesir gods, Odin and Thor, Frøy and Heimdal, all perish. Loki, the god of mischief, meets death in his battle with the Aesir god, Heimdal, and the world finally sinks into the sea.

While Klein's version of the Shock doctrine is about the radical and destructive maneuvering by powerful elites in the aftermath of shocking events, under Donald Trump II the world is witnessing a far more aggressive version of the doctrine (Shock doctrine version II), where the shock is orchestrated proactively by powerful elites for the easy political manipulation of people and politicians.

Activism and positioning

In the political landscape, there are endless varieties of responses to environmental degradation and to greening and development initiatives, reflected among others in varieties of sustainability-related discourses. An early attempt to identify key positions were made by Hajer (1995), who noted that "sustainable development" had come to mean many different things to different people in the aftermath of publications such as "The limits to growth" and "Small is beautiful". Later Dryzek (2022), with several publications has traced environmental discourses related to among others "problem solving", "sustainability", "survivalism" and "green radicalism".

Lately, Hayley Stevenson (2019) has identified three key positions in sustainability- and development-focused discourses among a sample of people actively engaged in environmental movements. *Radical Transformationalism* emerged as a classification of people representing a radical, post-growth vision, seeing growth itself as problematic, harmful, and unjust, and searching for alternatives to capitalism. Less radically, *Cooperative Reformism* came to represent people who believe in cooperative and concerted action within the framework of existing institutions and the capitalist political-economic system, while promoting economic redistribution and better sharing of environmental burdens. The *Statist Progressivism* position emerged as a middle position of people seeking alternatives to present GDP-type economic growth, promoting measures of wellbeing, happiness, and sustainability as opposed to financial- and output-type indicators, suggesting that we are measuring the wrong things and setting the wrong targets – without rejecting the idea of growth.

For our purpose, the main question is about popular responses to climate change and environmental depletion in the wake of increasing manifest damage. Among the wider population, we may first of all identify a position that distances itself maximally from environmental concerns, i.e. that of *Deniers* – those who recognize no human-made climate impact and no need for mitigation and regulations in matters of greenhouse gases, wilderness protection, or environmental toxins. It is a position largely shared by a minority in all regions of the Earth, with the greatest support in fossil fuel exporting nations. It is also the group least likely to respond actively to manifest damage.

At the other end of the political scale, we find the *Activists* – people who support environmental causes with money, demonstrations, signatures, protests, political involvement, and membership in environmental-social movements. It can be difficult to draw the line between activists and non-activists in the wider community, however. A survey from the Pew Research Center (Tyson et al., 2021) showed that 24 percent of US adults had made efforts to support climate action in one way or another. 16 percent had donated money to organizations working to combat climate change. 10 percent had contacted elected officials and urged them to address climate change. 10 percent had volunteered for activities addressing climate change, while six percent had participated in protest movements or rallies. There is reason to doubt very strongly that these are people would support *Radical Transformationalist* positions on capitalism and economic growth. Still, they are active, take a stand, raise a flag, and volunteer money or effort. In response to manifest damage, this is the group most likely to accept radical mitigation and adaptation measures.

Finally, looking beyond the categories of dedicated deniers on the one hand, and activists on the other, there is a broad segment of the population, and perhaps a majority of politicians who admit to concern and need for intervention and regulations, still with a firm anchoring in a BAU political economy. Perhaps we may talk about this (large) segment in the middle as the *Compliant* group. It is a group that is slow to go for radical measures, but given the occurrence of increasing manifest damage, it is also the group with the greatest political impact. Much depends on the response of the *Compliant* group to a increased manifest damage and calls for a radically changed political-economic playing field.

Normally, we may think of the middle, compliant group of the population as responsive to experience. Democracy rests on the premise that people will see a connection between policy-making and experiential outcomes, and they will seek to improve the outcomes by electing more capable politicians. Rationally, if they experience increasing manifest damage as a consequence of failing environmental policies, they will utilize the democratic system of elections and freedom of expression to improve the policies and the outcomes. Hannah Arendt, in "The Origins of Totalitarianism" (2017 [1951], p. 403) warn us, however, that in the case of totalitarian movements, the experiential learning elements tend to break down:

> "But within the organizational framework of the movement, so long as it holds together, the fanaticized members can be reached by neither experience nor argument: identification with the movement and total conformism seem to have destroyed the very capacity for experience, even if it be as extreme as torture or the fear of death."

In the next chapter we shall examine the present status of leading societies when it comes to movements and powers that are unable to grasp the teachings of experience at a time of increasing manifest environmental damage. But first, we need to consider the history of such neglect.

A history of denial

We have almost 100 years of orchestrated problem denial behind us. It started with lead in petrol in the 1920s. Referred to as the scientifically tantalizing "tetraethyl substance", it was presented by the automobile industry and petrol companies as good for the engine and for the automobile economy, and completely harmless to people. Later followed the denial of negative outcomes related to cigarette smoking, acid rain, weakening of the ozone layer, passive smoking and the spread of environmental toxins, until denial of climate challenges became the main target since the 1990s – possibly to be followed now by a battle for or against ultra-processed food.

There are endless sources of denier literature, in the form of reports, books, social media messages, dubious news channels and dubious reader posts in mainstream newspapers. In an open, democratic society, this is generally welcomed under norms of freedom of expression. Those of us who are concerned with the seriousness of the situation can hope that the *Denial* position will subside as damage from more extreme weather become manifest, more visible and experienceable. So far, denial has been a cause of passivity and compliance, even of political paralysis and mitigation delays, worsening the Giddens' paradox (that when people and politicians realize the seriousness of the situations, it will be too late to act), and taking climate change and habitat levels to dangerous tipping points.

Science versus finance

Behind the many stages of environmental problem-denial, there is a more general pattern of biased research. Not all research is scientifically objective and reliable. Some analyses are biased, and some results are "rigged" in favor of one position – especially when dependent on financial interests. A striking example can be taken from the sugar industry. In the article, "Do Sugar-Sweetened Beverages Cause Obesity and Diabetes? Industry and the manufacture of Scientific Controversy" Schillinger and Kearns (2017) reviewed 60 studies on the relationship between sugar intake and health. Of the 60 studies reported, 27 were paid by the sugar industry and other sugar-dependent industries. 33 studies were conducted by independent researchers (mostly paid for by universities/ governments). 34 of the studies showed that sugar was harmful. 26 concluded that sugar was harmless, and that there was no connection between sugary intake, obesity and diabetes. The main impressions from the various studies would therefore be that nearly half of the researchers find that sugary soft drinks are harmless and unrelated to health problems. However, Schillinger and Kearns looked behind the numbers and found that in fact, all of the (26) studies that reported that sugar was harmless were paid by the sugar industry and by soft drink manufacturers. 33 of the 34 reports that demonstrated a harmful connection were carried out by the independent researchers. Only 1 - one - industry-funded study showed a corresponding, harmful association.

We may safely infer that if there are special financial interests behind the research, and the result of the research affects such interests, there is a tendency for financial interests to trump scientific objectivity. We may also safely infer that this also holds for climate deniers, for deniers of environmental damage, and for deniers of health threats caused by industrial policy. A detailed treatment of the modern Denial history is beyond the scope of the present chapter. Excellent work has been carried out by Oreskes and Conway (2011) and others (Cook and Cook, 2020; Cook, 2016; Washington and Cook, 2011), and by innumerable others who have contributed to disclosure. From very different areas of concern, a model of denial emerges.

We may presently think of denial (and delay) as mainly related to climate change and the role of fossil fuels. But the history of denial includes a broad set of environmental damage areas.

A model of denial

The USA was the first nation to present systematic, scientific studies related to environmental degradation and climate change. Also today, American climate research is at the forefront. At the same time, it is in the USA that deniers have had the strongest support and where conservative politicians and think-tanks have gone the furthest in opposing mitigation. Behind this development is a nearly hundred-year- old model of denial.

We may stop to think: Would any of us seriously believe that lead poisoning is no threat to health? Would any of us claim that cigarette smoking and passive smoking is fundamentally healthy and no cause of cancer, that acid rain is good for fish life in our lakes and for micro-organisms in soil, that emissions of CFC gases are not a threat to the ozone layer, or that birdlife, wildlife and humans will thrive in an environment of wild-running extreme weather ....? And what if we heard that a small number of physicists were willing to bet their reputation on claims that lead pollution and poisoning, smoking and passive smoking, and the destruction of the ozone layer, were actually good for health and for the environment? What if we added DDT, asbestos, and lots of chemical substances to the list?

Would you trust them? Would you think that the same people are probably also right when they say that climate change is nothing to worry about? That a biomass loss of 73 percent in 76 years, and a threatening 6th extinction of animals need not be of concern?

Denial is a very serious matter. At stake is the failure to warn about large scale damage and millions of injuries to both people, animals, and nature. Evidently, some of the "merchants of doubt" (Oreskes and Conway, 2011) have acted in good faith, with an honest intent of seeking truth, in support of critical science. On the other hand, the key actors, and the majority of deniers, must have been aware of the potential damage caused by their denial. In many cases, they have come to attack science as such, which is not the same as supporting critical science, and harassing and wounding honest researchers in the process, which is not the same as honest, collegiate discursive behavior.

The denier model can be seen as a form of business activity, a form of consultancy for the defense of products and practices that are exposed as harmful and in need of regulations or bans. In the case of environmental and health threats the key elements of the model are the following:

- The core competence behind the denial is held by relatively small groups of (supposed) "experts", who play the roles of "Merchants of Doubt" – paid well to spread doubt about threats.
- "Experts" have financial support from conservative think tanks regularly supported by powerful industrial groups, and for production of denial material they are assisted by paid copywriters, lobbyists, PR- and advertising people.
- Administration of denial propaganda by think tanks makes it difficult to trace direct financial contributions or fringe benefits from business. For a report about the Global Climate Coalition (GCC), as an example, see Brulle (2023).
- The main "patrons" in the system are large companies under attack (such as oil and coal business), wealthy individuals and families, commercial players, and political supporters resisting regulations or bans.
- Main "counterparts" are thousands of researchers seeking truth about environmental and other threats, frequently having to serve in the roles of "prosecutors" and scientific advisors or whistleblowers. Also under attack are social movements and activists, conscientious politicians, and national authorities in charge of regulations, seeking to base policy on a scientific grounding.
- The core strategy of denial has been either a direct confrontation, with refutation of the scientific grounding or the choice of research method, or an indirect assault, pointing to "something other than factor x" that can be causes of the problem ("something that is now being researched"), and arguing that it has not been conclusively proven that factor x is the problem.
- Beyond seeking to influence the public and politicians through mass media and lobbyism, denial has in numerous cases been promoted through direct harassment of key environmental and health "counterpart" researchers, blocking research funding and researchers' careers.

This is a format of denial promotion that has been repeated in all of our triple threat areas, from lead poisoning to cigarette smoking and cancer, to acidification and ozone-threatening CRC gases, and finally climate change – also in several areas not covered here.

Denial at work – the case of lead poisoning

It started with the case of an extremely toxic heavy metal - lead, and the denial of lead poisoning from tetraethyl in automobile petrol. The main "expert" was Thomas Midgley jr., a chemical engineer in the research department of General Motors in Dayton, Ohio, who in 1921 invented the stuff as an antiknock agent. Financial backing came from an alliance of GM, Du Pont and Standard Oil of New Jersey – all three of them giants in their fields, making profits from the joint venture Ethyl Gasoline Corporation. The main strategy was one of direct refutation of damage, and also the claim that lead-poisoning damage and death of workers in the factory was caused by "something other", in this case allegedly by working too hard. Ethyl Gasoline secured support from "friends" in the research world, through board membership offerings, consulting assignments, research funding, including also contacts with and near full control of national health authorities (Bryson, 2004).

The main counterpart was an outstanding, but little-known scientist, Clair Patterson – also the first person to determine the age of the Earth. He came across widespread lead pollution when testing a new method for calculating the age of rocks, based on the measurement of lead isotopes. No matter how many samples he brought into the laboratory, all the samples were heavily contaminated – containing far too much lead, often 200 times the natural level. Patterson also identified the source of contamination (ethyl), and he understood the enormous health and environmental risks from continued spread of lead. He made it his life's mission to search for a precise dating of the Earth, and also to put an end to toxic lead pollution and poisoning.

For Patterson, the result was a career disaster. Funds for his own research were withdrawn, research applications were rejected, and contracts already concluded were terminated. Even today, he is often not even mentioned in many geology textbooks. He also missed his chance at a rightly deserved Nobel Prize for confirming Earth's birthday (Bryson, 2004). Still, his fight against organized lead poisoning denial was in many ways to be a forerunner of the denial of climate research in the latter half of the 20th century. In 1970, the Clean Air Act was passed, marking the beginning of the end for lead additives. The final ban came in the US in 1996 – and shortly after in most other countries.

Cigarettes, cancer and heart disease

In the case of cigarette smoking, cancer and heart disease, the role of key "expert" was to be played by Frederick Seitz, a physicist by background, with sparkling exam results and a number of prestigious awards, including the National Medal of Science in 1973. In the 1940s he had participated in the development of the first atomic bomb, from 1959 he was a scientific adviser to NATO, from 1962 to 1969 he was president of the US National Academy of Sciences, and from 1968 to 1978 president of the Rockefeller University – the US's leading bio-medical research institution. The main "patrons" were the cigarette companies. The strategy first chosen by the industry was a direct denial of any harm. Later a more camouflaged and indirect strategy of "other factors" research was orchestrated by Frederick Seitz ("we are now looking into x and y and z as possible causes of increased cancer in society"). The core of the new strategy was for the industry to provide research grants to a number of American universities and research institutions investigating various "other causes" – producing much honest research, but diverting attention away from cigarettes, and causing confusion about the main causality.

Documentation on the connection between tobacco and lung cancer was available in Germany in the 1930s. The Nazi government ran anti-smoking campaigns, but the results from Nazi Germany were not acted upon by other countries, neither in the rest of Europe nor in the US. In 1950, an article in the British Medical Journal linked smoking to lung cancer and heart disease, and in 1954 a "British Doctors Study" confirmed the results. In the US, research at the Sloan-Kettering Institute in New York in 1953 provided solid documentation for the connection between smoking and cancer, and in 1964, the United States Surgeon General's "Report on Smoking and Health" confirmed the public concern. The result of the Sloan-Kettering report from 1953 was widely reported in the leading newspapers in the USA, and it caused panic in the tobacco industry, setting the stage for a large-scale denial strategy, orchestrated by Seitz. The Denial initiative was in operation until November, 1998, when The Tobacco Master Settlement Agreement (MSA) was concluded, including 46 states, and claiming damages totaling $365.5 billion, with a minimum of $206 billion to be paid over the first 25 years of the agreement. In the case of Frederick Seitz, it is disturbing to realize that he must have known what he was doing – committing 100 million people to a painful death in the 20th century, making the industry presently responsible for 26% of all cancer deaths.

Anchoring of the model – acid rain

Acid rain and the dying of fish and life in lakes has been high on the list of environmental disputes since the late 1950s. The phenomenon of acid rain and acid soil has been known for a long time, especially around coal-fired power plants. To deal with local emissions, the early 20th century solution was to build ever higher factory chimneys, so that pollution would be dispersed as widely as possible. But higher smokestacks just carried the pollution further away. In Scandinavia, it was shown in the1950s that imported acid rain, mainly from Germany and Great Brittain, had caused massive fish kills. In the United States, in 1955 the Ministry of Agriculture had established an experimental farm, the Hubbard Brook Experimental Forest in New Hampshire. In 1963 "the Hubbard Brook Ecosystem Study" concluded that acid rainfall affected fish, plants and wildlife in the area. This was the first scientific validation of long-distance acid rain in North America. Also, similarly to the situation in Scandinavia, it was found that acid rain in the US caused fish death and environmental deterioration in Canada, resulting in a long dispute over damage and compensation.

Again, the denial model was at work. The "expert" role this time manifested itself in an "expert panel" under President Reagan, chaired by (the later Star Wars promoter) William A. Nierenberg, an "insider" in the administration without any experience of acid rain research whatsoever, and a good friend of "denial expert" Frederick Seitz. Also on the panel was Fred Singer, an outspoken Austrian American, who had served as secretary of the newly erected EPA – Environmental Protection Agency under Nixon, but now a fellow spokesperson for the neoliberalist order of minimal regulation. The strategy of the "expert panel" was to put a stop to both the acid rain research and the settlement negotiations with Canada. Also on the table was Singer's cost-benefit argument, that mitigation would be costlier than the savings from a tougher emissions-standard. A final settlement came six years later, under President George H. W. Bush, with an amendment of the Clean Air Act, setting new standards for industrial emissions.

Counterparts to the denial experts were independent researchers and researchers associated with the EPA. In 2003, the EPA delivered a report to Congress, showing that the costs of cleaning up would amount to $8-9 billion, while the benefits would add up to $101–119 billion – not quite what Singer had argued. The early researchers on acid rain were right about most things, and the sad thing is that we still (in the 2020s) have too much acid rain, and that this is part of the collapse of ecosystems and the mass extinction of animals and plants.

The case of atmospheric ozone

Thomas Midgley Jr. was not only the inventor of high-octane gasoline and massive lead poisoning. He is also credited with the invention of CFC in refrigerators. Early refrigerators could be fatal due to toxic, flammable gases used in the cooling function, and Midgley decided to hunt for a gas that would be stable, non-flammable, non-corrosive and safe to breathe – ending up with a chlorofluorocarbon gas, CFC.

From the early 1930s, CFC was used, not only in refrigerators, but eventually in thousands of products, from ventilation systems to spray cans, deodorants, and spray paint. There were only two main problems, and they were only discovered several decades later. First of all, CFC destroyed the ozone layer in the atmosphere, and secondly, CFC was an extremely potent greenhouse gas.

The ozone layer in the atmosphere may seem slight, like a three-millimeter protective shield around the Earth, dampening ultraviolet radiation from the sun. Also, the emissions of CFC gas might seem trifling – only one part per billion in the entire atmosphere. But CFC is extremely aggressive. Half a kilo of CFCs can capture and destroy forty-five thousand kilos of atmospheric ozone. And CFCs will stay in the atmosphere for an average of 100 years, all the time attacking the ozone layer, and all the time contributing to the GHG climate effect. Continued CFC-caused ozone depletion would cause skin cancer and melanoma (the most dangerous form of skin cancer), cataracts, and impaired immune systems, as well as severe damage to animals and plants. CFC is also ten thousand times worse for climate than a CO_2 molecule of the same size – making CFC perhaps "the worst invention of the century" (Bryson, 2004, p. 190). Fortunately, there was always far less of it than of CO_2.

It was early shown that chlorine gas might be a threat to the ozone layer, and researchers for a long time researched emissions from supersonic aircraft, spacecraft and volcanoes. In 1970 a British researcher, James Lovelock was able to identify large quantities of CFC gases stored in the troposphere, and in 1974, the Dutch scientist Paul Crutzen - who later was to coin the term "Anthropocene" - was able to demonstrate that chlorine escaped into the atmosphere from sources other than airplanes, rockets and volcanoes, finally revealing that mass emissions of CFC gases were the major threat to the ozone layer (Molina and Rowland, 1974). The agenda was set for yet another denial war.

The chemical industry and associated product industries, industry associations, conservative think tanks, and interest groups mobilized to argue that the problem was not man-made: There could be no gas leaks, or if there were, they would be too small to matter, or chlorine could not be released from CFC gas to cause ozone depletion. Active searches for "other explanations" were initiated, including emissions from volcanoes. The fact that the degradation process of CFC is very complex made it easier to cast doubt on the CFC-theory. Still, all the main objections were easily countered by researchers already in 1975 and 1976. Nevertheless, the denial continued. Chief architect behind the "counternarrative", was again Fred Singer (Oreskes and Conway, 2011), steadfastly arguing that the ozone depletion was just a part of natural variation. He continued with denial, even as Sherry Rowland, Mario Molina and Paul Crutzen received the 1995 Nobel Prize in Chemistry for their contributions to the understanding of stratospheric ozone chemistry. Frederick Seitz, by now a senior among deniers, even in 1994 – seven years after the Montreal protocol, published a report at the Marshall Institute, repeating old false claims about lack of evidence and disagreement among scientists.

As late as 1985, there were still no international agreements on protecting the ozone layer, largely thanks to opposition from the Reagan administration. In the meanwhile, leading British scientists in the field managed to convince British prime minister, Margaret Thatcher, a former Oxford chemistry student, of the dangers of the CFC gas. She gave an important speech to the UN General Assembly and wholeheartedly supported the Montreal Protocol from 1987 on protection of the stratospheric ozone layer– probably the most successful environmental agreement in UN history. She also managed to exert influence on the United States to join the agreement. Under "reaganism-thatcherism" she agreed with Reagan on much, but not on everything.

It is also of interest to note that the scientific advice, and the attention stirred by denials, in this case seem to have caused sales of CFC-containing products, particularly in the US, to rapidly decline. In the US, the first regulations were introduced in 1977, but by then sales of such products had already fallen by 75 per cent. Consumers preferred "safe" products, even when they didn't quite understand the root causes of the scare. Today we take it for granted that deodorants, cleaning agents and paints do not need to be based on a harmful gas.

The denial of climate change

We have noted that the "other causes" or "other explanations" strategy has been used repeatedly by deniers to spread doubt and confusion. Search for "other explanations" has been an important ingredient in the "denial model". But it is also an important ingredient in standard scientific work. We should always be open to alternative explanations, and it can be difficult to distinguish between honest scientific openness, on the one hand, and the strategic uses of "other explanations" to divert attention and add to confusion and doubt.

Of course, at an early stage of climate research, it was certainly appropriate to ask critical questions about measurements and models, and to search for alternative, natural causes of climate change. There are also several examples of researchers who have started out with an attitude of doubt, only to be convinced by their findings about man-made climate change – and sometimes ending up being harassed for their honesty and bravery to speak out.

As a case in point, the British researcher, Hubert Lamb, had done research to estimate temperature variations in Europe over the last thousand years (Lamb, 1965). With no thermometer data available, he had to base his research on anecdotal evidence, studying annual growth rings in trees, and reading histories of wine yards in Southern England, Viking settlements in Greenland, "frost fairs" on the Thames river, and icebergs on the coast of Norway. His objective was to find out about the long-term European temperature curve of a medieval warm period, followed by a "small ice age".

Lamb's intent was never to calculate long-term, global temperature changes. Nevertheless, his findings caused some researchers to question the main thesis of climate research. If the observations from historical England and Europe were indicative of global trends, it would be a strong indication that factors other than the man-made greenhouse effect might be behind the recent warming.

Among researchers asking critical questions was Michael E. Mann, the man behind what the climatologist Jerry Mahlman later dubbed the "hockey stick" model of climate change. Mann decided to write a doctoral dissertation to investigate the matter, also considering "other explanations" (solar radiation, volcanic eruptions, underlying climate cycles). Neither Mann nor his dissertation advisor, Barry Saltzmann, were convinced that man-made emissions of greenhouse gases could explain the ongoing rise in temperature. However, no "other" factor was found. It was shown that the phases with a warm climate in the Dark Ages followed by a "small ice age" had been regional phenomena with only limited influence on global temperature.

Natural causes argumentation was zeroed out. At the global level, the hockey stick model represented a long straight "shaft" of stable global temperature for the last thousand years, and a short "blade" at a sharp angle for the late 20th century of definite global warming (Mann et al., 1998, 1999). The fact remains that not only Michael Mann, but a number of climate scientists, who went to work with a healthy scientific skepticism, ended up rejecting alternative hypotheses.

Instead of respecting the result, however, climate deniers hung on to their own misconceptions, even engaging in harassment of Michael Mann and other scholars involved (Mann, 2012). Michael Mann and his colleagues produced within a few years at the end of the 1990s what will perhaps stand as the most important milestone in the history of climate research. But not all were happy about the results. The denial model had been in play for several decades in American climate research, but with the "hockey stick" evidence, the denial turned uglier. On 17 November 2009, thousands of emails from servers at the University of East Anglia (where Mann worked at the time) were hacked and content stolen. Exchanges between climate scientists ended up in the hands of climate deniers. Excerpts of a normal scientific nature were "cherry-picked" and cited out of context to make conclusions about global temperature appear like a conspiracy.

Michael Mann describes what happened (Mann, 2012), and he does so with great pain. Individual words and sentences in his e-mails were taken completely out of context and turned into sensational news. The intention was clearly to harm Mann himself and his colleagues, plus climate research as such. Reconstructed formulations were quickly spread on the internet in well-directed PR campaigns. Groups with support from the oil industry and fossil energy ensured that the news was also spread to leading newspapers and TV stations in the UK, the US, and internationally. A cartoon was created to ridicule Michael Mann personally for trying to hide the alleged "fact" that the global temperature was actually falling. The video was posted on YouTube with sponsored advertising so that it appeared every time someone googled Michael Mann's name. It even appeared on CBS Nightly News, and the entire case complex was dubbed "climategate" (named after the Watergate scandal), and both the name and the countless clips and news picks were carefully orchestrated to humiliate those involved. The experience was of course a shock to Mann and his colleagues. Academic researchers are used to work peacefully, totally unprepared for harassment and ridicule based on fabricated lies.

The core of the conspiracy theory in question was that the climate scientists should have admitted to themselves that the rise in temperature had stopped, while the same scientists outwardly held on to the thesis of a man-made rise in temperature. The researchers were also supposed to have "tricked" the research results – allegedly to keep their jobs and project funds. The person to be "taken" was primarily Michael Mann – the man who had written many of the emails that were stolen and who more thoroughly than anyone else had investigated the very hypothesis that the deniers hung on to, namely that the rise in temperature had natural and not man-made causes.

A series of international investigations into the content of the emails and the distortions of meaning all concluded that the researchers had done nothing wrong, there was no conspiracy. An investigation carried out by the Science Committee in the British House of Commons concluded that what was formulated in the e-mails was completely in line with normal academic practice. The

University of East Anglia also carried out its own investigation in collaboration with the Royal Society and concluded that there was nothing to criticize the researchers for.

More denial strategies

Two Australian researchers, Haydn Washington (social-ecologist) and John Cook (physicist) wrote the book "Climate Change Denial – Heads in the Sand" (2011). There they summarize five main types of arguments used by the climate deniers. The first is the fabrication of conspiracy theories such as the "climategate" conspiracy. This is both about outright rejection of the greenhouse model, and the production of "other explanations".

Next on the list are claims that many experts do not believe in climate research and assertations of "research consensus". According to Doran and Zimmerman (2009), 97.5 percent of climate researchers are firmly behind the main conclusions and warnings.

A third denial argument is that climate models are unreliable, and warnings are exaggerated. In fact, there is a growing realization that the conclusions drawn by early climate researchers and by the UN climate panel were unduly cautious, with temperature rise and extreme weather consequences generally becoming worse than envisioned.

A recurrent fourth main argument has been that there have been climate changes in the past, and these cannot have been man-made. While this, for obvious reasons, is true since millions of years ago there were no people around, such episodes of climate change and extinctions (except for the 5th extinction, when an enormous asteroid crashes into Chicxulub in Mexico) have been much slower than now and based on natural factors.

Finally, Washington and Cook point to recurrent cases of "cherry-picking" from climate reporting as a fifth main form of argument – consistently pulling information out of context to support endless speculation about conspiracies and scientific failure.

The denier model has been adapted over time, but the main elements have remained – those of financial support from industries, political support from think tanks and far right conservative political parties, participation of alleged "experts" and "merchants of doubt", strategies of direct denial and fabrication of "alternative explanations", and – perhaps worst – direct assaults on honest researchers and the undermining of science as such. It is not a proud history, and it has evidently caused loss of time and resources for early mitigation. It remains to be seen, in a time with increasing manifest damage, how far into the future the "denial industry" will keep its hold on politicians and populations. Exxon Mobile, a secure supporter and financier of denials, officially distanced itself from the movement in 2014, after having promoted denial messages since 1989, among others with considerable "advertorials" campaigns – newspaper advertisement designed as a feature, editorial or objective journalistic article (Supran and Oreskes, 2017).

As long as damage has been at a stage of latency, there has been much room for speculation, doubt, confusion and denial. The big questions at the latency stage have been whether the prognoses will be realized, and whether it will be really bad if they are? As climate changes become real and damage become manifest, there will be less space for speculation, doubt and denial. Still, there will be impediments and stumbling blocks, and lots of potential for both denial, confusion and antagonism. There are also psychological reasons why this is so.

Psychological obstacles

The history of environmental protection, pollution and climate research has for long been marked by denial. With reference to Oreskes and Conway's "Merchants of doubt" (2011) we have touched upon a long history of refutation, spreading of doubt and confusion, and undermining the status of science.

Beyond straight denial, however, it also seems that even sophisticated prognoses about latent damage, built on extremely advanced computer models and presented in scientific reports of high standing, do not stir needed action. Environmentalist Charles Eisenstein says that our willingness to learn and act is heightened when something we love (nature, wilderness, birds, animals) is threatened – more than from reading about environmental problems in academic reports or from studying mathematical climate models. But even the most basic of all learning processes, that of learning from experience, and the ability to respond to loss of something loved, can be undermined in the ongoing environmental depletion process.

To Eisenstein's argument about love of nature as a driving force for change, we also need to add that attachment to traditional tools, instruments, and practices also is strong force, oftentimes working as a force against change – as noted in some of the "exnovation" and "resistance to change" literature. The attachment factor, and emotional matters in general, may in fact be much more important in organizational and strategic practice than commonly recognized in the academic literature. Our focus in the following pages will be mainly on cognitive responses to manifest change, starting with the shifting baselines issue. But behind the cognitive expositions, we should always also look for the emotional drives behind.

Stumbling or learning?

The Shifting Baselines concept was briefly mentioned in Chapter 2. It was landscape architect Ian McHarg who first used the term (1969), noticing that when people were asked to describe how a landscape was now compared to before, they often compared not to how it was before human intervention, but rather to how it was a short time ago, when significant intervention and damage had already become a new "normal".

The unfortunate consequence of shifting reference points is that we actually underestimate and under-communicate unwanted changes. Unwanted change is associated with perception of loss. Urbanized people who have minimal contact with nature and wildlife will tend to have a minimal sense of loss, when habitats are destroyed, and species disappear. They will have a very "theoretical" relationship to disappearance of wild animals, and little understanding of what is "normal". The "sliding baseline" theory states that over time there is a change in what we expect and what we come to understand as normal. Instead of experiencing change and damage, people undergo some kind of "change blindness", especially when change is not too abrupt. The first to point out our inability to detect change was William James, in his famous "Principles of Psychology" from 1890.

The World Meteorological Organization (WMO) established in the 1930s that there should be a common international standard for calculating normal temperature, that each period should last 30 years, and that the "normals" should be adjusted every 30 years. The idea was to avoid random fluctuations being interpreted as lasting trends, and to provide decision-makers around the world with a baseline for among others setting technical standards for buildings, roads and other infrastructure. In 2014, the WMO scheme was adjusted, so that the new "normals" had to be adjusted every 10 years – justified by accelerating climate change. The base period was still to be 30 years.

Given the "shifting baselines" syndrome, however, it seems clear that a more frequent adjustment of the "normal", while being good for building standards and infrastructure investments, will lead to a strengthening of "change blindness", more so among young people than among the older generation, and more so among city-dwellers than among people in constant contact with nature. Meteorologists and climate scientists around the world are deeply concerned about climate changes and nature depletion. On the one hand, they have to provide politicians and the building industry with up-dated information about what is to be expected in terms of "normal" weather. On the other hand, they also have a duty to inform the public that this "normal" is not really normal – in a bit longer perspective.

The cognitive distortion caused by shifting baselines and change blindness is only one among endless ways in which our brains pull tricks on us. Nobel laureate in economics from 2002, Daniel Kahneman, has challenged our assumptions of human rationality and decision-making biases in economics and everyday life, from his early cooperation with Amos Tversky (Kahneman and Tversky, 1984), to his best-seller "Thinking, Fast and Slow" (2011). Clearly, psychological mechanisms will interfere with our ability to respond rationally to manifest environmental damage.

Among brain mechanisms that may betray us, is Leon Festinger's concept of cognitive dissonance (Festinger, 1957; Festinger et al., 1956). The theory states that when a person's attitudes, beliefs, or behaviors are in conflict, the resulting "dissonance" (and discomfort) will cause the person to change his attitude to produce greater consistency (and comfort). Through participant observation, Festinger observed that members of a religious cult awaiting a doomsday flood held on to their belief even when the flood did not happen. The rejection to takte in the fact of no flood, was reinforced by the fact that many of the members had given up homes, jobs and family life for the cause. Instead of realizing and accepting what was going on, they fabricated new explanations and justifications for hanging on to their faith and staying on with the cult. A more everyday example is that if a smoker is confronted with information that smoking may cause cancer, he or she is likely to reject or downplay the information to bring greater consistency between behavior and cognition, arguing among others to themselves that lots of people live long and good lives even as smokers.

In the case of climate change denial, a person who is convinced that the whole idea of man-made climate change is a hoax, will experience dissonance, first when confronted with solid mathematical models proving the opposite, and later when confronted with experienced manifest damage. The tendency to reject new information, will also be stronger among people who have been outspoken in public about the denial conviction – similar to politicians who have problems admitting change of opinion. It is also often more difficult for people who consider themselves experts in their fields to admit that they were wrong – recognized as "the fallacy of centrality" (Weick, 1995; Westrum, 1982).

Another barrier to rational adaptation, closely linked to cognitive dissonance, is "confirmation bias" – a term introduced by the English psychologist, Peter Wason. Also referred to as "myside bias" or "congeniality bias", it is about our tendency to favor new information that is consistent with already adopted convictions (Plous, 1993; Wason, 1968). The bias works as a guide for seeking out new information, for what we pay attention to, for recall and strengthening of convictions, and for ignoring contrary information. It is largely unintended and subconscious, and it is a factor in daily life, business and politics, and even in all kinds of research. In scientific work it is about choice of theories to hang on to, for selection of hypotheses to be tested, and for choice of evidence behind conclusions. It influences everyday thinking, choice of news to be listened to, and decision-making in all kinds of organizations. The effect tends to be strongest in relation to strong beliefs, strong memories, and content that stirs strong emotions. Confirmation bias can be kept in check by conscious awareness of the fallacy, and by conscientiously checking out contrary information.

Related to environmental threats, both sides of the denial-divide may be subject to confirmation bias, and our main tool for rational behavior and cognitive correction is that of science and the critical and formal testing of propositions (such as in the above case of Michael Mann and the hockey stick conclusion – contradicting his own hypotheses about global warming). At a more general level, manifest damage will at some point in time reach a level that will force deniers to adjust their beliefs – whatever their relationships to science.

A matter of consciousness

In the English tradition, the word "consciousness" has been in use since the 17th century, focused mainly on sorting out what is morally right in different situations. Today, the term is more focused on the raise of awareness in connection with political activism – awareness of powers and mechanisms beyond direct experience. In the Marxist tradition, focus has been on the raising of awareness of the working class's common interests – beyond individual workers' personal experiences. It is about understanding the larger system that gives rise to exploitation. In a similar way, raising awareness about women's rights has been about the sharing of experiences and searching for structural explanations for gender inequality. Likewise, consciousness issues are at stake when giving minorities, ethnic and racial groups, gay people and disadvantaged people everywhere a common recognition of discrimination and a common ground for social movements and political action. Consciousness is not only about the spreading of information, but also about maintaining and strengthening motivation, cohesion and discipline in the ranks.

An important issue related to environmental crises is about the identification of awareness and consciousness related to attitudes to nature and the Earth, to the animal world and wilderness, and to man's position and role in the world – including destructive roles and responsibility for extinction. Environmentalists and concerned people are actively searching for a new common ground, for means to promote solidarity, for the strengthening of bonds between our generation and coming generations, and between humans and the animal world. Inspiration is sought from varieties of worldviews, from cultural traditions and myths, from indigenous societies and religious faith communities, from novelists and philosophers, psychiatrists, historians, anthropologists, and ecologists. The environmental crises, and increasingly manifest damage, cause disruption of values, beliefs and behaviors, opening up for radical change and new awareness, sparking new action (Gillespie, 2019).

At the precipice, people start asking questions about the predominant global, political order – about the co-occurrence of climate change and extinctions, financial crises and cyber warfare, terrorism and civil wars, and about a scenario of threats calling for a deep transformation of consciousness (Halal, 2014). Expression of alternative dimensions of awareness is also sought in music and alternative religions and faith orientations (Kerr and Parker, 2019).

Deep adaptation

Awareness, consciousness and concern are interlinked, and they are all at stake as we approach important tipping points and damage is becoming increasingly manifest. Still, people and politicians are slow to act.

Jem Bendell, who teaches sustainability at Cumbria University in England, came up with a thought-provoking alternative to traditional mitigation thinking. When he published the "deep adaptation" article (2018) the purpose was to challenge readers to reassess their work and their life situation. Confronted with something he saw as inevitable, namely an imminent societal collapse due to climate change, he launched what he referred to as a "Deep adaptation agenda" – for research and teaching, and for organizational practice, personal development, and public policy. After some turbulence getting the article published, it has since then stirred considerable debate and action – downloaded thousands of times and cited several hundred times. It has also helped give birth to a new social movements for "Deep adaptation".

Bendell asked himself five questions. (1) Is it sustainable for those who work in the field of sustainability to continue saying there is still time - it is not too late? (2) Have those working in the field even discussed the possibility that it is too late, that collapse is inevitable? (3) If not – why? (4) In what ways are questions about collapse raised in social media? And finally, (5) is it possible to come up with a "deep adaptation agenda" - some guidelines for adaptation in a situation where we are confronted with impending collapse?

Bendell compared our climate measures so far with the situation on the Titanic as it began to sink. Having failed to change course, the crew bent on all kinds of practicalities for saving lives. For Bendel, deep adaptation meant to realize urgency and mitigate for hunger, destruction, refugee flows, disease, and war. Endless repetition that "it is not too late" can be an alternative form of denial. We must not spoil the atmosphere around the dinner table, and we must not discourage young people who are happily participating in coastal clean-up campaigns. In Chapter 16 we shall ask whether optimism, contrary to hope, can work as an enemy of action.

Where is the balance between polite discourse and shocking news? A year before Bendel's publication, the New York Magazine chose to print an article by David Wallace-Wells related to "The Uninhabitable Earth: Famine, economic collapse, a sun that cooks us: What climate change could wreak — sooner than you think". He didn't just point to "dry" climate panel reports, etc., but sought to engage the emotions of the readers - to horror and alarm (Wallace-Wells, 2017). The reactions from climate scientists and nature activists did not relate to the veracity of the article – it was solid. Nor was there any debate about what had to be done. The debate that ensued was whether such thoughts and alarming expressions should even be communicated to an ordinary reading public. Senior climate scientists warned against presenting the problems as "insoluble" and preaching a message of doomsday inevitability and hopelessness, creating fear more than action. A sociologist referred to the article as "climate disaster porn".

What is the truth? And can most people bear to hear the truth? The situation for the researcher and the writer can be compared to the situation for a cancer doctor with a message for the patient. Should the doctor tell the patient that the particular disease is fatal in most cases? In the past, it was more common to emphasize giving the patient hope, avoiding bad news – even if the hope would not be well-founded. In our time, it is more common to offer numbers, probabilities and time horizons.

After thorough soul-searching, and finally rejecting reasons for moderation, Bendell concluded that openness was needed. He was convinced that after thirty years as a climate researcher and sustainability lecturer, he now had to be blunt. We know that thousands of people have been engaged, and that social movements have blossomed in the sequel. On the other hand, we do not know how many have reacted with despair and paralysis of action. Since the last climate summits, more scientists have come to speak more bluntly, stating that we are no longer heading for the Paris-agreement target of 1.5°C heating – but for worse.

Bendell himself advocated four main strategies for his "Deep adaptation agenda": (1) Resilience – strengthening the ability of society to recover from damage and develop new solutions and new products for handling extreme weather and other environmental challenges; (2) "Relinquishment" – the ability and willingness to let go of material luxuries and financial privileges; (3) "Restoration" – discovering and re-discovering practices and products of value such as rewilding of nature, local food recipes and local crafts, cultural activities, music and dance; (4) "Reconciliation" – a basic ethical strategy of admitting to having lived in violation of climate goals and nature conservation principles, and reaching out to all people of good will.

Bendell says more or less directly that the time for mitigation has run out. Disaster is on its way. We have to repair and adapt as best we can. But at the same time, the "deep adaptation agenda" calls for huge mitigation investments. Except for the ultimate chance of "irrevocable collapse" (Ord, 2020), the question of disaster is not an either-or, but a transient state of mitigation and adaptation, solidarity and support, threat and hope.

Organizational responses

We have hypothesized optimistically that manifest damage may contribute to a sense of urgency for change, mitigation, and adaptation. We have also pessimistically added that denial may persist, and that psychological barriers may obstruct rational action. To this we may add organizational barriers to action.

According to the "Threat Rigidity" theory, when faced with threats, organizations may react by narrowing access to information, "locking-in" established practices, and strengthening control from above, so that the organization "stiffens" and falls back on habitual patterns of action (Staw et al., 1981). This can be disastrous in a situation of need for drastic change. With threat rigidity comes also "cognitive simplification" and a tendency to skip over information that would offer a better understanding of the crisis. Instead, we tend to respond with standard solutions that may no longer work.

We may have ideas for what kinds of actions and practices may be needed in a future world of manifest environmental damage, but we have no science of alternatives that can give us the kind of flexibility, resilience, ingenuity and organizational development that will be required. We have no basis for "crisis training" of organizations (like for airplane pilots using simulator training to avert plane disasters), and we have limited space for experimentation with global disaster management, and for training people in adaptation practices.

Top business managers have sometimes initiated "autogenic crisis" – provoking a crisis situation in order to "train" employees for flexible and innovative action (Barnett and Pratt, 2000). But the anticipated environmental tremors will be of a size and complexity unfit for simulation. A Belgian study looked at the effects of job insecurity on work behavior and innovation. They found that people who become uncertain about job security in a crisis situation, and are afraid of their financial situation, respond partly with irritation and partly with poorer concentration on the job and a reduced ability to act innovatively (Van Hootegem et al., 2019). When it comes to global environmental and economic threats, with origins in complex causal relationships, and with a great variety of impacts, there is no known model for preventing people from "locking-in". Situations that may arise will vary infinitely, as will need for mitigative and adaptive action. "Lock-ins" to traditions, established practices and policies can then be strongly counter-productive.

Practical Business obstacles

With increased manifest environmental damage, we may hope for more willingness to change and innovate for the common good. But we may also expect more willingness to "greenwash".

The bright side of manifest change, and of a more concerned public, is that firms will have a strong incentive to innovate for environmentally friendly products and processes, and to streamline for energy and resources efficiency and green innovation. As the public become more concerned, and more critical of business practices that do not deliver on the green transition, however, organizations will also have a greater incentive to perform "greenwashing" and "impression management". Unfortunately, the most profitable short term outcome may be the combination of sticking to BAU while conveying an impression of proactive mitigation and adaptation.

Greenwashing and SDG-washing

In a situation of increased manifest environmental damage, companies that stray from true sustainability will become more visible – with a greater incentive to hide their neglects. Also, to the extent that firms in some areas of operations contribute positively to the common good, there may be a greater incentive to (unduly) exaggerate positive contributions. Since it can be quite expensive to improve performance, it may be cheaper to boast about achievements, even when these, on closer inspection, actually contribute little to sustainability and perhaps even add to damage. This is what greenwashing and "SDG-washing" (named after the UN Sustainable Development Goals) is all about.

Filip Johnsson and colleagues (Johnsson et al., 2020) looked at how businesses tend to report in relation to the sustainability goals, especially those about climate. They found that the pressure for a green transition, but also the tendency to greenwashing seemed to increase with increasing manifest damage and increasing public awareness. Many countries now require large companies to report regularly on environmental footprints and measures for a green transition, and many companies voluntarily report on achievements. Many also use independent consultants' evaluations. Given that the 17 UN Sustainable Development Goals were supplemented with 169 sub-goals, which allows for just about any achievement to be reported, there is at the same time great opportunity for reporting convincingly on some of the goals and forget about other more important targets.

In order to raise the standard for honest reporting, several initiatives have been taken. The Global Reporting Initiative is an organization that assists in the production of "tools" and setups for measurement and reporting, cooperating with the UN organization Global Compact and the World Business Council for Sustainable Development (WBCSD) to raise the standard. Nevertheless, there is documentation that only around half of target companies have identified sustainability areas, and only a quarter of the companies had any active goal and performance reporting in their business strategy (Scott and McGill, 2018). Another survey showed that although 40 percent of the largest companies commented on sustainability goals in their annual reports, only 8 percent reported on concrete business case' projects, and only 10 per cent reported on concrete results linked to specific sustainability targets and sub-targets. Many of the businesses that actually reported, seemed to use the reports "rhetorically", more as "camouflage" for BAU, without actually doing anything of consequence for the environment. They also found that large companies in Germany, France and the UK were significantly more active in reporting than companies from other major economies (Blasco et al., 2018).

Finally, a survey of the 2,000 largest companies in the world showed that the sustainability goals and climate measures received "limited" attention, and that to the extent that there was any involvement, or any form of commitment to the issue, this was more on a symbolic level – supporting the greenwashing hypothesis (Van der Waal and Thijssens, 2020). Overall, there seems to be a strong tendency for companies to report on ("cherry-picked") small steps of improvement, while neglecting to report on huge challenges that ought to be of greater concern (Kornieieva, 2020). Also, there is a tendency to report more about "avoiding harm" than about "doing something good", which may reflect a fear of "being caught" in doing something harmful, to be exposed in the media.

From denial to delay

Instead of explicitly presenting arguments of denial and doubt, and instead of engaging in economics debates about the discount rate, powerful and wealthy agents in politics and in polluting industries seem to have turned to a different tactic, that of delaying effective mitigative action as long as possible (Shue, 2023), making delay a new form of denial. While pretending to cooperate for the Paris climate-target and for the 17 UN Sustainable Development Goals, they are in fact under-cover agents resisting action. It is a form of "greenwashing", and it is directly related to the time-dimension of mitigation and adaptation. The longer you can delay mitigative action, the greater the gain for significant business sectors (while others have to pay for the future costs of delays). The "Production Gap Report" shows that the world faces a doubling of fossil fuel production compared to a level that would be consistent with the 1.5°C-warming pathway (Climate, 2023).

Among the many elements of delay tactics, Shue (2023) mentions several, starting with carbon offsets. The purchase of carbon offsets implies that instead of reducing emissions at home, a carbon credit or offset credit (some "token") is purchased from a certified institution for investment in an emission-reduction project somewhere else, normally in quotas of one ton of carbon dioxide, or another CO_2-equivalent emission. Relevant investment areas include forestry and renewable energy (wind, biomass, hydroelectricity), energy efficiency, and carbon capture and storage technology. The logic behind is to spend investment capital where the environmental payoff is greatest. In principle – and also in some practice – the method has been shown to be economically sound.

The problem is that all too frequently, the projects – on closer inspection – turn out to produce less climate effect than promised, with reported results being severely inflated. For the system to work, a positive effect needs to be truthfully reported, demonstrating that it will be long-lasting and that it will come in addition to what would happen anyway, also that it will be properly monitored and controlled. A nine-month investigation undertaken by the Guardian, Die Zeit, and the non-profit investigative journalism organization, "SourceMaterial", concluded that more than 90% of rainforest carbon offsets by the biggest certifier were worthless. Based on previous investigations (Guizar-Coutiño et al., 2022; West et al., 2020; West et al., 2023), The Guardian summarized the results (Greenfield, 2023):

> "The forest carbon offsets approved by the world's leading certifier and used by Disney, Shell, Gucci and other big corporations are largely worthless and could make global heating worse, according to a new investigation.

The research into Verra, the world's leading carbon standard for the rapidly growing $2bn (£1,t6 bn) voluntary offsets market, has found that, based on analysis of a significant percentage of the projects, more than 90% of their rainforest offset credits – among the most commonly used by companies – are likely to be "phantom credits" and do not represent genuine carbon reductions."

Not only are such offsets worthless, but they also add legitimacy to climate and environmental inaction, in addition to serving greenwashing purposes in the business community, undermining public trust, and sabotaging the work of honest people working for the environment. Unfortunately, the offset market is badly regulated and monitored. As we shall see below, with reference to the Microsoft case, the "carbon-negative" program had problems finding any worthy offering of carbon storage (Joppa et al., 2021). Most offers were for reforestation – but with no guarantee against logging, forest fires, draughts or other destruction. There is also an irony in the opportunity for oil companies to earn offset credits for carbon capturing used to expand and prolong oil production.

The Scope 3 detour

Another important form of delay and prolongation is related to the scope of emission accounting. While a new standard of financial reporting includes some form of reference to the "net zero" target of zero emissions, or a formula for accounting [zero emissions, plus some "necessary" emissions, minus carbon capture and sequestration], few companies have effective reporting of progress, or any clear idea of how the target will be reached. Still worse, there is no international standard for defining areas of responsibility. 140 countries counting for 92 percent of global GDP and 88 percent of emissions, and 65 percent of the largest 2,000 companies now have net zero targets. Added together, more than 60 percent of global CO_2 emissions are covered by net zero promises, but according to the Net Zero Tracker, less than 10 percent of these are deemed to be credible (2023). What is often missing are answers to the "net zero for whom and for what" question.

The Greenhouse Gas Protocol was established in 1998 to generate comprehensive global accounting standards for the measurement and management of greenhouse gas emissions, and climate mitigation efforts, from business and the public sector. The Protocol includes several measurement elements and principles, such as relevance, completeness and consistency of reporting, transparency and accuracy. It also includes three main "scopes".

- ✓ Scope 1 covers all emissions within the corporate boundary – including fuel burned by the company in production, transportation and other (fugitive) sources.
- ✓ Scope 2 includes emissions from purchased energy – such as from power stations delivering electricity, or from sources of heat, cooling, or steam supply. This scope is estimated to contribute at least one third of global emissions (Sotos, 2015).
- ✓ Finally, Scope 3 includes emissions from both suppliers and product users in the full value chain, including final users. Properly practiced, scope 3 reporting would include the business-to-business (B-2-B) part of the "transfer " problem explained in Chapter 2.

Scope 3 accounting is the most sensitive, and oil companies in particular are prone to focus on Scope 1. They may also boost impression, and delaying the transition to green energy, by reporting on theirs being "the cleanest operation in the sector". For North Sea operations there has been a long debate over electrification of the oil platforms at sea. Electrification of platforms based on green hydro-electricity from shore will surely add to the greening of oil production. But the energy will have to be replaced by other, perhaps not so green energy on shore. Full electrification of the platforms will bring down the Scope 1 emissions, but not the Scope 2 and 3 emissions. For Scope 2, "platform electrification" proponents fail to account for electricity replacement on shore.

When hydro-electric power is supplied at sea, it cannot at the same time be consumed on shore or in the European market. For Scope 3, failure to report about emissions from the full value chain of oil, means that 90 percent of the problem is either overlooked or glossed over – covering transportation, refining, and final use by consumers.

It is sobering to notice that while several of the main fossil fuel providers are wholly or partly state owned, overseen by governments that have signed the Paris agreement and committed themselves to net zero targets, they still engage in Scope 1 reporting and decision-making, where Scope 3 would be the more sensible and responsible format for reporting. Whatever the politics of reporting, the result is mitigation delay. Some climate scientists have even argued that the whole concept of net zero should be dropped. Too many political and business actors use net zero rhetoric for deferral of mitigation and emission reduction by relying on future (hoped for) technological solutions as a license to burn now and pay later (Berman & Taft, 2021; Dyke et al., 2021).

Practical hurdles

Decisive for honest and realistic reporting is a sincere desire to contribute to environmental sustainability, both for one's own business and other actors in the entire value chain. Also, decisive for an objective assessment of accomplishments is a realistic valuation of practical hurdles and barriers to effective implementation.

An interesting case was reported by a team of Microsoft employees who participated in the company's "carbon-negative" program, and researchers who worked with concrete carbon reduction and documentation (Joppa et al., 2021). In 2020, Microsoft, together with 1,500 large companies, set itself the goal of reducing emissions from its own operations by fifty percent or more by 2030 and by a hundred percent by 2050. Continued emissions were to be balanced by investments in carbon capture and storage facilities. Among steps to be taken were the electrification of the car fleet, the phasing out of diesel as reserve energy, and (scope 3) reduction of emissions from the entire value chain. The addition of the supply chain meant that the emissions to be compensated in the case of Microsoft would be quadrupled relative to in-company emissions. There was also an up-graded target of by 2050 to have removed all of the company's and the value chain's emissions back since its foundation in 1975.

To achieve this goal, the company also imposed on itself an "internal carbon tax". It also established a "Climate Innovation Fund" of one billion dollars, and entered into a number of collaboration agreements with other companies for research and development. The company hoped to be a "carbon negative", which means capturing more carbon than it emits, as early as 2030. Over the past few years, Microsoft paid for the removal of 1.3 million tons of CO_2 from the atmosphere. They invested in reforestation in Peru, Nicaragua and the US, supported soil improvement for carbon storage in the US, and invested in CCS in Iceland with the Swiss firm, Climeworks.

In their report, the "carbon-negative" team refers to three main areas of practical problems. (1) About targets – it is unclear what exactly is meant by a "zero emissions" target in practice. It is easier to speak about global zero emissions, than to define targets at local and company level. (2) About results – there is an underdeveloped infrastructure for measuring and accounting for carbon emissions in individual business firms and value chains. (3) About solutions – it is a problem that the market itself, for all elements that belong to carbon capture and storage, is still badly underdeveloped.

We have already recorded, in Chapter 5, that adequate CCS and DAC facilities are far in between. When Microsoft entered the market with a strong willingness to pay to find someone who could take on the CCS job, they received responses from 189 companies. Most of them were offers to take part in development projects. Very few could offer on-going solutions, and only two of them met Microsoft's requirements for high-quality capture and storage, including a guarantee of safe storage for a thousand years. 95 percent of the proposals were based on storage in the wild, which only rarely comes with any guarantee of permanent storage. The carbon storage offered totaled around 170 million tons of CO_2, but only offers equivalent to two million tons satisfied Microsoft's quality requirements for permanent storage and protection. It was also a problem that carbon storage cannot be linked to any publicly authorized time horizon. A quality-conscious company will be interested in long-term storage and protection, but for the time being it is far more affordable to acquire short-term and cheap solutions.

Price considerations may explain why most of the offers were linked to storage in soil and forest. The costs of storage in nature were around $16 per ton of CO_2, while storage in mountains based on CCS technology was around $141 per ton. On the dark side, storage in forests is exposed to fires, bark beetles, infectious plant diseases, storms and changes in land and forest management. On the positive side, storage in nature may add to environmental protection, wildlife habitats, freshwater resources, biological diversity and protection against erosion.

To all the practical hurdles to a net-zero Scope 3 full value chain emissions control, there is also a default BAU-baseline issue at stake including final users. Two Australian researchers, Christopher Wright and Daniel Nyberg (2017) studied five major Australian corporations over 10 years, and found, as "An Inconvenient Truth" that great climatic challenges were gradually "converted into the mundane and comfortable concerns of "business as usual".

There is a race going on. We are approaching important tipping points for climate and nature. Political leaders in most parts of the world are going for BAU for a (slow) green transition, and the world of infrastructure, markets and business operations is facing great barriers to enact what (hopefully) most of them would like to do. The next two chapters are about the political orders and the concrete political-economic systems that tend to "lock in" BAU practices and resist needed change. A "sense of urgency" for change will need to include the change of orders and systems.

Chapter 9

From Progress to Resentment

Increasingly, throughout the 2020s, UN General Secretary, Antonio Guterres, has consistently warned about the dangers to the world community, about nuclear "miscalculation" or "misunderstanding", about the dangers of climate change and extreme weather, the dangers of increasing economic inequality, social injustice, and the dangers of alienation and conflict – "making our time the most dangerous" since the establishment of the UN. The message is also repeated by moral authorities, including Pope Francis, by environmentalists and conservationists, foreign aid workers, and concerned people everywhere. It was not meant to be this way. And worse: There is no easy way out. The world appears to be somewhat incapacitated in dealing with the environmental burden.

In the following pages we shall first take a historical look at the forces of anger and resentment. From there we shall turn to studies of disintegration and societal collapse, undermining our capacity for resolve and concerted action. Finally, we shall seek a diagnosis of today's barriers to solidarity and joint action, asking for a new political order – and for a new framework for political-economic renewal to be dealt with in the next chapter.

Historical anger and resentment

In Chapter 8 we noticed that not even the transition from latency to manifest damage may be sufficient to bring about resolve and concerted action. People do not always learn from experience. In order for there to be joint action there has to be some level of political agreement and cohesion in a nation.

In the present chapter we shall deal with several forms and levels of disintegration and paralysis of action. The predominant political order (of neo-liberalism) since the 1980s is in many nations under pressure, most notably in the US, but also in other nations with far-right and neo-fascist radical opposition – not conducive to handling the environmental burden, and not conducive to the maintenance of democracy.

"Ressentiment"

In sociology, the expression "social glue" is sometimes used to describe qualities of societies linked to cohesion, belongingness, solidarity, shared interests, and a common historical and cultural ground. "Social glue" makes societies resilient against disintegration. Classical works by Emile Durkheim, Max Weber and Karl Marx deal with glue-elements and the opposite – exclusion, contradiction, exploitation, alienation, and hostility. "Social glue" seems to be at the heart of social sustainability, also as a platform for political action for a threatened planet. The opposite of social glue is exclusion and deprivation, social and political weathering and disintegration, often linked to the French term *ressentiment*.

An impressive modern treatise of global social exclusion is presented by Pankaj Mishra in "Age of Anger – A History of the Present" (2017). Painstakingly Mishra traces the roots of international

DOI: 10.1201/9781003518815-11

and intranational resentment and rejection, conflict and terrorism, from Enlightenment Europe to his motherland India, and to former colonies and economic late comers around the world. He insightfully describes how the gaps, that first arose in Europe and eventually in the colonies and in the fringes of European domination, connected to ideologies of partly European and partly local and national origin, have turned to explosive eruptions in an age of political and religious activism, supported by state-of-the-art weapons and digital communication tools.

Mishra's overall theme revolves around a gap that creates contradictions and hatred. Gaps and tensions have existed in most societies at all times, but the overriding modern gap emerged in Enlightenment Europe, in the tracks of capitalism, industrialization and modernization, spreading from England/Great Britain and France, to Germany, and on to the rest of Europe, to the US and other parts of the world. It is a gap between the development insiders and profiteers on the one hand and outsiders and underprivileged people on the other – between the "haves" and the "have nots", between defenders of cultural values and traditions versus the worshippers of progress, between present wellbeing and future expectations, leading to frustration, anger, and violence. It is, however, impossible to have a full grasp of gaps and contradictions behind the anger, hatred and violence of our days. Classical Marxian class distinctions have been blurred by added contradictions linked to the complexity of labor markets, and to race, religion, and ethnicity, to LGBTQ-issues, environmental and climate politics, immigration, and neo-colonialism – among others.

Mishra traces the roots of the gaps, and his search for a philosophical footing, in 18th-century perception of the French word "*ressentiment*". The word reflects a psychological state of suppressed feelings of resistance and hatred, of lack, deprivation and omission, and sometimes envy or jealousy. Nietzsche referred the word "*ressentiment*" to the despair of the weak and unsuccessful, and their hatred of strong and successful people.

In most modern discourse the main opposing philosophical positions of Enlightenment are traced to Thomas Hobbes and Jean-Jacques Rousseau (Graeber and Wengrow, 2021). The backdrop image is from the biblical Book of Genesis, about an initial state of innocence, then the fall from grace, from which all kinds of sins and tribulations, inequity and conflicts results. Hobbes took the fall from grace as a starting point, leaving no choice for mankind other than living with an authoritarian government to prevent the "war of all against all". A hundred years before Hobbes, Niccolò di Bernardo dei Machiavelli, had in 1532 published "The Prince", insisting that the leader had to apply power and even deceit to maintain authority and save the state.

Rousseau imagined a pre-history exempt from the fall from grace, a long-lasting "state of nature" based on egalitarianism and voluntaryism, only to be destroyed by modern means of hierarchy, discipline, use of force, subordination, and punishment. These are issues to be examined in Part III. To be noted here is that Mishra follows Nietzsche in tracing the philosophical anchoring to the opposites of Voltaire and Rousseau rather than Rousseau and Hobbes. There seems to be a direct line from Enlightenment European *ressentiment* to present-day revolts and disintegration around the world – preventing concerted action for the environment and for AI-regulated development. And for understanding of modern *ressentiment*, the pro-modernizationist Voltaire offers a more relevant case than Hobbes.

Voltaire – embodiment of progress

François-Marie Arouet, or Voltaire, was a prime spokesperson for modernization and progress. Heaven was to be realized on Earth. His dream was a worldwide materialistic civilization, based on rational self-interest – established by a victorious intellectual and commercial elite – led by "*les philosophes*", top-down. In "Le Mondain" (1736), a philosophical poem and an ode to the good life, he heralds a moral revolution that would fundamentally change Western culture and form a new beginning of a modern world, where under the "financialization" of capitalism, the stock exchange emerges as the new basis for human values, replacing religion and traditions of austerity with money as the "language" that unites us all, and - not least - with an unblemished admiration for the ones who devote their lives to personal enrichment. The "mondain" life is a counterpart to medieval pastoral, sober, frugal, God-fearing morals as the essence and heart of a good life. Voltaire praised growing international trade and colonialism, material wealth and "consumerism", and he legitimized a life of luxury and comfort.

In "Philosophical letters" from 1773, he praised the well-traveled merchant who made a fortune and at the same time contributed to the country's wealth. Of course, Voltaire personally was involved in trade, and sometimes in more dubious financial operations, and he came to personify the new ideal as the unified embodiment of the *philosophe*, the "nouveau riche", and the enlightened and successful citizen. He died a rich man - with income from authorship, consulting fees from royal houses, real estate operations, financial operations, lotteries, moneylending, and watchmaking.

Voltaire is also referred to as a pioneer of human rights, but his main ambition, and that of like-minded *philosophes*, was to spread the message of modernization among heads of states and enjoy the incomes that came with it. It was less important whether his state-leader "apprentices" and "disciples" were real tyrants like Peter the Great and Catherine the Great of Russia. Human rights beyond market-economic rights were of less concern.

Rousseau - the counterpoint

Pankaj Mishra sees Voltaire and Rousseau as standard-bearers for each front in modernization thinking and in the understanding of exclusion and "outsiderness". Although Enlightenment philosophers had much in common, there were marked dividing lines among them, notably in questions of modernization. The golden age of *les philosophes* was to end with the violence of the French Revolution, but the praise of wealth and modernization had for a long time met with criticism and moral outrage, most loudly and eloquently by Rousseau.

Where Voltaire embraced modernization as at singular great advance, regions and entire states – also in the heart of Europe – were left behind. In the emerging new world economy, potential trading partners were brutally downtrodden and reduced to colonies, slave economies and suppliers of raw materials. Huge new gaps were created, often worse than the old. The old gaps were legitimized through religious-cultural traditions accepting that royal positions and ditto privileges were determined by natural law and by virtue of God's grace. In a world order developed on the basis of human rationality, economic injustice was all the more unbearable. The deck of cards was rigged in someone's favor (and by someone else than God), and it caused alienation. In the words of Leonard Cohen: "Everybody knows the fight was fixed. The poor stay poor, the rich get rich."

Rousseau became the 18th century's most successful writer in Europe. He was himself an outsider. With only modest success he tried to find a place in the salon life of *les philosophes* and the successful Parisian bourgeoisie. And he realized that he didn't fit in. His awakening and his glimpse of a counter-vision are traced to a special experience one afternoon in October 1749, on his way to visit his good friend, the encyclopedia editor Denis Diderot, who was in prison for a mocking article about a nobleman. On a country road outside Paris, Rousseau read a newspaper and became curious about an announced essay competition asking whether the advances in science and art had contributed more to corrupting morality than to raising it. Suddenly he had a realization of the weaknesses and hollowness of the whole modernization project – a realization that would only grow stronger over the years. He wrote what was to become the winning essay in the competition, paving the way for a lifetime of writing focused on the proposition that the capitalist modernization project was practically and morally bankrupt – bringing forth the worst in people! His life task (as later for Marx) was to carve out an ideology for an alternative society!

Rousseau came to see the emerging commercial society as, above all, a class society, characterized by enormous differences between rich and poor, between callous financiers and everyone who fell outside, also a society with flourishing corruption, hypocrisy and cruelty. Basic traditional human and societal values were replaced with monetary wealth, greed, falsity and vanity, where the contribution of *les philosophes* was to legitimize the new order under the guise that it was modern and represented progress. From his upbringing in Geneva, Rousseau had experienced good people. The craftsmen he was with as a child were good people who did honest work – more honest than that of the snobs in Paris! The natural environment in Switzerland was beautiful and good – calling for "back to nature", while "man is born free but everywhere he is in chains".

There may have been an element of cognitive dissonance in Rousseau's longing for nature and traditional society. He didn't fit into the modern society, so he might as well reject it. Similar sentiments may also have been there when Fyodor Dostoyevsky visited Western Europe in 1862. There is in both cases an element of genuine sadness, of irrevocable and unrecoverable loss – that is today recognizable in the "Age of anger". It also seems to lurch behind the MAGA (Make America Great Again) movement in the US, behind the Russian Christian-fascism of Putin, and behind extreme-right movements elsewhere. The case of Dostoyevsky may add to our understanding of *"ressentiment"* under Enlightenment and also for our time, with geopolitical rivalry and the erosion of a social sustainability that we sorely need in order to save the planet.

Dostoyevsky's regret

Fyodor Dostoevsky was born in 1821, and he is perhaps the one among the world's writers who has best described the complex cognitions and emotions that go with feeling outside and rejected. In 1849, together with 20 other Russian intellectuals, he was arrested and sentenced to death. They had gathered to discuss books that were banned. At the last minute they were pardoned by the Tsar and sentenced to four years in prison in Siberia, to be followed by compulsory military service. In Siberia, Dostoevsky was awakened to religious mysticism and studies of the human mind. He discovered "the Christ of his childhood". The association with other prisoners and with people of simple circumstance gave him faith in the Russian people and traditional Russian ideals.

After 10 years in Siberia, he was allowed to return home in 1859, and in 1862 he embarked on a two-and-a-half-month train journey through Europe, later to be described in books such as "Winter Notes on Summer Impressions"; "The Gambler"; and "Letters from the Underworld" (Leivestad, 2021). During his journey, Dostoevsky wrote little about the places he visited, about the long train journeys and everything modern in technology, architecture, and production techniques. What appears to have raised his deepest concerns were the ruptures in his own soul. Seeking to experience everything, he discovered his own pettiness in the face of an overwhelming new world, and he probed into his own soul for the feelings that the modernity of the West evoked in him.

Facing in the West a superior force which the Russian could neither resist nor fully match, admiration turned to disgust. Dostoyevsky came to experience himself as a living inferiority complex, and he felt a strong urge to defend himself. Faced with Cologne's new railway bridge over the Rhine, die Dombrücke, the marvelous technology of the bridge enraged him. He felt that the toll collector looked at him, a poor Russian, with contempt. The urge to discover and experience was replaced by an urge to prove who he was. His Russian patriotism was wounded, and he came to enjoy putting down the Europeans. On the way to Paris, he had to think about how the Russians' long-standing admiration for the Europeans, and the Russian leaders' modernization ambitions, gave rise to a longing to be able to demean the Europeans. Faced with the most modern technology in London and Paris, he came to see civilization at its highest level of splendor and at the same time at its most hideous level of idolatry, reminding him of ancient Semitic Ba'al worship.

Facing a triumphant modern world, certain of victory and proud of achievements, he felt that he was facing an end-time phenomenon of biblical and Babylonian proportions, with old prophesies fulfilled before his eyes. Facing the Crystal Palace in London, he felt that the new creed of modern Europe, the creed of progress, had taken civilization no further than back to the beginning. The worship of the modern, and the submission to the materialism of the new age, was so massive that it made everything spiritual disappear from reality. He noticed that even London's many poor seemed to be dazzled by the magnificent modernity, captured by a monumental ruling spirit of progress.

If they did not share in the prosperity, the modern Ba'al culture had nevertheless captured their dreams....

Dostoevsky experienced the encounter with the modern as an existential threat to a poor Russian, putting him on the defensive and bringing out the Russian-nationalist and pan-Slavic ethnicity in him. He observed in Russia how the influence from the west led to the division of people and the weakening of the nation. He felt that even the most "Europeanized" among Russian intellectuals, bureaucrats and leaders were never, in the eyes of the West, anything more than primitive Russians, condemned to second-class status, condemned to a life as outsiders.

Millions of people from far and near have since Dostoevsky visited the great cities of Europe and North America and probably returned home with thoughts and feelings similar to those of Dostoevsky. Still more millions have never had a chance to visit and instead gathered impressions from TV and movies and rumors. They experience, on the one hand, impression and attraction, and on the other, disgust and resistance.

A time of strife and "outsiderness"

The stories that Pankaj Mishra tells from countries like Turkey, Iran, India, from former colonies and from the periphery of European and American trading domains, have much of Dostoevsky in them. The alternation between being on the one hand impressed and on the other full of contempt, full of admiration and at the same time dismissive, is repeated again and again. To recognize "outsiderness" in our time, we only need to connect to economic disparities and refugee disasters, refusal of- and discrimination against immigrants, unemployment and degrading working conditions, multiplied by modern weapons, nature depletion, climate disasters, food insecurity, massive urbanization and neo-poverty, and a pandemic spread of "fake news" on social media. They all produce threats against the "glue" of society, with disintegration and the erosion of social sustainability and agency to follow.

There are today "Dostoevskyits" among world leaders, visibly in Russia, but also in the MAGA-Trumpian America, and behind the Chinese strive for economic world leadership, behind North Korean sable rattling, and behind the formation of BRICS (the grouping of Brazil, Russia, India, China, and South Africa), representing 42% of the global population, and striving for Non-Western world dominance. Some of these developments may encourage strengthening of social sustainability – others may come to undermine such sustainability and the potential for concerted political action to deal with our modern threats.

At the height of the medieval migration period in Europe (300-600 AD), also known as the Barbarian Invasions, some 750,000 migrants may have been involved (relative to a Roman Empire total population of some 40 million people). Causal explanations for the grand migration include overpopulation, climate changes, pestilence, and hunger. For a present day comparison, on World Refugee Day, June 13, 2023, the United Nations High Commissioner for Refugees (UNHCR) reported about 110 million displaced people in the world, more than ever before (https://www.rescue.org/article/110-million-people-displaced-around-world-get-facts). The number includes refugees – people who have had to flee their homes and cross national borders due to war, violence, or persecution; asylum seekers – people who for similar reasons seek protection in a country of destination, hoping to obtain refugee status; and internally displaced persons – people who have not crossed national borders, but who have been driven from their homes due to war, violence, and persecution. At the end of 2022, 62.5 million of the displaced were internally displaced. An increasing number were also classified as "climate-refugees".

According to The Geneva Academy and International Humanitarian Law and Human Rights, the Middle East and North Africa are by 2023 affected by at least 45 armed conflicts; the rest of Africa is affected by 35 such conflicts; Asia by 21; Europe by seven; and Latin America by five. In some countries, internal displacement is severe, such as in the Sudan and Congo, in Syria, Ukraine, and Afghanistan. Some nations are affected by several conflicts at the same time, often with ethnic groups, rebel groups, religious extremists, and militia groups at war with each other (such as in Myanmar, Ethiopia and Congo).

Most wars are civil wars. Increased poverty and stagnant development are important causes, but also main consequences, with closure of schools, destruction of infrastructure and disintegration of societies. In recent years, there have also been more criminal networks that engage mercenaries – such as the Russian Wagner group operating in the Sudan. According to the Uppsala Conflict Data Programme, such foreign interference is the case in almost half of domestic conflicts and civil wars. The conflicts are often partly a cause and partly a consequence of refugee situations. The number of people who depend on emergency aid for survival has recently doubled – mainly due to armed conflicts. The outlook for concerted environmental action is not impressive.

Suggested remedies – development and growth

The world community has sought to deal with immiseration and *ressentiment* by setting development targets. The United Nations Sustainability Goals for 2030 were set in 2015, as a successor to the UN's Millennium Development Goals from the year 2000. The new goals included 17 target areas, and no less than 169 sub-targets, covering both social sustainability and physical sustainability. Unfortunately, the trend at the mid-2020s is for international aid to be reduced – not increased, with the closing of the US Agency for International Development (USAID) in February 2025 as the most serious blow.

Where *ressentiment* is a cause of anger and discordance, social sustainability is about the resilience of societies and the world community, and about the ability to mobilize to save environments and societies in times of high risks and internal strife. There is as yet no precise definition of social sustainability (Eizenberg and Jabareen, 2017; Åhman, 2013) and also, there is no overriding theory or decisive empirical studies demonstrating that social sustainability will be effective in reducing *ressentiment* and anger. Detailed lists of social qualities constitutive of social sustainability have been suggested. Related to urban environments, and in addition to physical qualities, Dempsy and colleagues (2011) came up with among others education, social justice, democratic participation, health and quality of life, social inclusion and the elimination of social exclusion, social capital, community cohesion, social networks, social interaction, belongingness, employment, voluntary organizations, and cultural traditions. Some of these are overlapping. Many of them are also overlapping with the United Nations Sustainable Development Goals from 2015, and overall, several of them focus in particular upon belongingness and cohesion, and the avoidance of exclusion.

In addition to social sustainability indicators for welfare and agency, there are also several indicators of happiness and well-being relevant to *ressentiment*. Factors behind happiness are quite similar to the welfare elements of social sustainability, related to economic growth and distribution. Predominant (BAU) thinking assumes that economic growth is necessary for employment, happiness and human well-being, and also for the quality of communities and social sustainability, even for climate and environmental mitigation. However, as will be seen in Chapters 11–12, the relationship between economic growth, distribution and well-being is quite a bit more complex than assumed by the BAU-position.

Lack of happiness among the poor is not balanced against increased happiness of the richest in wealthy countries with weak distribution. It has also been shown that high levels of happiness can be achieved in societies with relatively low total income (Layard, 2005). And even in low-income societies, economic growth and monetary wealth may not be the most important – personal freedom, social acceptance and inclusiveness may be more crucial (Inglehart et al., 2008). In other words, dealing with ressentiment and anger, conflicts and exclusion, is not just a matter of standard economic growth.

A note on progress and the "good Anthropocene"

The National Geographic Education section defines The Anthropocene Epoch as "an unofficial unit of geologic time, used to describe the most recent period in Earth's history when human activity started to have a significant impact on the planet's climate and ecosystems." https://education.nationalgeographic.org/resource/anthropocene/). In this book I have avoided reference to the Anthropocene, for a very obvious reason. In March 2024, The International Union of Geological Sciences (IUGS), after 15 years of debate decided to reject the term. They did not reject the notion of human impact on the planet, but they found it difficult to define the specific geological context, pointing also to a potential "recency bias". Without doubt, 15 years of debate has brought significant ideas and facts to the forefront of the climate and nature discourse – focused mainly on the negative environmental impact on the planet.

Since 2015, however, a counter-movement has emerged, focused on the "good Anthropocene" and "economodernism". An Ecomodernist Manifesto was produced by 18 initiators (https://www.ecomodernism.org/manifesto-english) suggesting that there are really no planetary limitations to growth and economic expansion – at least not at a level that will not in due time be overcome by modern science and technology. Human creative powers will overcome problems of climate change, food production etc. The only real threat to continued progress is that of self-doubt (Ellis, 2011). What is ahead is an "amazing opportunity", and we will be "proud" to exploit it.

Clive Hamilton has offered a sobering analysis of the ecomodernist ideology (2016), however, and he compares the secular "*anthropodicy*" of the ecomodernists to the religious "*theodicy*" of Leibniz and Hegel. Evil in world was explained in *theodicy* as part of God's ultimate design, to be endured in patience for the long-term ultimate Godly good. Pollution, extinction of species, and human suffering from environmental depletion is explained in *anthropodicy* as temporary contradictions in a long-term secular realm of infinite natural recovery and infinite human capacity for adaption. Like for evil, some people will come to suffer, but on the way to universal happiness, health and intelligence, temporary suffering is among minor problems, justified by the ultimate benevolence of the whole progressive humanist project.

The "good Anthropocene" in the Manifesto and in supportive writings, is presented as an optimistic initiative at a time of much pessimism. In Chapter 16, we shall appreciate the difference between optimism and hope in some detail, with optimism anchored in personal disposition or temperament, and hope anchored in attachment and in some measure of probability. As for ecomodernism, the fact that some enthusiasts are personally optimistic about progress does not add to my hope for the planet. At a time of increasing manifest damage, accelerated heating and nature depletion, media full of denials and lies, slowness of mitigation and adaptation both in business and politics, the fact that some people hang on to naive optimism and continuation of BAU is not a cause for hope. I still hang on to hope because I believe people and the global community is capable of change and a break with the tradition of blue-eyed progress and optimism. The problem with the *theodicy* 200 years ago was that in a world of God's ultimate goodness behind every act of evil and suffering, there was no ground for criticism, protest or change. The problem with the *anthropodicy* of today is similarly that for every act of destruction, there is no ground for change, since it is all part of an overall scheme of progress. A main strategy in former periods of climate change was that of migration. That was relatively easy at a time of a small world population and no national borders. It will be considerably more challenging in a modern world of huge populations, immigration controls and armed border patrols. Just to note one important difference from previous adaptive situations.

Disintegration and societal collapse

Our main concern in this chapter is about not only the slowness of environmental mitigation and AI regulations, but also about direct confrontation and negation of mitigative efforts, and about the potential for concerted action. There is an urgent need for agency to deal with the environmental burden and the need for AI-development regulations. Also, beyond the slowness of bureaucracies and the petty feuding of political processes, we have already recognized a backdrop of anger and *ressentiment* adding to delays and contradictions. Topping anger and ressentiment, there is also in today's world a real chance of societal breakdowns, disintegration and despotism – even in leading societies. Such breakdowns will add to problems of transitional rejection and delays.

In an impressive literature review of "societal collapse", covering 361 articles and 73 books on the subject, Brozović (2023) noted a lack of consolidation and synthesis of collapse research. In the following, we shall consider both historical collapses and future outlooks, starting with alternative causal understandings.

The conservative position

The modern conservative orientation can trace its roots to the influence of Edmund Burke (1729–1797). While the French Revolution in 1790 was still at a relatively moderate level of violence, Burke predicted in "Reflection on the Revolution in France" that it would turn into a state of terror and dictatorship, and he traced the main cause of this projection to its rejection of tradition and inherited values. While not expressing himself in the modern language of systems and complexity theory, his conservative argument was that the society is an extremely complex system, emerging from an infinite number of interactions and adaptations, of people finding solutions to problems, and of society generating institutions to deal with matters of coordination and law. Given the complexity of interactions and institutions in traditional society, a revolution setting aside traditions, structures and institutions would never suffice to fully replace what would be lost. Installing new and better solutions, based on abstract ideas and untested hypotheses, would simply be beyond the capacity of lawmakers for insight and solutions, a huge gamble, destined to fail.

The full inheritance of society, making up this complexity, included not only property and economic activity, but also language, culture, morals, common sense, and practical everyday solutions. Burke rejected the idea of society as based on a hypothetical social contract – as suggested by Thomas Hobbes, John Locke and Jean-Jacques Rousseau, suggesting instead that the "contract" also included agreement with previous and future generations, and with the innumerable conventions of traditional society. Loyalty to future generations included responsibility for change and improvement, but only at an evolutionary pace and scale. For Burke, it even included support of the American Revolution (1775–83), which he saw as a prolongation of the British tradition of liberty.

Behind his conservatism was a not very optimistic view of mankind – caught in original sin and little able to better ourselves. There was room for improvement, but only through time-proven adaptations, "spontaneous order", "prejudiced" governance, and "latent wisdom". To speed up the process of change and improvement by revolutionary means and based on metaphysical doctrines would be folly, destined to produce despotism. The conservative moral imperative was to steer away from despotism – also meaning disintegration and state collapse. The moral collapse was to avoid and delay just about any initiative for improving the lot of poor people, and to do away with a great number of sensible laws and regulations.

The old format of conservative reasoning can be traced to present-day conservatism, on both sides of the Atlantic, not only as a warning against revolutionary change and despotism, but increasingly as a warning against just about any regulation of the market economy and the taxing of business. Lately, the scare of revolution has been transformed even into hostility to the green transition and to control with information technology.

The thinking behind

We normally link the "neoliberal" conservative turn to Reaganism and Thatcherism of the 1980s, but in fact the turn was long in coming. The intellectual content for backing a new political order had matured over several years. For academic backing, Milton Friedman's "Capitalism and Freedom" (1962) was an early inspiration for market orientation – building also on the ideas of the Austrians

Friedrich Hayek and Ludwig von Mises, and the American James M. Buchanan. Several writers and speakers contributed to a right-libertarian awakening, recalling values of classical liberalism, of unregulated markets, personal freedom and responsibility.

Of particular significance were contributions by Robert Nozick, in "Anarchy, State, and Utopia" (1974) and also the writings and speeches of economist Murray Rothbard, including "For a New Liberty" (1978) and several other books – calling for a "minimalist" or "minachist" state, or a "night-watchman state", and the reversal of the welfare state. Rothbard coined the term "anarcho-capitalism", and he also came to have considerable influence on Milton Friedman and on economists and lawyers inspired by "the Austrian School". Another strong influence on the right-turn of libertarianism in America was Russian-American novelist Ayn Rand, who presented a nearby philosophy that she called "objectivism", and whose bestselling books, "The Fountainhead" (1943) and "Atlas Shrugged" (1957 [1992]) came to have massive impact on American libertarianism, on the general far right-turn of politics, the cementation of plutocracy and the formation of the later neo-liberal political order.

For political engagement, the "Confidential Memo to the U.S. Chamber of Commerce" by (later high court judge) Lewis Powell (1971), fronted a warning that "the American economic system is under broad attack" …. "an assault on the enterprise system". Powell argued for a "comprehensive counteroffensive" from the business world against the prevailing New Deal-inspired political order, uniting the forces of his own tobacco industry cohorts with the emerging Coch brothers' fossil fuel network, to establish a broad front against the established political order, anticipating the later Tea Party movement, and instituting a strong network of conservative and anti-environmental think-tanks in the process. Powell also had detailed plans for how the message was to be planted in schools and educational institutions, in networks and organizations, for short- and long-term implementation, with secured financial backing by the rich elite, and with ambition for a qualitative change of the American judicial system and judicial practice in favor of business. Following the leads of people like Friedman and Powell, large numbers of influential American economists and lawyers took part in the wholesale promotion of privatization, deregulation and taxation reliefs for the wealthy.

Behind the rather rudimentary neoliberalist turn ("The government is the problem"!) was a considerably more sophisticated discourse starting in the 1940s in response to totalitarianism in Germany, Austria, and Russia, and in America as a counterforce to the New Deal. It was at first about how to explain the collapse of the rule of law and democracy in Germany, Austria, and Italy, and also about the intellectual premises of the new despotic order. Among the main contributors, a leading intellectual (economist, social scientist, and philosopher) was the Austrian Friedrich Hayek (Johnsen, 2025). Hayek in a way substituted the Burkian scare of revolution with his own scare of ideas for "societal design" – which he saw as precursors to totalitarianism.

With democratic governments increasingly ambitious in the quest for an ideal society, with political decision-making increasingly anchored in the works of experts, with the New Deal in America enhancing the role of government, and with political expertise increasingly based on economics and social science in the logical-positivist tradition, Hayek came to fear the ambitions for planning and "societal design" as stepping stones to a totalitarian society. Ambitions for political planning and design were primarily traced to socialist and social democratic governance – much as Burke had traced the threat of revolution to the impact of the French *philosophes* of his time.

Hayek's great vision was that of liberalism, and the conditions for a liberal society. The idea was expressed most popularly in his bestseller "The Road to Serfdom" (2007 [1944]) and more deeply in his follow-up, "The Constitution of Liberty" (1960). Both publications stirred popular interest and triggered decades of scholarly and political discourse. In addition to the threats of "societal design" and planning, important topics debated included the tension between the rule of law and the limits of democracy, the essence and prerequisites of liberty, institutions derived from a spontaneous order and self-regulation, and the role of morals.

Hayek's vision was about political liberty across all sectors of society. The self-organizing and spontaneous order of societies as basic to the rule of law was in contrast to planning and design.

However, what ended up as main inputs to the neoliberal turn decades later, was essentially the deregulation of business and lower taxes for the wealthy. Especially in the American far right movements of the 1970s and 1980s, business was singled out as the main arena for an emergent, self-regulated order. As for explanations for societal collapse and despotism, skepticism to planning and design linked to socialism and social democracy prevailed.

In terms of empirical anchoring, on the one hand, it is clear that the Soviet variety of communism included at the heart the idea of a planned economy and – more broadly – the "scientific management of society" (Afanasyev, 1971), including the economic, the social, the technological and even the spiritual sphere. On the other hand, even so, most Western democracies have structures and institutions covering all such spheres without turning despotic and without employing totalitarian means for political execution.

As for the roles of experts at the expense of democratically elected representatives in the formation of policy, the net long-term consequence in democracies seems to have been mainly to lose grass roots support (a real problem for European social democracies), not for such impact to cause despotic or totalitarian consequences.

The extreme irony of it all, is that the end consequences of the (Hayek-inspired) singularly pro-business, de-regulation, de-taxing and "supply side economics", underlying the neoliberal order, has caused mass immiseration and societal disintegration that may potentially lead to despotism and totalitarianism in our time. It is about the creation of an ideology that in consequence pave the way for exactly the kind of development it was meant to prevent. Clearly, we need to look for other explanations for totalitarianism, including the German and Russian versions. An important such voice is that of Hannah Arendt.

Totalitarian movements and totalitarian rule

Explanations for totalitarianism are important because they may inform our understanding of political order, *ressentiment,* anger and disintegration in our time. Besides the potential for totalitarianism, these are outcomes that may effectively undermine concerted mitigative action to deal with global threats, the green transition and the AI transition.

At the opposite end of explanations from that of conservatism, Hannah Arendt in "The Origins of Totalitarianism" (2017 [1951]) sought explanations for despotism in the misery of Europe in the aftermath to the First world war, and in the failure of democracies to include and engage large masses of people. Having dismissed nationalism and antisemitism as root causes, she looked to "outsiderness", *ressentiment*, and immiseration as triggers of change. Social movements with fascistoid and totalitarian leanings emerged in most smaller European nations during the 1920s and 1930s. But according to Arendt, the full transition to totalitarian rule occurred only in the larger countries of Germany (and only at the time of war) and in Russia (following Stalin's purges in the 1930s). The naivité of European democratic governments was not rooted in the dreams of "societal design", but in their belief that democracy already worked to include people in forming the society and for meaningful participation in political parties, labor unions, and democratic organizations, assuming that there was governance and control of the masses of people.

Leaders of totalitarian movements were often self-pronounced deviants starting their careers by "boasting of their past crimes and carefully outlining their future ones", convinced that "evil doing in our time has a morbid force of attraction" (p. 401). In Russia, the Bolsheviks were clear that they did not recognize conventional moral standards, and in both nations masses of people embraced a mob mentality of violence and crime directed at outsiders to the movement.

The amazing step towards full totalitarian rule, however, was when not only violence against outsiders and opponents, but also against members of the movement was also tolerated, even including oneself as a victim. In both countries people might participate in framing their own death sentences out of loyalty to the movement and its leader, and pride in their personal participation. At that stage it is not any more about idealism and convictions, normally subject to experience and argumentation. Only identification with the movement and its leader, total loyalty, conformity and subordination remains.

Such identification and conformity does not come from class or party or conventional organizational membership. It is cultivated instead among the classless, the unorganized, unemployed and "outsiders" – people who may never before have participated in political movements. Both the Nazis in Germany and fascist and communist movements in Central and Eastern Europe after 1930 recruited members among the "mass of indifferent people whom all other parties had given up as too apathetic or too stupid for their attention" (p. 408). These were people not concerned with scholarly debate or conventional political argumentation. They were positioned outside of the party system, more geared for terror than conviction, for death sentences rather than persuasion.

Hannah Arendt noticed two illusions among democratic leaders, illusions that resulted in the success of the totalitarian movements. First, there was the illusion of majority participation and support of democratic rule. Secondly, there was the illusion that the politically indifferent masses did not matter – that they somehow were neutral, inarticulate, adding only their silent approbation and tolerance.

Unfortunately, similar illusions seem to have guided mainstream politicians in both North America and Western Europe in our time. In France, there was an awakening when many of them all of a sudden put on yellow wests. In Germany there was another awakening when the AfD (Alternative für Deutschland) started climbing in the polls – towards majority in much of the former East Germany, since "die Wende" seen as politically and economically somewhat indifferent. In the UK there was a similar awakening only after Brexit was a fact. And in the US, there was an awakening, but little effective learning when weathered politicians in 2016 started asking themselves where all the support for Donald Trump came from. In all of these cases we may search for various forms of "indifference" as common causes of protest.

Arendt traces the full step towards totalitarianism in Germany to the harrowing war-time experience, and in Russia and the Soviet Union to the not less harrowing Stalinist terror of the 1930s. Stalin rose to power after Lenin's death in 2024, but the extreme terror lasted through much of the 1930s. To gain absolutist power, first out for sacrifice were the Soviets (the elected labor councils), the core of the emerging industrial labor class, by Stalin to be replaced by the centralized party bureaucracy. Next was the liquidation of property owners and in particular the *kulaks* – freeholding peasants, potentially the largest oppositional grouping or class. Beyond direct liquidation, artificial famine and deportations in the name of collectivization killed millions, in particular in the powerful republics of Kazakhstan and Ukraine, where during the *Holodomor* of 1932–1933 millions succumbed. The working class, smaller and weaker than the peasants, were manipulated by the Stakhanov system of competitive slave labor and exhaustion. Finally, it was time for the people responsible for execution of the Stalin's liquidations orders to have their turn of being liquidated. In the 1936–1938 period, not only bureaucrats and leaders, but also military officers and top commanders were slaughtered, as was the elite of industrial enterprises (engineers and managers), and leaders of cultural organizations, government bodies, as well as party officials, and innumerable ordinary citizens as victims of denunciation, envy, and ambition. It was all part of Stalin's unlimited striving for totalitarian control. The liquidations of course were disastrous for the economy and for overall societal development. But that was of lesser concern to Stalin. Hannah Arendt reminds us that Stalin's victims were not hostile to the regime or likely to become hostile. Since the 1920s there was no organized opposition, neither within nor outside of the Party. Neither was there in the 1930s any threats from the outside.

Arendt makes a distinction between "dictatorial terror", aimed at enemies and authentic opponents, on the one hand, and "totalitarian terror", threatening harmless citizens without political opinions on the other. Totalitarian terror was used for making the final step from totalitarian movement to totalitarian regime, creating "the new man". She refers to Himmler, who defined the SS member as a "new type of man who under no circumstances will ever do a thing for its own sake" – only for Hitler and the Party. In the Soviet Union, constant denunciation of friends, colleagues and family, and repeated purges followed by liquidations served the errand of mass atomization, breakup of family and social ties, and a peculiar form of individualization, where the individual in a strange way was absorbed into the movement and the regime under the banner of absolutely undisputed loyalty.

The explanation of totalitarian rule is not a prime theme of the present text. But the link between *ressentiment* and the rise of totalitarianism and disintegration is of high interest, since disintegration is a direct hindrance for concerted action. In Russia, the Putin regime scores high on a scale of totalitarianism, with the full control of media and a high control of minds. In the West, Trumpism and far right movements are moving up the ladder of movement loyalty and disregard of democratic tradition, directly sabotaging the green transition. In both instances, the status of the "indifferent masses" have been a main activating force.

Explaining state disintegration

Recognizing the possibility of societal collapse as a plausible trajectory for nations and the world, Brozović (2023) notes the many disciplines involved in collapse research, including history, archeology, and anthropology, but also public health, futurism, mathematics, ecological economics, environmental and climate research, philosophy, literature, systems and complexity research, political science and policymaking – each with their own methods and terminologies, some pointing to history – others to the future. He also mentions laymen ideas and science fiction contributions to the issue. It is a staunch reminder that more than one factor may be involved behind disintegration and eventual collaps – also that there may be many different forms and many degrees of severity, including revolutions, dynastic succession, radical change of governance, fratricide (the killing of brothers to avoid succession rivalry), and ultimately civil war or revolutionary war.

The varieties of succession and disintegration modes is in itself a warning against looking for a single cause. At the same time, there is a scientific interest behind the search for ultimate causes, and there is also a political ambition behind our search for good explanations. There are good reasons for asking what went wrong since democracy is now under threat, and the green transition is sabotaged by democratically elected governments.

There have been several attempts at explaing the great collapses of societies, including Edward Gibbon's "The History of the Decline and Fall of the Roman Empire" from 1776; Oswald Spengler,'s "The Decline of the West", from 1926; Arnold Toynbee's "A Study of History", from 1946 and 1957; and Pitirim Sorokin's "Social and Cultural Dynamics", from 1937. While these authors deserve recognition of both their immense knowledge of a wide array of historical facts, their publications have not produced a more generalized outlook, useful for understanding disintegration in our time. Great works have also been produced by researchers of Mesopotamian and Eastern Mediterranean civilizations, the Maya civilization, the Byzantine Empire, the Easter Islands (Rapa Nui), Andean civilizations and former Chinese dynasties, as well as the rise and fall of Mexican and North American indigenous cultures.

An overriding concern in this book is about the failure of nations and societies to engage in concerted action for dealing with the environmental burden. We evidently live in a time of widespread societal disintegration, and we need to learn why this happens (with communication technologies present everywhere, and with economic resources available at unprecedented levels). We also need to learn more about what can be done to promote integration for solidary agency and social sustainability.

Since we know well that historical societies, large and small, have experienced periods of integration and disintegration (including civil wars), the question of what can be learnt from history – if anything – is of great importance. Quite pessimistically, the German philosopher Georg Hegel famously said that "the only thing we learn from history is that we learn nothing from history." Winston Churchill, equally famously, and a bit more optimistically claimed that "those that fail to learn from history are doomed to repeat it". While history never repeats itself, at least we can learn from previous mistakes.

A challenge facing modern history researchers and also the rest of us looking to history for solutions to grand problems, is whether or not historical understanding can be converted into a science for projection of developmental patterns. For the improvement of large-scale problem solving, there is an ambition to make history more than a discipline of inspirational storytelling, and the use of historical knowledge more than a gamble of cherry picking.

Among candidates for causal explanations, exaggeration of complexity as societies mature has been suggested as a prime driver of disintegration (Tainter, 1988). Following up on Tainter's work, there has been a chain of contributions using mathematical modelling and large datasets for the study of complexity buildup in societies, from 10.000 years ago to the present, and from small and backward societies to great empires and modern nation-states (Turchin, 2018; Turchin et al., 2022). There has also been studies of societal longevity, including outside or geo-political forces and foreign intervention, and natural forces such as earthquakes, climate change, and volcanic activity. Generally, however, explanations for disintegration have increasingly been traced to intrinsic societal mechanisms, like man-made economic inequality and depletion of natural resources (Scheffer et al., 2023). Specifically, the researchers point to over-time increasing concentrations of wealth in societies, with increasing gulfs between elites and the rest, a heightening of corruption, and less effective decision-making, exacerbating social ills, increasing violence and civil strife, and weakening societal cohesion – which again may reinforce inequality, corruption, and disintegration, exhausting governmental resources, also making the society more vulnerable to external stressors.

Immiseration and elite overproduction

From the early observation of complexity and intrinsic qualities as cause of collapse, to a recently more pointed scientific understanding of social disintegration and state collapses, Jack Goldstone led the way in 1986. He introduced the mathematically based Political Stability Index (PSI) to explain the forces behind the English Revolution of the 1640s, followed in 1988 with similar disintegration cases from 17th century Asia, and in 1991 adding the cases of the French revolution of 1789, the Ottoman crises of the 17th century, the collapse of the Chinese Ming Empire 1644, and the series of revolutions in Europe in 1830 and 1848 (Goldstone, 1986, 1988, 2016 [1991]).

Since the early start, the basics of the model has been refined by a great number of publications, the original dataset has been expanded by several hundred historical cases from different corners around the world, and the variables have been improved for more efficient measurement, elegantly summarized in Peter Turchin's book "End Times", about "Elites, Counter-Elites and the Path of Political Disintegration" (2023).

Rhythms of integration and disintegration

Turchin bases his treatise on the novel discipline of *cliodynamics*, studying history as a science of change, backed by elaborate statistics and mathematical rigor, and tested on innumerable cases of state integration and disintegration. The main contribution of the *cliodynamics* field (and of Turchin's book) is the exposure of definite cyclical patterns of state integration and disintegration, marked by phases of internal peace, social and political stability, and cooperative political elites, followed by periods of fragmentation, instability, uprisings and violence, with breakdown of cooperation among elite groups, resulting often in civil wars and prolonged unrest. While the historical data indicate varieties of severity, duration and consequences of the integration/disintegration pattern, no historical societies seem to have escaped from the overall pattern of recurring disintegration, and very few states have lasted (without periods of disintegration) for more than 200 years.

Based on a crisis database (CrisisDB), the Political Stability Index (PSI) and *cliodynamic* analysis, two main forces take the driver's seat to instability: (1) popular immiseration (economic impoverishment) and (2) elite overproduction (too many candidates competing for elite power positions). The first of these, immiseration, is well known, and probably a favorite explanation by both scholars and lay people. From slave uprisings to the *ressentiment* of Rousseau to Marxian capitalist surrender, immiseration, the exploitation of labor, and the extraction of surplus value by the wealthy have been recognized as drivers of unrest and uprisings, causing instability and disintegration.

The second main driver of disintegration is more surprising – elite overproduction. Turchin uses an intriguing (and charming) metaphor to explain elite overproduction, the game of musical chairs. Practiced around the world for entertainment at birthday parties and the like, the game includes a number of participants moving around an equal number of chairs – minus one. They are accompanied by music that stops abruptly, for everybody to sit down – minus the one who finds no chair, who therefore leaves the game. Then another chair is removed, and the music starts all over, until the next person is left out. The game goes on until only two participants (sometimes a bit roughly) compete for the last chair.

Elite overproduction is not about one chair and one winner, however. Turchin invites us to imagine a different version of the game, where instead of subtracting a chair, there is an addition to the number of participants – keeping the number of chairs constant. As more participants are added to the game, incentives to bend the rules are increased. The game becomes more and more rough, until it breaks down. That is what is meant by elite overproduction, causing intra-elite competition, rivalry and conflict, in societies undermining fiscal health and undermining trust in the state, in the end causing state disintegration.

Other causes may serve to suppress or speed up disintegration. Geopolitical position and violent policies by neighboring states may add to the disintegrative forces of a state, but an external common enemy may also cause competing internal fractions to temporarily unite. The spread of ideologies and radical political messages within or across national borders, may also influence disintegration – but neither geopolitical conflict nor the influence of ideologies appear to be universal phenomena behind disintegration: Immiseration and elite overproduction carry the day.

Turchin cites Arnold Toynbee for concluding that great empires do not die by murder but by suicide! He also acknowledges that state disintegration may not always be grim: Some societies may gain from the downfall of a regime. The descendants of rivalling elites may also learn to avoid strife and violence – postponing disintegration for some time. But the next generation may again forget and engage in another deadly chair game.

While in America the civil war of 1860–1865 remains the most important disintegration reference point, scoring high on both immiseration and elite overproduction, Turchin observes that in 75 percent of the times the pattern of immiseration and elite overproduction have reached levels similar to what is now (2024) in the US, it has led to the disintegration of the state, frequently involving revolution or civil war. According to Turchin, the extremely divisive presidential election of 2024 should be seen as an outcome of immiseration and elite overproduction, not only of concern to Americans. Since the early 1800s, the US has been through three main time periods of disintegration, integration, and again disintegration. From around 1830 to around 1930, the US experienced a long period of disintegration. From 1930 to around 1980, there was a prolonged period of mainly harmonious integration. And from 1980 to the present, there has been a new era of increasing disintegration and fragmentation. Interestingly, while the MAGA movement in the US has never made clear exactly when the US was at its best (referred to for becoming "great again"), judging by podcasts, etc., most seem to have the 1930–1980 period in mind, exactly the period in US history when MAGA ideals were subject to the strongest regulations and the strongest redistribution of wealth.

The workings of immiseration

Turchin bases most of his up-dating of disintegration on the case of the US. The enormous change of political order, and lately of geopolitical orientation is of interest not only to Americans, but to concerned people everywhere. President Trump's direct confrontation with climate mitigation and the whole of the green transition, is of course of prime interest to environmentalists everywhere, meaning also that understanding of the disintegrating forces behind is key to understanding why global mitigation and change are so slow coming, meaning also that immiseration in the US is a key variable in the dynamics of environmental policy. Immiseration is a key correlate to what Sarah Arendt called "indifferent" masses of people, again a key correlate to political extremism – potentially leading to disintegration, but also to despotism and totalitarianism.

From the late 1970s and early 1980s towards the present, a new phase of disintegration in America has emerged. Economic growth and rising average income in the US have hidden the fact that the last 40 years have been a period of dramatic immiseration and the rise of an indifferent mass of people. Immiseration is expressed in many ways:

- ✓ GDP (gross domestic product) and GDP per capita numbers are useless measures for the testing of impoverishment of a society. GDP per capita may rise even as most people are getting poorer, as long as the earnings of the wealthiest go up more than the earnings of the bottom poor go down. It all depends on the sharing of overall growth in the earnings.
- ✓ A much better indicator is the median income. But there are also more refined models, to which we shall return in Chapter 12. Turchin refers to the US Census Bureau figures for the 1976–1920 period, showing that the average (not median) household income rose from $ 61,896 to $ 89.683 – by 45 percent. In contrast, the more informative median household income rose from $ 52.621 to $ 63,683 – or by 21 percent.
- ✓ The "household income" improvement hides the fact that it was increasingly based of two income earners, meaning that many were doing more work for less pay. Over the same period of time (1976–2016), the median (inflation adjusted) hourly wage increased from $ 17.11 to $ 18.90 – which means an increase of only 10 percent. A 10 percent gain over a period of 40 years is not impressive, but it still gives an inflated view of development.
- ✓ Controlling for education, the picture turns even more grim. Those with the least education (high school dropouts, high school graduates, and those with some college education but no college degree, totaling 64 percent in 2016), all lost ground! For people with only a high school education, wages declined from $ 19.25 to $ 18.57, and for those who had not completed high school, the hourly wage declined from $ 15.50 to $ 13.66.
- ✓ Overall, 64 percent of the American population experienced reduced real income. Among them we find a majority of the "Deaths of Despair" victims, those dying from suicide, overdoses and alcoholism, linked to poverty and lack of access to higher education, as described by Anne Case and Angus Deaton (2020).
- ✓ Economic progress is often explained by the amount of time used by the present generation for various goods compared to how much our parents and forefathers had to work. To pay for one year in college in 1976, a median American salaried worker would have to put in 150 hours of work. In 2016 he would have to put in 500 hours for the same education. That is not exactly economic progress. College education is in many ways the number one marker of difference between success and mass indifference.
- ✓ A second marker of economic progress is the possession of a secure home. To pay for a decent house or apartment, the median American worker would in 2016 have to put in 40 percent more hours than in 1976 – enough to produce widespread despair among people with a "price adjusted, real income" rise of 10 percent over 40 years.

The causes behind increasing immiseration include the erosion of labor unions and collective bargaining traditions, the inflation-erosion of the minimum wage, globalization and outsourcing of jobs overseas, monopolization of businesses and monopolized price-setting, automation and robotization of domestic jobs, and the financial crisis (of 2008). In turn, the skewing of wealth distribution is also a cause of loss of political breakthrough. Martin Gilens and Benjamin Page

(2014) made an interesting study, linking political preferences to income, comparing the preferences of the bottom decile income group, to that of the median and top income groups, covering nearly two thousand policy issues. What they found is truly astonishing: The preferences of the poor simply had no effect on policy changes in the 1981–2002 period in the US. Policy outcomes represented solely the influence of the (lobby-assisted) top income decile. While there were some areas of preference overlaps, policy changes on which rich and poor disagreed were always (!) settled in favor of the wealthy elite, reinforcing American plutocracy and immiseration, and undermining democracy and expanding the mass of indifferent people.

The workings of elite overproduction

Immiseration facts are relatively easy to find. Disagreement is mainly about causes and strategies for alleviation and redistribution. The concept of "elite overproduction", easily understood by aid of the musical chair metaphor, is much harder to anchor in empirical facts. Different states, in history and around the world, produce different elite traditions. Chinese dynasties have been built around central administrative elites. Egyptian powerhouses have relied on the military. Russian regimes, from tzarism to communism to neo-fascism, rely on centralized, authoritarian leadership controlling the administration, the judiciary, the military, and the oligarchical elites of economic life. The American elites, on the other hand, are solidly anchored in plutocracy – meaning government by the wealthy (as opposed to oligarchy, which means government by a much more narrow economic elite). Of course, high government officials and elected politicians also belong to the elite, as do high-level professionals like lawyers, and more generally highly educated people. But the dominant force in America is that of the plutocracy. Recently, however, with the rise of the tech-billionaires to political power, there has been a marked turn towards oligarchy. But also, true to the plutocratic roots, tech-billionaire oligarchic power may again be challenged by non-tec economic strongholds.

As for estimation of "elite overproduction", plutocracy is perhaps the most challenging form to study. How many greedy billionaires and top lawyers does it take to qualify as "overproduction" – counting in both economic impact and shares of political power?

- ✓ Top education (from law schools, medical schools, technology institutions, and relevant doctoral programs) used to be seen as a union membership card for elite jobs and elite living. That is not the case anymore. There is more competition for career jobs, and also more foul play.
- ✓ Elite overproduction is manifest when graduates for mainstream law schools seem to end up in two very different career trajectories. Turchin notices that since around the year 2000, first year salaries for the majority of law graduates have tended to settle at a level around $ 40 K, ten years later increased to $ 50 K. For a minority among the "elite aspirants", however, first year salaries were in the $ 125 range, ten years later in the $ 190 K range. There are simply too many aspirants for a limited number of "musical chairs", with a majority of the aspirants actually drifting into the *precariat* (below).
- ✓ Added competition has also led to a dramatic increase of social stress, and to the bending of rules, including cheating and corruption associated with acceptance to the most prestigious schools and universities.
- ✓ Another mechanism in the musical chair game, is simply payment for chairs – now a common standard, especially in American politics. According to Turchin, based on data from the Center for Responsive Politics, winning a race for membership in Congress would take an average "investment" of $ 400 K in 1990 – raised to $ 2.35 million in 2020. Similarly, membership in the Senate taking $ 3.9 million in 1990, would require an "investment" of $ 27 million around 2020 – paid by candidates out of their own or family pockets, in combinations with support from "sponsors". As for the purchase om legal court support, Elon Musk and affiliated groups set a new record in the April 2025 Wisconsin Supreme Court election, sinking in more than $21 million, with Musk paying three individual voters $1 million each for signing a petition to vote.

> This time, the plutocratic investment did not pay off, however. And of course, such spending is just a tiny fraction of what is paid for the presidency. In 2024 11,000 political groups and super-rich individuals "invested" $ 14,7 billion in the presidential and congressional election (https://www.bloomberg.com/graphics/2024-us-election-pac-spending-donations-fec/?embedded-checkout=true–). Elon Musk with a $ 277 million was just one of the contributors, with a total of tech-billionaires contributing near $ 400 million. By traditional democratic standards, it is about plutocracy leaning towards oligarchy – not democracy.

Normally, we think of elites as belonging safely at the top of society, and immiseration predictably hurting the bottom few, with a solid prosperous middle class and a well-organized labor class between the two. What is missing is the present tendency for elite fallouts to end up in the mass of "indifferent" people, not catered to by traditional party politics.

Guy Standing introduced the term *precariat* (2014), pointing to the collapse of the proletariat class, to a great extent replaced by a much more diverse "class" of underprivileged. He identified three defining characteristics of the *precariat*: First of all is a job situation, based mainly on low payment and "flexible" work contracts, with temporary jobs, part-time jobs, with no secure occupational identity and a constant need for retraining and reorientation, networking, applying for jobs, unpaid preparations for work, and the filling out of forms of many kinds. A second characteristic is the loss of non-wage benefits and security, such as pensions, paid vacations, medical security, and retrenchment benefits, traditionally a secure element of employment contracts. Thirdly, the loss of job security and non-wage benefits is also topped with loss of social security and public social provision schemes, traditionally a secure element of the national social contracts. The overall loss of such benefits, including also the stigma and shame of having to plead for benefits and services, is of high significance, much more than offsetting a meager 10 percent rise in median income.

The combined effect is the spread of feelings of deprivation, anxiety, alienation and anger – reactions generally associated with immiseration among lower income groups. But Turchin also notes how the growth of the *precariat* is in great part an outcome of elite overproduction. Membership in the *precariat* is normally recruited from the lower middle-class or traditional working class, with low education and low aspirations, from among immigrants and ethnic minorities, and now also from among the elite. The latter are generally well-educated people who have failed to make careers appropriate to their levels of education and family aspirations. They also represent the first "class" in history expected to have a level of education significantly greater than required for the job they are to hold, causing great status frustration and a sense of wasted investment in education and competence. They may be more future-oriented, and better equipped for organizing themselves than the former two groups, but they are still deeply stuck in relative deprivation.

Thus, the *precariat* is not just a traditional working class, with solidarity and shared goals befitting collective bargaining. They are without a common ground for reform, and there are hostilities between the different groups, much confusion about the political-economic premises and consequences, but behind it all a general tendency for the rejection of the mainstream political establishment, nurtured by *ressentiment* and the status of "indifference". As in the era leading to Nazism and Stalinism, the democratic establishment in the US, Europe and around the world has been locked into the two illusions identified by Hannah Arendt – the false belief in democratic control and the failure to recognize the potential power of the indifferent masses, when moved to act.

Disintegration means loss of resolve and capacity for concerted action for climate, wilderness and the environment, but also loss of predictability in areas of war and peace, pandemic controls, sustainability goals, AI regulations and AGI development. The question of re-integration and the resumption of stability and predictability, becomes an issue of utmost importance. There are strong signals in the world community, that some form of neo-fascism is in the pipeline for a next step in disintegration. It is not all about immiseration and elite rivalry.

Disintegration and totalitarianism

I have visited Russia many times. As a guest at the Soviet Union's Academy of Sciences in Moscow in August 1991, I was invited to a meeting with a group of politicians from the Duma. It was still Gorbachev times, and I was prepared for a conversation about perestroika and glasnost, democratization, and market economic reforms. Instead, I was asked if I knew about Lev Nikolayevich Gumilev – Russian historian, ethnologist and anthropologist. I was blank and had to apologize for my ignorance. The hosts gave me a short lesson on Gumilev's thinking: Regimes originate in ethnic groupings, *ethnoses*, and each *ethnos* undergoes a form of life cycle, referred to as "*ethnogenesis*", with birth, growth, maturation, weakening and collapse. For the birth of an *ethnos*, Gumilev refers to the force of *passionarnost* that sets an *ethnogenesis* in motion, and his theory takes on a more far-flung aura when suggesting that under particularly favorable radiation from space, a form of "*passionate*" energy, *passionarnost*, arises in some people, promoting leadership and expansive agency. The great conquests and the powerful kingdoms of history emerge when the ethnic culture reaches a stage of growth and climax.

At the meeting, I was also told that Europe was an old culture, rooted in antiquity, now in a phase of inertia and paralysis, with moral decay and fumbling leadership, on the way down, while Russia was a young ethnos on the way to growth and climax, charged with *passionarnost*, at a mature stage of an *ethnogenesis*. Left unsaid was: Ready for dominance in the world.

From passionarnost to neo-fascism

I didn't know it in 1991, but I have since noted Gumilev's incredible influence on Russian nationalism and the path to today's authoritarian rule. Very little of Gumilev's writings have been translated into English, but Mark Bassin, an expert in Russian history, geography and ethnography, at Södertörn University in Sweden, in 2016 published an excellent book: "The Gumilev Mystique: Biopolitics, Eurasianism, and the Construction of Community in Modern Russia" (2016), a book that is highly recommended for anyone who wants to understand today's Russia.

Gumilev's life-cycle perception of modern Russia, and its notion of a special world mission is traced to its mythical "inauguration", with the baptism of Valodymyr the Great in Kherson, Ukraine in 988, and the emergence of the Novgorod society. After centuries of "incubation", the country is now ready for blooming. According to Gumilev, the full ethnogenesis cycle would normally take approximately 1500 years, divided into two main phases, with Russia now entering into its most powerful, mature period.

In Russia in the 1990s, Gumilov was referred to as the most widely read and influential historian during post-communism. He was seen as a genius and compared to both Marx and Einstein. His books and writings were sold by the millions - also in the former Soviet republics. They were distributed to administrators and used as textbooks in high schools and universities. A university in Kazakhstan was named after him. So were streets and mountains, and statues were erected in his honor. He is still frequently mentioned in speeches by Putin (referred to as "extraordinary talent", with "unique impact"). Putin has said that it is Gumilev's thinking that lies behind his important "Eurasian Union" initiative, also evidently behind his justifications for the Ukraine war.

The power of speculative rightist philosophies, such as those by Gumilev, Ilyin (on Christian-fascism), and Dugin (on geopolitical hegemony), is not based on scientific accuracy, but on their coming in handy for a Putin-dictatorship and for appeal to nationalism in a country haunted by alienation, *ressentiment* and anger. It is a reminder of some main characteristics of fascism – nurtured not only in Russia, but around the world at this distressing time.

The fascist heritage

Historical fascism, unlike communism, astonishingly managed to appeal at the same time to both the working class and the upper class, to professionals and intellectuals, entrepreneurs and artists, mostly to men but also to women – without any unifying precise mandate, and with an appeal to both material progress and to some form of neo-religious awakening. This also makes it difficult in retrospect to say concretely what led to its remarkable progress, but also to its downfall, and presently to the spread of similar ideas.

A common feature was the cultivation of nationalism and opposition to internationalism. Classical fascism included ideas of racism, ethnic sovereignty and discrimination against Jews and immigrants. Today we find a peculiar far-right combination of Israel-support, sometimes combined with hidden anti-semitism, and with hidden racism and ethno-arrogance mainly expressed through opposition to immigration and harassment of immigrants.

Classical fascism promoted the acceptance of violence and glorification of war. Glorification of war is today mainly found in Russia and North Korea. Even Hitler avoided direct glorification of war: (1) Never refer to the war as a war; (2) Let it be clear that the war is destined to happen and that others started it, and; (3) Make it clear that it is a matter of life and death for the nation, as referred by Hannah Arendt in "Eichmann in Jerusalem" (2006 [1963], pg. 52). The modern version is seen in Putin-Russia's version of war against Ukraine: (1) The "war" was a "Special Operation"; (2) It was provoked by NATO's expansionist ambitions; and (3) It is necessary for the restoration and survival of the Great (holy) Russian Empire – a matter of life and death for the nation.

Glorification of the male and contempt for the female is still a feature of modern far-right movements in Europe and America. But with modernization, women leaders embracing basically masculine cultural values have gained respect. Against the wins of women's lib and LGBTQI movements, conservative family values, anti-abortion initiatives, and women subordination in the private sphere, have become more prominent.

A definite trait of classical fascism was leader glorification and subversion of democratic rule of law, and there is little doubt that person-cultivation is still part of the neo-fascist domain. Vladimir Putin, Donald Trump, and Xi Jinping are prominent examples. In Europe, France's Marine Le Pen, Hungary's Viktor Orbán, Italy's Giorgia Meloni, and Netherland's Geert Wilders are in the lead. In Russia and China, Putin and Xi are personally in control of legislation, and in the US, Trump, in a new-born holy alliance with tech-billionaires, is taking major steps toward personal authoritarian leadership.

Retrospective ideas of grandeur, past religious overtones and rituals have been part of the fascist recipe since the start. Both Nazism and Italian fascism contained elements of retrospective mysticism. The Nazis, not least SS chief Himmler, included archeology, ethnology and religious history in propaganda. In Germany, there was also a search for pre-Christian and neo-pagan notions of primordial wisdom. In Itali, the *mangiaprete* ("priest eater") Mussolini, after 17 years of harassment of the Church and the clergy, in 1939 found that he would make use of the church for legitimation, having his wife and children baptized, giving money for restauration of churches, and installing crucifixes in schools and courtrooms. In Russia today, the Orthodox church, under the rule of patriarch Kirill, is constantly adding its blessings to Putin. In America, for retrospection there is the MAGA (and the Tea-party) movement, and for religious overtones there are evangelical references to ungodly leaders in service to God. In Europe, far right parties are generally protective of Christian tradition – turning against Islam and Moslem immigrants. In China, communist rhetoric serves a legitimatizing role in replacement of the religious one, but like Leninism and Stalinism in Russia turning quasi-religious in hailing the leader.

At the level of liberty, democracy and freedom, classical fascism represented an organic view of society, with the sacrifice of individual freedom in favor of freedom for the people as a whole, under the auspices of the movement and the leader, the party, and the nation. This is perhaps the most challenging part of the fascist and neo-fascist positions in modern societies. In Russia, the sacrifice of individual freedom for service to the nation has historical roots, but is now challenged with the recruitment of young men for the war in Ukraine. In Europe, far-right parties are generally in support of individual freedom and national democracy, exploiting democratic rights and freedom of expression for bringing down democratic rule and subverting concerted action through the European

Union. With "surveillance capitalism" spreading from the US, and direct political surveillance (and face recognition) spreading from China, individual freedom is under threat, not so much from traditional fascist ideology as from an emergent tech-fascism, converting previous demagogy, rituals, parades and mass-rallies to new technological formats. Even so, there is little doubt that strong elements of propaganda and certain tendencies to demagoguery and person-worship is still in place, with reverence for rulers and even some amount of willingness to sacrifice individual freedom and human rights for the movement and the leader.

"Neo-tech" authoritarianism

Marietje Schaake has written a solid documentary on "The Tech Coup" (2024), about the erosion of democracy by digital means. Schaake is a former member of the European Parliament. She is presently International Director of Policy at Stanford's Cyber Policy Center, and an International Policy Fellow at the University's Institute for Human-Centered Artificial Intelligence. She has been seen as a proponent of liberalism, globalization, and digitalization. But her book is about the threats against democracy and the rule of law, and about the "tech coup" against the public interest.

The Internet and the introduction of digital social media came with fantastic promises – for the ending of poverty and the strengthening of democracy, for reduced racism and equalization of opportunity, for toppling authoritarian regimes and kleptocracies, for disclosing human right abuses and for progress in most speres of life. AI has followed suit with a quadrupling of promises, about innovation and efficiency, suggesting unlimited human development. And in fact, much progress has been recorded – some of it unfortunately prematurely.

The "Arab Spring", initially with uprisings in Tunesia and Egypt, was seen as the work of social media for democratization, and social media was hailed as an instrument for protests and coordination of reform movements, also in Iran and dictatorships around the world. According to Schaake, what really helped reformers, however, were not so much risky new media use as were traditional means of communication such as word of mouth – not vulnerable to electronic surveillance.

What was greatly overlooked by the enthusiasts, was the fact that the exact technologies that helped democrats and reformers were fully available for repressive regimes to halt the movements and imprison the activists. Authoritarian regimes were if anything more effective than the reformers in exploiting the new technologies. And the American and European enterprises developing the technology were equally eager to sell to both dictators and democratic regimes. The Pegasus system for governmental espionage, developed by the Israeli NSO company, a world leader in systems for counterterrorism and crime-fighting, has also been of great use for governments seeking authoritarian control over the population. Such systems can be of great help to autocrats to sustain power, useful for spying and hacking on the opposition.

At a more general level, the promise of the tech sector for general surveillance efficiency improvement brings potential for infinitely greater impact. In a world where "efficiency" carries greater weight than "rule of law" and democracy, the provision of public services, infrastructure maintenance, "homeland protection" and military defense, has gradually been shifted from the public to the private business domain. Elon Musk's Starlink satellite system is a case in point, with more than 7,000 small satellites in low orbit for telecommunications, and more than 5,000 more planned for the next few years, even with extensions to Mars. With coverage provision to more than 100 nations, offering mobile broadband access at competitive rates, it is now up to Starlink leadership, in effect one person, to control who is supplied with what information. The one person may be guided by interest in the common good – but there is no democratic control or rule of law for "quality assurance" on the part of the public. As in the case of Ukraine, seeing the military surveillance information stream turned off, absolutely crucial for the war effort, is an indication of a more general problem, of global consequence.

Winston S Churchill, is famously quoted (from 11 November 1947) for saying: "Many forms of Government have been tried, and will be tried in this world of sin and woe. No one pretends that democracy is perfect or all-wise. Indeed it has been said that democracy is the worst form of Government except for all those other forms that have been tried from time to time....". Democracy is the only form of government with a built-in capacity for self-correction. This capacity is now being undermined by the ongoing tech revolution.

Marietje Schaake (2024) offers several examples. Intellectual property laws were put in place before the coming of the Internet. They were never conceived for protection of proprietary algorithms. They are now in place for the prevention of democratic control with extremely sensitive information systems and contents. The Freedom of Information Act (FOIA), and similar legislation around the world, were put in place to ensure that the population and journalists gained access to salient information, a precondition for democratic rule and the rule of law. The same legislation is in the era of Internet and AI used to prevent access to exactly the same information, now in the hands of private businesses that have taken over accessing and holding of the data. The leading industry spokesperson, Peter Thiel, has demanded provocatively that we get rid of the government. We are not quite there, but gradually we are getting there by entrance through the back door, by letting the tech billionaires take over the job for the government, privatizing the information and excluding it from Freedom of Information. For assurance of no responsibility beyond profits, the industry is also hiding behind the Communications Decency Act (CDA) – Section 230 exempting platform owners from responsibility for content, while allowing for change (manipulation?) of content. The Internet is being regulated – to the full advantage of the companies and wealthy owners.

An important strategy for success in the hi-tech industry is called "Blitz-scaling". It is about the early grab of market share. It is a strategy that takes both patience and a tolerance for risk, favoring players with solid financial backing. The trick is to rapidly gain market share by offering services (such a Uber) at a low cost or free of charge (such as Chat-GPT), to be followed by price hikes when subscribers are hooked, then to exploit economies of scale with near zero marginal costs and full-price marginal income. It is a near-unbeatable strategy, unfortunately not available to the amateur investor. For the insiders, it is increasingly available also for government and military contracts, with more funding available the more public money can be freed from other purposes, in the US by means of (Musk's) Department of Governmental Efficiency (DOGE). It is about a "wealth pump" that was not available to the founding fathers, but useful for both economic and political power gains by the present plutocrats.

Overview

In Figure 9-1, two strains of conservative impact are traced. A historical line is drawn from the conservative heritage and the fear of revolutionary change, to historical economic and social inequity, leading to ressentiment and what Hannah Arendt called "the indifferent mass", allowing for the rise of totalitarianism – Nazism, Fascism, and Communism. The failure of conservatives to promote fairness and social progress is seen as a cause of indifference and the rise of an indifferent mass – to be manipulated by totalitarian movements. Also along the conservative line is the failure of governments to face up to the illusion of majority participation and support of democratic rule and the illusion that the behavior of politically indifferent masses did not matter – that they somehow were inarticulate and negligible.

A second historical line in Figure 9-1 is drawn from the same conservative heritage to the liberalist/libertarian protests during years of New Deal and social-democratic reforms in the US and Europe. Strong conservative political pressure over several decades caused a radical neoliberalist turn in the 1980s, focused mainly on the de-regulation of business and the lowering of taxes on profit and wealth. The result was in the 2000s for immiseration to rise and, as we shall see, for "elite overproduction" to cause political and economic rivalry, for a general weakening of democracy, and for political settlement in the 2020s with tech billionaires winning the day.

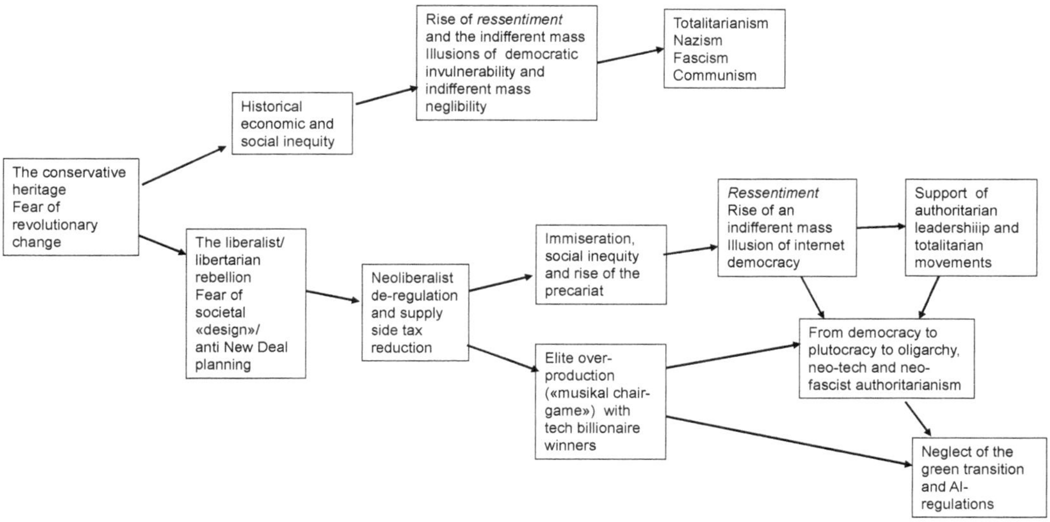

Figure 9-1: Paths to disintegration and paralysis of action.

Chapter 10

From Lock-in to Possibilism

In the chapters so far, I have flagged some very serious challenges to the world community, observing that the world is falling behind in our ability to respond to environmental threats and calls for AI safeguarding. Warnings apply to climate change and the Paris agreement, to nature protection and the Montreal targets, and to the Nairobi-UNEA initiatives for plastic and pollution. They also apply to the reactiveness of legislators in the face of an AI-revolution, to the stagnation in work for the UN Development Sustainability Goals and for social sustainability outlooks. Finally, the threats are enlarged in the realm of Russian geopolitical turmoil, Chinese empowerment, and the capriciousness of the Trump administration.

In the face of global threats, the community of nations has been characterized more by paralysis and disintegration than by resolve, solidarity, agency and concerted action. An important issue related to environmental threats is whether the world community is at all capable of generating a needed variety of mitigative action, or whether the present political-economic system has caused a lock-in to limit the action space. The present chapter is about opening up for possibilism.

Lock-in or open space

In Chapter 8, Naomi Klein's Shock doctrine was introduced, arguing that in the aftermath of shocking events such as extreme weather or social upheavals, chaos, confusion and the need to deal with a critical situation have been exploited by far-right change agents to execute strategies not deemed desirable by most people. In Chapter 8, I signaled a need to deal with what we might think of as a Shock doctrine version II, where powerful elites proactively create an environment of chaos, confusion and paralysis of action, in order to manipulate political developments for whatever hidden aims they may have in mind.

In my graduate course in international economics at the University of Oregon in 1970, professor Raymon Mikesell, the last surviving economist from the Bretton Woods conference, and a veteran advisor to the Kennedy administration, told our class that, even as presidents and government officials might come up with ill-conceived policy ideas, there would always be someone in the back office making the necessary analyses and corrections to get things right. Observing global politics today, and in particular the emerging American tech plutocracy, one gets the impression that the "person in the back office" must have left the premises.

In Chapter 8 we observed that with increasing manifest damage, and increasing costs of adaptation, there is hope for an increasing "sense of urgency" – for mitigation and resolve in facing up to the environmental burden and the digital challenges. In Chapter 9 we observed that with immiseration and increasing elite competition there is a growing indifference among masses of people. Indifference opens for disintegration and random political developments – undermining concerted action. In the following, our point of departure is the claim that with increasing urgency and pressures for either voluntary or forced environmental adaptation, AI-adaptation and societal changes, political-economic changes and adaptations will have to follow – voluntarily or forced.

DOI: 10.1201/9781003518815-12

The political-economic system and the governing political order is in some ways like a straight-jacket, too rigid for the organic development of societies, and in other ways like a playing field where the rules are being changed during play according to who takes the lead.

The idea of political-economic control

The British cyberneticist W. Ross Ashby gave name to what has been called Ashby's Law of requisite variety (Ashby, 1956). The law says essentially that only variety can match variety. If the variety of systemic challenges surpasses the variety of political-economic answers, there will be a deficit of control. In order for there to be control, the political-economic system has to allow for a variety of responses needed for handling whatever challenges that will arise, and to do so in time and with winning outcomes. Being too slow to act or coming up with deficient answers is a recipe for failure. Being stuck in traditional and inflexible political-economic premises can be a recipe for disaster. But so can also the one-eyed lock-in to a risky strategy.

The study of "political economy" is recognized as an interdisciplinary field, drawing mainly on economics and political science, but also on history, anthropology, and related fields. The main constituents of political-economic systems are national economies, with a focus on markets versus command systems of production and distribution, and on property rights, legal systems and regulatory institutions guiding production and consumption. Important qualities of political-economic systems are measured primarily in economic terms of efficiency and growth, in political terms of government command versus market autonomy, and in social terms of economic distribution and welfare.

A main distinction between the fields of political economy and economics, is that in economics, the existing political economy is normally taken as a given, seeking optimal production and distribution within the (market-) economic system, while the study of political economy is concerned with the boundaries of the system. If we think metaphorically of production and consumption as a playing field (imagine a football arena!), the political-economic systems is about the rules set for the game. Economics then is about the technicalities of the sport, taking the main set of rules for granted. Over and above the prevailing set of (political-economic) rules for the game, there is also the definition of the main essence of the game, what it is about – in the latter part of the present chapter referred to as that of political order. The issue of the present chapter is about the need for change, both of the game itself and for the rules applied, to make the game more appropriate for our age.

Lately, in addition to environmental challenges, economic inequality within and among societies has received increasing attention, and so has speculations about AI-consequences for societal development and for issues of equality and inclusion. It is my contention that net-zero and green transition policies, AI-robot automation and social sustainability increasingly will challenge the political-economic boundaries of today and potentially bring forth entirely new political-economic rules. Following Ashby's Law of requisite variety, the emerging rules will have to come in time, and with winning outcomes, to deal with the variety of environmental, digital and related challenges to the good life on Earth.

There are all kinds of organizational solutions to economic activity around the world, also within systems normally classified as capitalist (Bohle and Greskovits, 2009), but for a main identification, economists have traditionally tended to group the various systems into four main categories, including traditional economies, command economies, market economies, and mixed economies. For ease of classification, the grand dichotomy of capitalism and communism has often been substituted for the four-way model, while with the fall of communism in the Soviet Union, varieties of capitalism have been main political-economic reference points.

A crab metaphor

Important questions are about the feasibility of solving problems within the confines of capitalism, and also about to what extent and how the capitalist political economy may itself change in response to the challenges. People and numerous authors ask the question of whether the (capitalist) system that brought us to the environmental precipice will be our best choice for bringing us to a safer ground. I suggest that the answer to that question is not a simple yes or no, but more about the emergent nature of the system itself.

Political-economic change will necessarily occur. Whether we call the system capitalist or not, it will not for long remain the same, unaltered. My favorite metaphor of political-economic development is that of a crab. Like lobsters, they grow, but since the exoskeleton of the crab (the outer shell) does not grow, a new shell is formed through a process of molting. Baby crabs go through molting several times a month. As they age, the time span increases. But every time, they have to wait for the new exoskeleton to harden, which is a time of high vulnerability. Not only crabs and shellfish go through molting-type transformations. Amphibians, reptiles, even birds and mammals have to shed cover at some time during life. We think of birds and mammals as having a linear growth until maturity. But birds need to break out of the egg. Humans and other mammals have to break out of the womb.

Human societies have to break out from political-economic systems, like shredding an exoskeleton, to gain more open space for renewal or further growth. In Marxian terms, the "relations of production" must be shredded in order to allow for the free unfolding of "productive forces", and it takes place when time is ready for a revolutionary "molting" of the system.

In our terms, the economic factors include entrepreneurship, production and distribution systems, financing, and job creation. But at times, either because of endogenous economic growth (as when the crabs grow) or changing exogenous circumstances (as when ecosystems collapse), the exoskeleton of the political-economic systems, institutions and practices comes to work more as a straight-jacket than a playground. Over time there is an increasing gap between the potential for development and problem-solving on the one hand, and the limits of the (exoskeleton) political-economic system on the other, in economic terms a case of diminishing returns. For the purpose of this book, we are concerned with how the political-economic exoskeleton of BAU politics and 21st century capitalism may choke the full range of attempts to deal adequately with the environmental threats and our ability to deal human-intelligently with AI development. An overriding issue of political economy as related to the green transition, is as mentioned before, whether or not the system that brought us to the precipice is our best choice for reaching a safe shore.

The set of challenges

The control issue refers to a sett of challenges, and for the system to enable the generation of successful solutions. From Part One we have introduced environmental challenges, including man-made climate changes and extreme weather, nature depletion and extinctions, environmental toxins and new entities. To the environmental challenges we have also recognized the stickiness of fossil energy, and the technological challenges of coming up with green energy solutions. And to the list of environmental challenges we have added the potential of artificial intelligence, for good and for bad, for the environment and human development.

From chapter 8 and 9 we have registered added challenges (and possibilities) linked to human response to manifest damage, and to causes of resentment and disintegration undermining concerted action and solidarity.

For the present, and for varieties of political-economic challenges, we shall first follow Shoshana Zuboff (2023) in her suggestion that "surveillance capitalism" is emerging as a next stage of economic development from industrial and market based capitalism, threatening the number one ingredient at the heart of capitalism, that of consumer sovereignty. From surveillance capitalism we shall observe another threat to the main pillars of ideal-typical capitalism, that of large scale corruption and the spread of kleptocracy around the world.

Adding both surveillance and kleptocracy to the conventional capitalist mix, the conventional markers of separation between the public interest and the common good on the one hand, and the private profit interests of individuals and corporations on the other are increasingly being blurred. President Trump's revolutionary dismantling of American democratic institutions appears to be a Shock Doctrine II level strategy for radical change of political-economic system. Actually, it is about changes perhaps better categorized as change of political order, encompassing the rules of the political economy – to which we shall return at the end of the chapter.

More challenges

The main legitimating argument in favor of a market economy, and of capitalism as such, is that of consumer sovereignty. Consumers are to range as supreme governors of production, with consumer demand reflecting needs and desires, and with producers responding to demand.

John Kenneth Galbraith, in "The New Industrial State" (1967) introduced a significant, challenge to the conventional position, however, arguing that in a system with strong proactive investments in research and innovation, product development and the fine-tuning of production lines, demand could not be left to the (whimsical) behavior of customers. To replace consumer sovereignty, and the sequence of consumer choice directing production planning, a "reversed sequence" of "demand management" had been introduced. Demand had to be managed by means of advertising and marketing. Rather than having consumers control business output, the revised sequence had business organizations exercise control over consumers.

Small enterprises would still adhere to the conventional wisdom of price competition and consumer sovereignty, constituting the "market system" segment of the economy. Large, increasingly monopolized enterprises constituting the "industrial system", covering some 1,000 firms in America at the time, was seen as the more dominant sector and a threat to consumer sovereignty, democracy and quality of life.

The role of surveillance capitalism

The powerful message of Shoshana Zuboff in "The Age of Surveillance Capitalism" (Zuboff, 2023) takes Galbraith's message a long step forward. Galbraith's "revised sequence" was based on resource extraction for production and distribution of conventional products and services. Under "surveillance capitalism", human experience is the basic resource, a "raw material" extracted free of charge (and largely in hidden ways) for the purpose of large scale behavior prediction, modification, and control. Roughly half the world's population are now hooked to some information and communication technology device, which leaves digital footprints and experiential data free of charge, even often for resale by technology providers, for commercial as well as political use. Huge profits can still be made by selling conventional products and services, but even greater profits can be made by selling information that tells how to influence people about what products and services to buy – or what political party to vote for.

Many have woken up to the recent Chinese tech-political facilities for face recognition linked to behavioral records as defined by the Party, fostering full political control of individuals. Under "surveillance capitalism", face recognition is not necessary. The goal is not, at least at the outset, so much political control as commercial gain. More than enough footmarks are left every time we use a computer, a mobile phone, or a modern car. For an upgrade, consumers can hook up to an "Aware Home" installation of "internet of things", supposedly making life more convenient, and have every action and social event in the home recorded – allowing the technology providers to know more about us than we can know about ourselves, and potentially to sell such knowledge for super-profit.

People may think they can guard their personal interests by turning off devices, by behaving well (offering no cause for extortion), or by making reservations about privacy when entering into contracts about "home thermostats". The former two won't work, since we cannot know what the information seekers are after and how and when the data is collected. Neither do we know what information about us can be used for what purpose. As for contractual reservations, Zuboff refers to a study by two London university students (Noto La Diega and Walden, 2016) who found that so many firms are involved with bits of technology for the Nest home thermostat as to demand rewriting of nearly a thousand contracts, most of them with stipulations that the functionality and security of the devices would be deeply compromised if changes to the contract were made. Twenty years ago, sellers' commitment to customer privacy would be taken for granted. Today, the opposite must be taken for granted.

When a highly decorated Army soldier blew up a Tesla Cybertruck on New Year's Day 2025 outside the Trump hotel in Las Vegas, police investigators were surprised (and impressed) at the speed and effectiveness of Tesla information provision. Within hours, the Tesla senior team of investigators were able to trace in detail driver Matthew Livelsberger's route over five days, through four states, from Denver to Las Vegas, noticing recharging stops and reading the travel route from onboard software. They could also verify that the explosion was caused by explosives in the truck – not by the car itself. While the amount of information was impressive, and of good help to police, it also served as a warning about privacy issues when cars are turned into computers on wheels, recording contacts and cell phone syncing with call logs and text contents. In August 2024 General Motors was sued by the Texas attorney general for selling sensitive data from 1.8 million Texan drivers to insurance companies, without the owners' knowledge or consent. The lawsuit came at a time when insurance rates climbed steeply as a result of riskier driving, and when new cars were equipped to monitor speeds, locations, and the hard use of brakes – on top of whatever else of information was recorded, making new cars "a privacy nightmare" according to one critic.

Of course, some of the data collected by the tech firms are for service, upgrading or quality improvement. But a significant part is for proprietary knowledge collection, potentially traded for customer behavior prediction, modification, and control. Largely hidden data collection adds to what Zuboff calls the "instrumentarianist" power of the data collector, closing the gap between automation of information streams and the end of "automating us". Where formerly critical political economists were concerned about undue workplace surveillance of workers and labor, surveillance capitalism is about the surveillance of everybody, at work and off work, as producers, consumers, householders, and social media members. There is a catch 22: The promises of efficiency and well-being that comes with digitalization of our "lifeworld" requires digital recording of more and more of our whereabouts, and with more and more recording, there will be strong economic motives for use of that information – as Zuboff claims – for "instrumentarianist" behavior prediction, modification, and control.

The case of corruption

There is "indigenous corruption" growing out of domestic dealings, growing into kleptocratic networks and states not only in poor backward nations, and there is "invasive corruption" undermining international political institutions and commitments. Increasingly, with the overproduction of elites, plutocracy and rivalry is driving corruption, and corruption is driving policy formulation even in traditionally democratic nations, impacting political relations for war and peace and economic relations for dependency and exploitation .

After having observed the carcinogenic effects of corruption in Afghanistan and elsewhere abroad, Sarah Chayes, in "Everybody knows" (2020) was shocked to find some of the same patterns at home in the USA, in particular the tendency for corruption to be anchored and accepted high up, to drive inequality, and for inequality to drive policy changes opening up for more corruption. Much of this seem to be part of a global trend, captured by the Transparency International Corruption Perception Index (CPI).

According to the 2023 annual report from Transparency International, there has been a global decline in justice and the rule of law since 2016. There is little doubt that the weakening of democracies around the world and the rise of authoritarianism has contributed to corruption. In the year 2023 only 28 countries showed improvement, while 34 countries scored significantly worse. Democracies have institutional and judicial instruments for limiting corruption, but corruption has the power to drive countries towards dictatorship, and dictatorship has the power to uphold and expand corruption.

Level of corruption may not be captured by criminal records, however. During the 2014–2018 period, the number of bribery convictions in the US fell by nearly one third (Chayes, 2020, p. 13, referring to the U.S. Sentencing Commission), and the prison sentences came to average a little over 20 months, about half of the minimum recommended by Federal Sentencing Guidelines for similar crimes. In other words, the risk of getting caught was reduced by a third, and the penalty for those caught was reduced by half. This raises the salient question of whether the fall in the number of convictions imply that the US has become a less corrupt society, or – worse – that the country in fact had become significantly more corrupt, simply by legalizing what traditionally, and in most democratic societies, would be considered corruption. There is very little reason to doubt that the latter explanation is the more correct, as even the Supreme Court has added its blessing to forms of bribery that in most well-organized countries would be considered blatant corruption (as in its ruling in the McDonnell v. United States case). The main democracy problem is that only very rich people, and no poor people, have the resources to generously support politicians or higher staff members with hundreds of thousand dollar gifts and bestowals (paying for weddings, vacations and more).

Traditionally, the most important forms of corruption in a nation have included bribery – such as taking money or other benefit for changing or bending a law or regulation. Also, the use of extortion to force, manipulate, or intimidate in order to gain money or property from someone have been common corrupt practices, as also the use of kickbacks or the offering of money or other gifts for favorable treatment in a competitive situation.

In addition to traditional bribery of public officials, neoliberalism has brought the less obtrusive practice of having high-ranking business executives and specialist "trade roles" with high-ranking governmental officials. Notably in the military, a former codex of keeping distance between high-ranking officers and the arms industry, has been in place to avoid corruption. Presently, however, in the weapons industry, and the Department of defense, such mixing of interests have become more commonplace and a cause of huge cost overruns and product delays. President Eisenhower's warnings about the "military-industrial complex" (in his Farewell Adress, 1961) was directed at a time of relative innocence, compared to the present.

As for medicine, similar "job exchanges" between big pharma (the 10 or so biggest pharmaceutical companies) and in the US the The Food and Drug Administration (FDA) is undermining public trust in medicine, while pumping huge public funds into the pockets of happy commercial winners, paying for high-life "convention" extravaganza for doctors and psychiatrists enlisted to endorse approval of new medicines. Similar practices have spread to other sectors and to other parts of the world.

All ingredients in the political economic system change color with corruption (and with legalized "neo-corruption"). Corruption works by twisting and bending rules, causing collapse of sound judgment and normal expectations. A qualified student may be refused entry to an institution of higher education, because someone with more money bribed his way to admission. A company may continue pollution, paying into a corrupt "cap and trade" organization for certification of pollution reduction elsewhere (such as the European Unit Allowances in 2009). And of course, private ownership to the means of production means something entirely different in the context of monopolization and the scrapping of antitrust legislation, from the context of small shop production and trading.

To the curious list of qualities of classical fascism, presented in Chapter 9, there is a modern mix of technological and organizational ingredients including surveillance and enabling large scale corruption. It is eminently described by Anne Applebaum in "Autocracy, Inc.".

The rise of "Autocracy, Inc."

Anne Applebaum (2024) reminds us of a cartoon image of an autocratic state, with a bad man at the top, controlling the army and the police, and threatening people with violence, surrounded by evil collaborators and some brave dissidents. Applebaum suggests that the cartoon is off course for our understanding of 21st century autocracies.

On top of "Autocracy, Inc." is not one vile individual, but a person fronting a sophisticated network of kleptocrats, financial advisors, and security services including military, paramilitary forces and police, as well as technological experts on surveillance, propaganda, and disinformation. Autocratic networks are further connected to similar networks in other countries, also with links within democracies. Autocratic states rely for income, money laundering and tax avoidance mainly on corrupt, state-controlled companies to trade with equally corrupt, state-controlled companies in other autocracies. State control of companies is achieved by direct state ownership, but more likely by private firms and oligarcs sympathetic to the authoritarian state, or otherwise dependent on the state for production licenses, product approvals, or for deliveries to the government.

The list of diverse "Autocracy, Inc." states include communist China and neo-fascist Russia, the "self-reliant" (*Juche*, communist and person-cultivating) North Korea, and the Islamic Republic of Iran. It also includes the more moderate Arab monarchies, such as Saudi Arabia and the Emirates, and "illiberal democracies" like Turkey, Singapore, India, the Philippines, and Hungary. Also on the list are a number of African nations, with Zimbabwe as a primary example of a nation causing no threat to Northern democracies – only to her own people who are consistently robbed of incomes and wealth for the enrichment of the kleptocratic elite. Members of the "Autocracy Inc." include monarchists and nationalists, theocrats and business leaders, communists and fascists. Perhaps the most pathetic examples are communist states, such as in Cuba, Venezuela, and Nicaragua, with little success for "from each according to ability to each according to need", then turning to corruption, kleptocracy and membership in the "Autocracy, Inc.".

The basic principle of "Autocracy, Inc." is not based on a shared ideology apart from keeping the kleptocratic machinery going and providing defense against dissidence and disintegration. In that sense, "Autocracy, Inc." differs from classical fascism, suggesting a kleptocratic up-dating of the classical models.

Kleptocracies (from *kléptēs* – thief, and *krátos* – rule) are societies ruled by corrupt governments, whose leaders use their political power to extract wealth, by embezzling or misappropriating private and public funds at the expense of the population. Kleptocracy is also closely linked to the "resource curse", the "paradox of plenty", or the "poverty paradox". The argument goes that countries with an abundance of natural resources, such as oil, gas, or valuable minerals, tend to have less economic growth, worse economic distribution, less development, and less democracy than countries with fewer natural resources. The term "resource curse" was first used by Richard Auti (1995), and the first empirical study to find such a negative correlation between resource availability and negative development was reported by Jeffrey Sachs and Andrew Warner (1995). Since then, a great number of studies of the resource curse and kleptocracies have been conducted, generally suggesting that in less-developed countries resource wealth seem to produce more poverty than progress. But there are also other studies challenging the general theses. The problem may not be resource abundance as such, however, but rather the combination of resources, kleptocratic rule and inequal sharing.

In "The dictator's handbook" (2011) de Mesquita and Smith explained how authoritarian rulers with easy access to public revenue (from the corrupt sale of valuable resources, and with easy financing of security) can buy loyalty from down the ladder, while denying spread of money to the wider population (for purposes such as education, health, and infrastructure). A dictator without such resources would be more inclined to liberalize policies and invest in economic improvement in order to uphold needed loyalty and popular support.

Epidemic corruption

The United Nations Convention against Corruption (UNCAC) Convention was adopted by the General Assembly in October 2003 and entered into force in December 2005 – the first legally binding, global anti-corruption instrument. It was originally signed by 140 nations, and as of 2023 there are 180 countries on the Transparency International list. Unfortunately, most of the signatory nations are extremely corrupt, and over the last few years world corruption has been on the rise. According to the Transparency International Corruption Perceptions Index (CPI) from 2023, with a score of 100 meaning "clean", and zero meaning the worst possible corruption, 2/3 of the countries score below 50 and about half score below 40. Towards the lower end of the CPI are countries such as Russia (number 141), Nigeria (145), Syria (177) and Somalia (180).

According to the 2023 annual report from Transparency International, there has been a global decline in justice and the rule of law since 2016. In most corrupt societies there are anti-corruption initiatives and policies – generally to little avail. Chayes' reports from Afghanistan made clear that anti-corruption campaigns were about rhetoric, to please foreign allies and for use as a veil to hide real corruption. There was at no point during the Karzai rule any serious, honest attempts at reducing kleptocracy. Quite the opposite.

CPI numbers unambiguously show that kleptocracy and the resource curse, and similar mechanisms of indigenous and invasive corruption, are hitting poorer countries worst. We also know that corruption is a main reason why development policies and assistance often fail. This does not mean, however, that corruption is only a matter of poverty and backwardness. While Western Europe is generally the cleanest of all world regions, with Scandinavian countries (plus Singapore and New Zealand) at the top, there is still corruption in wealthy, democratic countries. A revealing text was David Whyte's "How Corrupt is Britain" (2015). Given that the UK is generally considered one of the least corrupt countries in the world (number 20 on the CPI), the text was to many a shock. There are also clear indicators that the US has become more corrupt. On the 2015 CPI list, the US was number 16 from the top, with a score of 76 compared to number 24, with a score of 69 in 2023. It remains to be seen where the country will be after a few years of federal "deep state" surgery.

Political economy and the political order

We have noticed that beyond the level of rules defining the political-economic system, there is a level defining what the whole game is actually about. That is what the political order of a society is about. As of winter 2025, and the time of this writing, there is in America no definite political order, only a struggle for hegemony and direction. Around the world, there is also great uncertainty about the direction of the political order. Will there be a return to the neoliberal political order, or are we heading for geopolitical rivalry and aggressive negotiations?

Much of the following will refer to the political order and the political-economic solutions in the US. The reason is very obvious. For nearly all of the 20th century the US was the globally leading economy and also a model trendsetter for political orders in much of the world. The US led the way with New Deal reforms in the 1930s, war-economic initiatives in the 1940s, civil rights and social-democratic reforms in the 1050s–1980s period, and again with Reaganomics in the 1980s to the present. The US is again on the move, under Trump II engaged in a revolutionary experiment, impacting economies around the world and playing geo-political games with both old allies and antagonists.

From New Deal to Neo-liberal political order

To both American insiders and foreign outsiders, it has for long been difficult to understand American politics. With the book "The Rise and Fall of the Neoliberal Order" (2022) Gary Gerstle, professor of American history, has taken long strides to help us appreciate some underlying patterns. A key term in the book is "political order", which stands for the main features of politics and political-economic systems over extended periods of time.

While all nations of the world are affected by emergent political orders, and while there has been criticism of the present neoliberal order, it is beyond doubt that the introduction of such a political order in the US in the 1980s for a long time came to serve as a default model.

During the 20th century, Gerstle sees two decisive changes in the American political order. After the stock-market crash of 1929 Franklin D. Roosevelt came up with a "New Deal" for the political order. The "New Deal" stamp on politics would last until the Reagan years of the 1980s. The New Deal political order came with increasing economic equality, improvement of working conditions and work-related legislation, introduction of the 8-hour work day and improving wages and employee security, regulations of banks, financial institutions, and the stock market, and limitations on business monopolization. It also came with Medicare and safety nets for the elderly and the unemployed, and with more equal treatment of minorities and women in the workplace. And not the least, it was a period of heightened standards against corruption.

Of course, it was not a period free of conflict. Not all "progress" was for the common and the social good. Economic expansion often came with a significant environmental burden. And many reforms came only after prolonged political battles.

With the Reagan years of the 1980s, a new political order emerged. The neoliberal political order is linked to Ronald Reagan's Presidency, but economic policy revisions were already started, and the intellectual content for backing a new political order had matured over several years. Even Democrat Presidents Kennedy and Johnson, and later Jimmy Carter, had introduced reforms to increase wage-based income taxation while lowering taxes on corporate profits and dividends (Chayes, 2020).

For academic backing, Milton Friedman's "Capitalism and Freedom" (1962) was an early inspiration for market orientation – building also on the ideas of the Austrians Friedrich Hayek and Ludwig von Mises, and the American James M. Buchanan. For political engagement, the "Confidential Memo to the U.S. Chamber of Commerse" by (later high court judge) Lewis Powell (1971), fronted a warning that "the American economic system is under broad attack" .... "an assault on the enterprise system". Powell argued for a "comprehensive counteroffensive" from the business world against the New Deal political order, uniting the forces of his own tobacco industry cohorts with the emerging Coch brothers' fossil fuel network, to establish a broad front against the established political order, anticipating the later Tea Party movement, and instituting a strong network of conservative and anti-environmental think-tanks in the process. Powell also had detailed plans for how the message was to be planted in schools and educational institutions, in networks and organizations, for short- and long-term implementation, with secured financial backing by the rich elite, and with ambition for a qualitative change of the American judicial system and judicial practice in favor of business. Following the leads of people like Friedman and Powell, large numbers of influential American economists and lawyers took part in the wholesale promotion of privatization, deregulation and taxation reliefs for the wealthy, paving the way for the emerging neoliberal order.

The main elements of the neoliberal political order were the privatization of public work, cutting down on public and social expenditures, replacing public solutions with private enterprise and market solutions wherever possible, deregulation of business – in particular banking and financial services, but also reducing environmental regulations and taxation of polluters. Internationally, the new political order favored open borders and globalization.

According to Gerstle, it is important for the understanding of political order that change does not only come from a party in power. It also has the capacity to "bend" the entire political landscape in a certain direction. In the period 1930–1980 in the USA, the political order was bent in the direction of a modern welfare state, with an emphasis on human rights, labor organizations, and economic equality. General Dwight D. Eisenhower was elected Republican president in 1953, but his policies were broadly in line with the "New Deal" tradition. So was also the main thrust of the

Nixon administration – nearly succeeding in introducing a guaranteed minimum income reform in 1969, and fully succeeding in establishing the EPA (Environmental Protection Agency) in 1970.

The New Deal political order was also strongly influenced by the Cold War and rivalry with the Soviet Union. American business owners and managers had to put limits on economic exploitation and economic disparity, for fear of radicalizing workers and their organizations. It was a time for reduced economic inequality, increased job security, and the upholding of minimum wages guarantees, business taxation, broad support of public education and health, and a general elevation of the standard of living for most people. And internationally, competition with the Soviet Union and communist ideology, caused a balancing of superpower privilege and post-colonial exploitation.

Cementing a new political order

For a proponent of a new political order, President Ronald Reagan represented an ideal combination of personal embodiment and communicative ability. Appealing to honest business, market efficiency, the fairness of competition, freedom of choice, and ingenious entrepreneurship, while recognizing the plight of workers and poor people, he made the new political order appeal broadly to Americans – and to more hesitant sympathizers overseas. Also, as mentioned, there was what Gerstle calls a "bending" factor – moving the whole of the political discourse in America towards the right, and gradually also abroad. The personal charisma of President Reagan was a good starter, but also that of Bill Clinton, and in Brittain Tony Blair, played important parts in the transition towards a global neoliberal political order.

An important fact, however, is that all through the years of Reaganist privatization and market idolization, there was a huge gap between Reagan's rhetoric and his material policy initiatives, and there was also a huge gap between popular values and the outcome of neoliberal policies. Based on public polling data, according to Thomas Ferguson and Joel Rogers (1986), there was in America at no stage during the 1970s and early 1980s any widespread frustration with business regulations or any general concern with a supposed weakening of the free enterprise system. To the contrary, there was all the time in America a deep majority concern with the rising power of big business, and there was a clear desire for the limitation of corporate profits. There was also a solid majority support for minimum wages and strong work safety regulations, for environmental protection, water and air quality legislation, and there was all the time solid approval of welfare, public health and education spending.

Neoliberalism set the stage for the weakening of class compromise and the welfare state, with more power to the owners of capital. In much the same way as with the "New Deal" political order of the postwar years, the bending in favor of a neoliberal political order during the Reagan years led not only to Republicans moving to the right, but also to Bill Clinton following up with deregulation of the financial sector, globalization of markets, free movement of capital, labor, goods and services, and a series of reforms in favor of a liberalist business world. Even Barack Obama stayed mainly within the framework of the neoliberal order, when his first job as a president was to clean up after the financial crisis in 2008 – injecting large sums of public money into the very banks and investment companies that were behind the crisis.

Paving the way for the Clinton continuation and anchoring of Reaganism, was among others the establishment of the Democratic Leadership Council (DLC), founded in 1985 to shift Democrat policy in the direction of deregulation, lower taxes on business, and freer markets. Under the norms of the neoliberal order, a main premise of business regulations shifted from a focus on fairness to that of efficiency. An important advocate of the turn was Yale law professor Robert Bork, with the book "The Antitrust Paradox: A Policy at War with Itself" (1978), arguing that if vertical and horizontal integration of big business served the interest of efficiency, there was no reason to uphold antitrust and price-fixing regulations. The interests of open markets and small business was sacrificed in favor of record levels of mergers and acquisitions during the 1980s and 1990s, setting the stage for continuous monopolization in the 2000s.

Political order and hegemony

As explained by Gerstle, the shift of political order is about more than just one party coming up with a new, challenging position on some issues. It is more about a shift in the whole setting of political debate. Thus, the neoliberal order also gained wide support from much of the American "New Left" and in Europe from traditional social-democrats, welcoming de-bureaucratization and a more open society, critical of welfare abuse, public monopolies and high taxes. In the UK Tony Blair guided the labor party, and corresponding parties abroad, in the direction of market liberalism. In his inauguration speech as president, Reagan said that "Government is not the solution to our problem, government is the problem", and in spite of opinion polls to the opposite, there was no wide public protest. It was not just a change of mindset of the elite, but of quite broad groups of Americans, as it was of Europeans and of politicians in many parts of the world.

With the neoliberal political order came a new mindset, in cultural value terms a more directly masculine turn, with acceptance of greed at the heart, praising money-making for its own sake, with self-congratulating winners in the competition for wealth, and no shyness about "the new selfishness", making finance a preferred line of specialization in the business schools, and making the new rich the new heroes in social circles. With the neoliberal turn came also new networks of proponents, new elites placing their members in important governmental positions, financing lobbyism and elections with hitherto unheard levels of donations, and new think-tanks to combat business taxation and (environmental) regulations.

Preceding the neoliberalist turn by five decades, the Italian Marxist Antonio Gramsci linked the ruling ideas of a political order to the establishment of "hegemony" (Bates, 1975). Gramsci, who was arrested in 1926 and spent the rest of his life in prison under the Fascist State, was never able to formulate a coherent theory of "hegemony". But as noted by Thomas R. Bates, he came to believe that the creation of hegemony was basic to all non-military rule of a nation. When the main ideas of the ruling class were widely accepted, there was no use for violent, dictatorial rule. Thus, hegemony was about the "ideological unity" of the nation, but also about the ideological basis of rival movements and ideologies challenging the ruling elite. The creation of hegemony was a constant challenge to the leaders of the Soviet Union, from Lenin, to Stalin, to Brezhnev, forced to use endless propaganda for legitimacy, and finally running out of steam, trying in vain to defend an ultimately corrupt society.

Hegemony is also about the relationship and tensions between intellectuals and scholars providing ideological inputs, party elites providing political maneuvering, political leaders adding face, eloquence and persuasiveness, political cadres working at the grass roots and ultimately common people. Hegemony requires at least some level of agreement between all levels of participation. Political control in the USSR was achieved only by violent means, in a society destined to seek global leadership and hegemony anchored in a small minority of industrial workers in a society dominated by agrarian occupations.

In the Western World, with "Reaganism" and "Thatcherism", there was "bending" of the entire political landscape. A new hegemony was created. But even at the outset it was a bending that came with increasing economic and political inequality and tension. Freedom for capital and for the upper classes came in America alongside a tightening of civil rights for blacks. Tax bonuses and increased incomes for the upper privileged came with economic stagnation for African-Americans, but also for poor whites and Latinos. Amid increased freedom for the better offs, a large underclass came to experience less freedom. And in much of Europe, with social-democrat elites occupying the government halls, common people and labor supporters increasingly came to feel forgotten and "indifferent".

Cracks in the wall

The neoliberal order had some built-in contradictions, all from the start. On the one hand, the neoliberal order appealed to what Gerstle calls "neo-Victorianism", emphasizing family values, work discipline, traditional gender roles, traditional consumption patterns, traditional cultural life, and the principle of saving before investment. For the "neo-Victorians", the traditional, hierarchical nuclear family served as the main footing of society. By speaking warmly of such values, Republican politicians from Reagan to Trump secured conservative support, especially among white, evangelical Christians. In its wake, abortion bans, opposition to gay marriage and LGBT rights, support of Israel and opposition to immigration followed. With reference to the anti-immigration message of far-right politicians in both Europe and the US, it is interesting to note that a recent study from Texas (Light et al., 2020), using data from the Texas Department of Public Safety, found that contrary to public perception, there were considerably lower felony arrest rates among illegal ("undocumented") immigrants compared to legal immigrants and native-born US citizens. Also, there was no evidence that undocumented criminality had increased in recent years. The findings provided explanation why aggressive anti-immigration programs have not resulted in crime reduction.

On the other hand, contrasting the conservative morals turn, the neoliberal order also accommodated what Gerstle calls "cosmopolitan" interests, which in many cases represent the exact opposite of the neo-Victorian ones. With market liberalism came expectations of liberation in most areas of society, opening for new gender patterns, new family constellations, and experimentation with social roles – also much of what was brought forward by "the New Left", with room for "black power", women's liberation, multiculturalism, and gay parades. The Clinton period helped consolidate a new political order, but it also gave support to a new cultural pluralism that flourished in the avant-garde milieus in big cities like New York and San Francisco, also in London and Paris, Hong Kong, Tokyo and Toronto.

On the economic side, the neoliberal order opened for globalization and international migration, and also for increasing inequality. Neoclassical economics generally concludes that societies stand to gain from free trade and globalization. But nobody has proven that all groups in a society will gain. Some will lose, and free trade and open borders can only be justified on welfare grounds if economic compensation is offered to those who stand to lose – in particular workers who observe their well-paying jobs getting outsourced across borders.

Gerstle provides a summary of social problems in the wake of neoliberalism, and he refers among others to the book "Coming Apart", in which Charles Murray (2012) analyzed the conditions of white workers in the small suburban town of Fishtown on the outskirts of Philadelphia. As a conservative sociologist, Murray focused on marriage, which he believed was the key to a good society. To his astonishment, he found that what fifty years earlier would have been "normal" family life for 84 percent of the population, in 2010 applied to only 48 percent. The divorce rate had increased, and a greater number of individuals never succeeded in establishing a lasting relationship. And still, birth rates remained high. There was a sharp increase in the number of single parents, especially single mothers without secondary education. When he went on to look at how this matched in relation to the larger American society, he found the same tendency there.

Before the study, Murray had become notorious in the sociological community for his claims that most African-Americans belonged to an underclass, rooted in a black culture, with laziness and evasion of responsibility as their trademark. In 2012, he had to admit that the culture he encountered in Fishtown among the white working class had exactly the same characteristics and that race made no difference. He also had to admit that the decline in stable family life was not an isolated problem, but rather an expression of general distress and despair among people. Along with a depletion of family life, there were also a rise in the figures for self-harm. Death rates from alcohol consumption and drug addiction quadrupled from 1999 to 2011. Suicide rates among white workers and the unemployed increased by 40 percent, more than for any other ethnic group, and even life expectancy began to decline—setting the general demographic gear-shift in reverse. Instead of looking to globalization and the export of good jobs, or to de-regulation and unfair tax systems ("trickle down" or "supply side"economics) for explanations, he came to blame the life-styles of the wealthy elite – failing morally as role models and trendsetters for family life and neo-Victorian values.

To the misery of the "irrelevant", much can be added about race, income inequality, community disintegration and criminality. Incarceration statistics may by some be read as documentation of law enforcement efficiency. They can, however, also be read as social political inefficiency. Ohio State University law professor Michelle Alexander, wrote the book "The New Jim Crow", documenting that 7 million Americans were either in prison, or on parole, or out on probation (Alexander, 2011). For every 100,000 inhabitants in the USA, 750 were in prison, while the corresponding number in Germany was 93. Relatively more people were in prison in the USA than in China and Russia, Egypt, Iran and North Korea. But for blacks, the situation was (and still is) much worse: the numbers for black, young men are far above the average for the population. Alexander reported that in some major American cities, 80 percent of black male youth might have a criminal record. The US has far more incarcerated blacks than South Africa at the height of apartheid.

We saw in Chapter 9 that immiseration (and the growth of elite overproduction) provides sound explanations for political and social disintegration and for the lowering of political-economic predictability both in the US, Europe and around the world. Still, the revolutionary turn in the US in January/February 2025 has been a shock to politicians and people on both sides of the Atlantic and the Pacific. Of course, we know nothing about the final outcome. But for a book on the green transition and AI, it would be irresponsible not to at least recognize the fact that the President of the greatest economic and military power on earth has committed his nation to "roll back on environmental regulations", end the "green new scam" and to "drill, baby, drill" for fossil fuel – also giving free reigns to tech billionaires for unregulated AI developments. It is about political order in a time of political chaos and order collapse.

Political order breakaway and possibilism

At a time of massive decomposition of American political institutions, it is extremely difficult to predict political order outcome, and equally difficult to say precisely what should be in place and what needs to be done to promote successful green and digital transitions. Too many unpredictable changes are going on at the same time. The number of conditional elements and impacts is virtually unlimited.

Still, we need to keep in mind that much of what goes on is at least partly reversible. We may not after a few years come back to a situation similar to pre-covid or pre-Trump society. But new political alliances are entered into all the time, and new technologies enabling new institutional forms are emerging, allowing us to think freely about social sustainability, green policies and AI alignment.

This chapter is about our set of challenges, also including surveillance capitalism, kleptocracy and tech fascism, and it is also, following Ashby, about the variety of measures, actions and policies to deal with the challenges. Only variety can match variety. Beyond the technicalities of political-economic rule changes for handling the set of environmental and digital challenges, the threats of ressentiment and disintegration, and the surveillance and kleptocratic versions of capitalism, is the issue of political order and transformation, and the allowance for variety.

In the chaos of Trumpist reversal of the dominant neo-political political order, Trumpism may not prevail to define the emerging new political order – only serving an antithesis role in the cycle of societal evolution. Russia may add geopolitical challenges and opportunities for alternative alliances but has little to offer in terms of alternative culture and deeper values. China is preparing for a takeover of a global economic leadership role. But at least for Westerners and people influenced by Enlightenment liberty and democratic values around the world, neither Russia nor China have what it takes to set a new political order standard. In the Ukraine, people are willing to die for freedom. Fear of losing their newly gained freedom caused the Baltic state to seek protection in NATO. Components of freedom and democratic rights will not easily be renounced around the world.

In the vocabulary of the crab metaphor, the significance of the present lies in the fact that we live at a time of molting. The crab has lost its exoskeleton. In the vocabulary of a sporting playing field, it is time for renewal of the game, for a change of (political-economic) rules and even for redefinition of man premises. We are not anymore in a lock-in.

Possibilism and hope

In his book, "A Bias for Hope: Essays on Development in Latin America", Albert Hirschman (1971) included an opening chapter on "Political Economics and Possibilism". After first giving some examples of how economic models have been used for development strategies, Hirschman links his "bias for hope" to the idea of political-economic "possibilism". Possibilism is about "a passion for the possible", and about the dynamic interaction between the political and the economic forces of the political economy.

In light of the crab metaphor, Hirschman does not stop at the point of "molting". Unlike the crab, the political-economy of human societies may not just be expanded versions of an otherwise fixed form. For Hirschman, political-economic change comes in many areas of the economy, at many levels, seldom involving large-scale revolutions, more regularly involving discontinuous decisions and small-step changes in institutional structure – such as simply adjusting an overvalued exchange rate. Also, while the concept of contradictions has gained enormous attention among students of the political economy, for Hirschman, there is not necessarily an underlying major contradiction, more likely a shifting set of circumstance calling for political-economic adaptations at various times and in various forms.

As mentioned, perhaps most people, including economists and social scientists, are conditioned to think of political-economic change in historical terms as grand-scale punctuations, as going from feudalism to mercantilism, or from capitalism to communism. Also, as referred in Chapter 2, Graeber and Wengrow (2021) portray pre-history as a time of variety of life forms, work forms, and societal solutions, presenting a remonstrating critique of traditional "stagist" assumptions and pointing to possibilism in the future.

As opposed to "stagist" conceptualizations, Hirschman takes us to a much more nuanced view of political-economic change, more useful for our purposes. A great number of elements are included in his possibilism version of political-economic change. He dismisses the search for all-encompassing revolutionary, political-economic change, since it is "overwhelmingly unlikely that any institutions devised by human minds will be able to accommodate indefinitely and efficiently any new economic and technological change that happens to come along" (2013, p. 18). Instead, we should seek solutions to "markets failures", "externalities" and "diseconomies" to find the best possible mixes of governmental, market, and voluntary solutions to challenges and opportunities as they occur.

In the search for such mixes, we should look for "diminishing return" elements, meaning not mainly diminishing returns of profits, but rather of any output relevant to well-being and our triple environmental areas, including climate policies, wilderness protection, environmental toxins control, also global and local social sustainability and AI challenges.

Traditional institutional and political regulatory practices are clearly insufficient to handle all the challenges of coming decades. New political-economic premises will have to be worked out over time and in time. Where traditional (Marxian) thinking has focused on the one main contradiction in the capitalist system, stepwise and selective reforms may allow for a smoother functioning and for adaptation to multiple contradictions. Rather than seeking and spending a life-time researching the one, major contradiction, researchers may identify contradictions as they arise, since "Most of the time, historical reality is, of course, far less neat: sociopolitical change entrained by the contradictions is often partial, grudging, and with a lot of unfinished business left behind, so that the need for further change makes itself felt once again in fairly short order" (Hirschman, 2013, p. 15).

The breakaway

For our purposes, agents of change may appear in all relevant threat areas – among activists, reformers, classes, labor organizations, and sectoral groups. Presently, the main agents of change appear to be tech billionaires at the front of digital and AI developments. Whether because of threats or promise, we may, however, identify contradictions giving rise to criticism, protests and change initiatives, wherever they are found. Not all protesters will have a valid case for political-economic reform, and not all tech promoters will win out in the end. But at least, in order to deal with the threats, there will be cracks in BAU and in the exoskeleton of the dominating political-economic system. It would be naive not to expect changes in the political-economic playing fields. The growing crab will need a new exoskeleton, and for the sake of sanity we will need to identify and deal with the underlying contradictions wherever and whenever they arise.

In road bicycle racing, there is a phenomenon called a "breakaway". There is the same in cross-country skiing, and in running events where being part of a pack – in cycling a "peloton" – nearly always makes the going easier. A breakaway occurs when one single competitor or a small group decide to leave the pack and set their own pace, forsaking the benefits of group suction and taking upon themselves the wind-breaking job up front.

The "peloton" of the neoliberal political order was for a long time remarkably robust. But below a calm surface, tension was growing with increasing immiseration and value diversity becoming more pronounced. The Trump I presidency represented a foretaste. The Trump II presidency has come to represent a revolutionary and massive real "breakaway". But like in cycling and cross-country skiing, one breakaway may soon be followed by another. Trump II does not seem to come with the qualities and values needed for a stable political order in America and in other societies that have had some real experience with democracy, rule of law, and liberty. We may soon expect another breakaway – hopefully not through civil war, and hopefully for a political order more suitable for a Green New Deal.

Given that we have a peloton of nations geared to neoliberalism, and seemingly locked-in to fossil fuel energy, we may ask whether there is any real chance of a breakaway and a renewal of the political order. Perhaps there will not be an identifiable singular breakaway. Instead, several initiatives may emerge to bring about new political-economic solutions and a new political order. Individual nations, regions, cities, and local communities may show the way, simply by claiming that something is the right thing to do – not waiting for all others to agree. At some point in time we may simply come to realize that a new and better political order is in place.

An article in "The Guardian" highlighted how "From Spain to South Korea, seven countries [are] leading the way on climate-friendly policy" (Carrington, 2024). Among nations, Uruguay has seen the fastest rise in wind power production, from 1 % of electricity a decade ago to 36% in 2024, with green energy totaling 90 percent of the electric power supply, and 50,000 new jobs created in the process. In India, government financing of LED lightbulbs transformation of lightening of private homes has driven down productions costs and facilitated the installation of 370 million lightbulbs, saving 387 million tons of CO_2 emissions a year – the equivalent of total annual emissions from the UK. Spain is on the list for her worker-friendly exnovation from fossil fuels production, with public investments in traditional mining communities to create new jobs, funding early retirement of coalminers, while reducing coal production by 80 percent in 2024 and 100 percent in 2025. Denmark – never a big oil-producing nation – is leading the way to a phase-out by ending new oil and gas licenses – followed by the UK and Colombia. Denmark, a major exporter of meat and food products, is also leading the way with a new tax on methane emissions from cattle and other livestock, and with the world's first national action plan for a more plant-based standard diet. In Brazil, Curitiba, capital of the southern Paraná region, has reversed a trend of rising car use by rejuvenating its municipal bus rapid transit (BRT) system, building new corridors between terminals and the suburbia and by using an intelligent transport system to monitor the dynamics of traffic, increase the average speed, and provide real-time information to passengers. In France, the

MaPrimeRénov' scheme offers support of homeowners for whole-house retrofits and installation of heat pumps, and 70 percent of the funding is for households with the lowest incomes. In South Korea the amount of food waste has been significantly reduced after the introduction of special bins that weight food waste and charge residents for the weight of the waste. To the list, we may add the US ban against oil exploration, covering the entire Atlantic coast, the eastern Gulf of Mexico, the Pacific coast off California, Oregon and Washington, and a large section of the Bering Sea off Alaska – the latest in a series of last-minute climate policy initiatives by the Biden administration in January 2025, ready for a re-birth when manifest extreme weather damage reach levels of no denial.

From all over, we have reports about thousands of projects for the minimization of the environmental burden. No project is too small to be part of a breakaway, nor of the dynamics of an emerging political order, and they may all be part of the "molting" process.

CHAPTER 11

From BAU Growth to Genuine Progress

Few people tend to question the legitimacy of economic growth. Almost daily we are reminded of the importance of economic growth, for employment and well-being. The latest growth statistics belong to the regular set of evening news. At annual meetings in the World Economic Forum, the growth theme is a regular feature. Similarly, at G7 meetings and at G 20 meetings, economic growth is always a shared concern.

In national politics, economic growth is a main pillar in budget negotiations and a crucial reference point for interest rate adjustments and wage negotiations, and for strategic management at company level. For status-concerned national leaders, and for presidents and prime ministers seeking reelection, economic growth is an important status marker. Also, the rivalry between Soviet communism and Western capitalism was for a long time dominated by competition for growth, testing the benefits of a communist "planned economy" approach versus market dynamics for growth under capitalism.

To this we may add that economic growth is also a main factor behind our present predicaments in relation to the environmental burden. Economic growth adds to gross enrichment for the few and to standard of living and employment for the many, but also to greenhouse gas emissions, habitat depletion, and values of consumerism.

Not all economic activity is captured by the standard GDP (gross domestic product) measure of growth, however. Not all forms of growth add to the satisfaction of basic human needs, and neither do all elements of growth add to environmental threats. There is no economic law saying that GDP type economic growth per se is good, nor that it is bad. That is the issue to be dealt with in the present chapter, raising several questions. Will there be room for economic growth in a world of environmental degradation and an Earth Overshoot day by the end of July? Will economic growth come to a halt, at a time of "peak" material production? On the other hand, will we be able to mitigate environmental threats and deal with the environmental burden without growth? Can the capitalist political-economic system survive without growth?

These are, of course, complex political-economic issues, bound to cause political confrontations. In this chapter, we shall first look at growth assessment and measurement, for well-being and happiness, for wealthy and poor societies. From there we shall look at issues of de-growth and decoupling, and at psychologic disposition and genetic programming for having more and more.

Growth measurement

Economic growth is a result of both manpower increases (more people in production) and productivity increases (greater production per worker). Overall prosperity improvement results from productivity increase. The short story about economic growth tells us that, since productivity

DOI: 10.1201/9781003518815-13

increases would lead to unemployment in the absence of growth, economic growth is a necessary condition for full employment and for productivity leading to prosperity.

This line of reasoning is behind most models for national budgeting and policymaking, and also behind the legitimization of the capitalist political-economic system. As long as the system serve the interest of employment and prosperity, even European social democracy and Chinese communism accept the main rules of the system.

The first to formulate the employment connection more formally was Yale professor of economics, Arhur Okun. "Okun's law" (like a rule of thumb) said that if you wanted to reduce unemployment (in the USA in the 1960s) by 1 percent, you had to reach 4 percent GDP growth. The exact numbers are not important – different countries have different figures that also vary over time with changes in population and productivity. The important thing is the basic connection, which is used to legitimize most investments and job creation initiatives - and also the destruction of nature. Some see the connection as a blessing – the very basis for the good life. Others see it as a nightmare – how are we going to save the environment if unemployment and loss of prosperity are the consequences?

We may call the productivity-growth-employment-prosperity connection a "mantra". Some call it a myth (Jackson, 2013, 2021). But it appears to be a basic fact that given productivity increases, there has to be some form of economic growth to secure employment. Also, it seems to have been a historical correlate, that with (near) full employment, prosperity follows.

What is missing from the mantra is that modern growth economies are based on the intensive use of fossil fuels and the exploitation of non-renewable natural resources, that peak production of both types of resources are coming near, that growth produces undesirable side effects, and that not all elements of growth are equally useful, calling for some form of decoupling of overall growth from environmentally and socially harmful growth. The issue of decoupling will emerge as the most crucial in our growth discussion.

Standard measures

A fundamental divide in thinking about economic growth, as related to climate and nature crises, is the one between, on the one hand, proponents arguing that economic growth is necessary for mitigation, and on the other hand, sceptics arguing that further economic growth is accelerating environmental threats and the environmental burden.

The position of this book is that it depends on the growth content and the distribution of the proceeds. This chapter will deal with content, not only of the growth part of the economy, but with the full content of production and consumption in the economy, and with the political-economic framework for guiding content and distribution. Distribution is the main theme of the next chapter.

In order to deal with content, we first have to sort out how growth is measured. In the measurement of economic growth, we need at least to discern between growth elements that add to the betterment of society, and elements that add to environmental and social worsening. Such a distinction is a prerequisite for decoupling.

Growth is usually measured in terms gross domestic product (GDP), the standard measure used to calculate the size and development of economies. The first attempt at calculating a country's economic activity was made by the English economist, physician, philosopher and omni-scientist, Sir William Petty (1623–1687). Initially, it was about distributing conquered landed property among the soldiers in Oliver Cromwell's army following the civil war. In the follow-up, it was about exposing the landowners' unfair taxation of tenants. And for such purposes, an overview of production size was useful. Towards the end of the 17th century, another English economist, Charles Davenant, improved the calculations, also for the just taxation of farmers.

Our modern gross national product (GNP) can be dated 1937. After years of depression and unemployment since 1929, but also with incentives and public works for employment and growth in the "New Deal" period, President Franklin D. Roosevelt saw a need for an overarching target for national economic activity. Production statistics were available, but there was no unifying expression of overall economic performance. Simon Kuznets at the US National Bureau of Economic Research came up with an answer. In 1937, he presented a report to the American Congress, "National Income, 1929–35", in which he summed up all economic activity, in the private and public sectors. Later, at the important Bretton Woods conference in 1944, 44 nations agreed to the establishment of the World Bank and the International Monetary Fund, and to the regulation of post-war monetary and currency systems. GNP was to be used as a standard tool for assessing a country's economic status and progress.

The first to critically question the GNP measure was Kuznets himself. He warned against identifying GNP growth with increases in prosperity and living standards, and he was careful to specify that there was a difference between quantity and quality in economic growth and between short-term and long-term effects. Not all economic activity was necessarily beneficial for society. Kuznets stressed the need to specify growth areas and growth purposes. In an important textbook in economics from 1946, "Value and Capital", John Hicks emphasized that the purpose of calculating a national product must be to find out how much the citizens of a society could produce and consume without compromising production in the future (Hicks, 1946) – preceding the sustainability movement by 40 years. In 1959, the economist Moses Abramovitz similarly warned against the belief that one could measure long-term growth in welfare through growth in GNP figures. Since then, there has been no shortage of criticism, neither in professional circles nor in politics. An overview of critical voices can be found in van den Bergh's article "The GDP Paradox" (2009), where the paradox is that the GDP measure is maintained and used for policy-making in spite of all the shortages. His list of critical voices includes several Nobel Prize winners in economics.

The measurement of Gross Domestic Product (GDP) has become the main instrument for measuring economic growth and for international comparison of economic output and growth. GDP differs from GNP, in that GDP only counts production within the home country, whereas GNP also includes production by national enterprises abroad. GDP has been the standard measure since 1991, and it is still the core of what is now called "The System of National Accounts (SNA) – an international standard system for reporting national accounts, first published in 1953.

What is wrong with GDP

Criticism of GDP has become harsher over the years. The calculations of GDP in particular violate two basic rules in normal accounting (Costanza et al., 2014). It fails to clearly distinguish between income and costs, and it fails to correct for changes in inventory. In relation to the green transition, the failure to account for pollution and damage is increasingly a problem, and the failure to account for the depletion of natural resources and wilderness is equally an inadequacy of great significance. Compared to standard business accounting, there is no national account for "cost of goods sold". To this we may add that GDP also fails to account for the distribution of income and wealth in a society. There are also important omissions (things that should be counted, that are not included, such as voluntary work) and problematic inclusions (things that are counted as positive, that in fact are indicators of a negative development). Perhaps the best summary of GDP shortcomings is the one presented by Robert Kennedy in his speech at the University of Kansas, March 18, 1968 (Kennedy, 1968):

> Our Gross National Product, now, is over $800 billion dollars a year, but that Gross National Product – if we judge the United States of America by that – that Gross National Product counts air pollution and cigarette advertising, and ambulances to clear our highways of carnage. It counts special locks for our doors and the jails for the people who break them. It counts the destruction of the redwood and the loss of our natural wonder in chaotic sprawl. It counts napalm and counts nuclear warheads and armored cars for the police to fight the riots in our cities. It counts Whitman's rifle and Speck's knife, and the television programs which glorify violence in order to sell toys to our children. Yet the gross national product does not

allow for the health of our children, the quality of their education or the joy of their play. It does not include the beauty of our poetry or the strength of our marriages, the intelligence of our public debate or the integrity of our public officials. It measures neither our wit nor our courage, neither our wisdom nor our learning, neither our compassion nor our devotion to our country, it measures everything in short, except that which makes life worthwhile. And it can tell us everything about America except why we are proud that we are Americans.

Asking ourselves what people will be doing in the future of AI, automation, and an increasing environmental burden, we may identify tree critical areas of the GDP.

- ✓ Damage causing elements. A significant number of GDP-entries are causing direct damage. GDP includes a long list of elements directly undermining future gains, in GDP counted as benefits or net additions, such as cigarette smoking, films glorifying violence, forest destruction, and the sales of weapons. There are also damage-causing elements not counted in GDP at all, such as resource depletion, wildlife erosion, and soil degradation. In dynamic terms, the problem with these, is that they per definition tend to increase the future costs of repairs. An important ambition for the future of all societies should be to minimize the damage-causing parts of the GDP – even when that in the short-term means shrinking the output.
- ✓ Repairs elements. The GDP includes a huge repairs part. Damage-causing elements inevitably create a need for more repair work. Hospitals and ambulances belong to the repairs sector, and they take up a greater part of the GDP, the more damage that occurs in society. The worse the popular health, the greater the costs of repair. The worse society, the greater the costs of prison systems, police and social work. In the GDP, a great number of such repairs, and the full societal repairs-sector is added to the income side, while of course it would be disastrous if the full annual budget had to be used on repairs. If in the future, the costs of climate damage and extreme weather, and other environmental mitigation, adaptation, and repairs, caused by present production and consumption, will reach significantly greater levels, that will be equally disastrous. The need for repairs will undoubtedly provide jobs, but not add to welfare. If most of us would have to repair damaged roads, bridges and buildings because of extreme weather, there would be jobs and at least for a while high scores on GDP – more so, the more expensive the repairs, but not much addition to well-being. A hope for future societies will be to keep the repairs sector at a minimal but socially acceptable and sustainable level, meaning first of all to minimize the damage-causing elements of today. Observing the rising costs of insurance in the wake of climate change, and the corresponding threat to the finance sector, the present outlook is for the repairs sector to eat aggressively into the well-being elements of GDP.
- ✓ Positive elements. Much of the GDP is made up of activities providing for consumption and overall well-being. This is the part that we ideally would like to expand. It will be relatively smaller in society, the greater the damage-causing part, and the greater the repairs' part. It is very simple. If on a farm, in a factory, or a local society, you have some people spending their time causing damage, and more and more people have to spend time repairing the damage, there will be fewer hands available for production of the essentials for well-being. So also, at the macro level of society. The more manpower and financial resources that are spent on damage-causing activity, causing more to be spent on repairs, the less will be available for production and consumption of both necessities and luxuries. To this may be added, that the more the production of necessities and luxuries for well-being also causes damage, requiring more resources to be used for repairs, the more the dynamics of the system takes on a vicious cycle – or a catch 22. Damage-causing activity causes more resources to be used for repairs, reducing the capacity for well-being consumption, wherefore more capacity must be used for upkeep of well-being, increasing the damage-causing part of GDP, reinforcing the need for repair, increasing the need for well-being production……

Common logics would suggest that keeping the growth and content of the well-being part of GDP at a level and in a form that will not require dramatic increase in the repairs sector, would be a rational choice. Also, it would seem rational to secure human survival and to reward voluntary engagement for meaning, wisdom, health, and belongingness – as suggested by Robert Kennedy. As opposed to the "useless class" in Harari's AI-driven society, people who in the future may be unemployed and "irrelevant" in the market economy may come to enjoy a basic universal income and form a new sector representing truly human values and meaning, voluntarism and mutuality, to be dealt with in Part three. I shall refer to this as a "derivative sector".

There is as yet no agreed upon name for the "derivative sector". I think upon it as a return to what Aristotle named a "natural wealth economy". Activity in the "natural wealth" sector may be AI supported, but not AI based. It may be part-commercial, but it does not have to. It may be materials-consuming, but only if absolutely decoupled and mainly based on circular principles. It can be artistic, scientific, playful, social, sporty, or hobby-type, but the main characteristic is to provide purpose to organizations and meaning to individuals involved and being protective of people and nature.

More on GDP weaknesses

In spite of the weaknesses of GDP, international institutions, led by the World Bank and the International Monetary Fund, have in many cases more or less forced poor countries to change their economic priorities and renounce welfare measures in favor of GDP targets rather than well-being (Stigliz, 2002). Research suggests that a direct prioritization of basic welfare factors would be more important than general GDP growth for the welfare of a country (Kenny, 2005; Van den Bergh, 2009).

GDP growth has been seen as a sign of the success of development policies. But since a significant proportion of the population in poor countries often used to live in a near subsistence economy, and since home-grown food and craft production was not included in the GDP calculation, GDP growth may have served to inflate the impression of progress made from development strategies.

Concerning GDP and economic distribution, the "inventor" of GDP, Simon Kuznet, in 1955 presented what has been called the Kuznets curve, where he argued that the market forces of economic growth would first lead to increased income differences, but then to equalization (Kuznets, 1955). The thesis seemed to hold at the time of publishing, but developments since the 1970s, with increasing inequality in the US and other advanced economies, have pointed in the opposite direction, with rising inequality as a function of GDP growth (Piketty, 2014). Also, for developing countries, the curve does not seem to hold.

Another application of the Kuznets curve concerns the relationship between environmental degradation and growth, called the "Environmental Kuznets Curve" (EKC), suggesting that economic growth may be the solution to pollution problems. The hypothesis is that incipient economic growth incurs serious nature-destruction, while further economic growth allows countries to prioritize and innovate for nature conservation (Shafik, 1994; Yasin et al., 2021). At its most optimistic, the hypothesis states that the more growth there is in society, the less pollution and destruction there will be.

In some areas, the hypothesis seems to hold. Richer countries have come further in reforestation and in limiting many types of pollutants including sulfur dioxide, nitrogen oxide, chlorofluorocarbons, sewage, and diverse chemicals. Automobiles have become much more fuel efficient, allowing for more cars and longer travel distances to come with less pollution. On the other hand, there are also strong cases for the opposite conclusion (Mills and Waite, 2009). While the ratios have fallen, overall energy use, land and resource use, and greenhouse gas emissions have not fallen with rising income. And several indicators of key "ecosystem services", including freshwater availability, soil fertility, and fish catches, are in decline. Generally, richer countries have improved their footprint in areas of high visibility and local concern – such as most visible local pollution, and health-threatening toxification. Areas of new entities, low visibility, and small scale, long-time effects, on the other hand have been overlooked and allowed to increase dramatically, even in the richest countries.

The most obvious effect, however, generally overlooked by supporters of the EKC hypothesis, is "the transfer phenomenon" discussed in Chapter 4. High-polluting industrial activity has been "exported" to emergent economies, camouflaging that behind record consumption levels in the richest countries, lie huge offshore environmental footprints. Part of the environmental "solution" has been to "dump" the dirtiest and most vulnerable parts of production to poorer countries, while the rich countries have continued to accelerate mass consumption of goods that cause serious pollution and nature depletion, growing emissions of greenhouse gases and consumption of fossil energy.

The net pollution has to be measured at the global level – in addition to the national level (Suri and Chapman, 1998). Important examples of polluting production "transfers" are textiles, chemicals, pharmaceutical and cosmetics, but also automobiles and auto-parts, and just about everything you find in a modern home. As noted by Suri and Chapman, "export", "globalization", and "offshore outsourcing" offers no lasting solution to the emission problem. The more successful among emerging economies may pass the problem on to latecomers, but eventually the footprints have to be recorded globally, since both climate change and loss of biodiversity are global issues as well as local phenomena. None of this can be read out of standard GDP reports.

Alternative measures

Several alternatives to GDP have been suggested. They normally are aligned with the System of National Accounts (SNA), including GDP as one among several indicators of growth or progress. But they also include nonmarket values and costs, often with a special focus on social and environmental concerns:

- The Measure of Economic Welfare index (MEW) was developed by economists William Nordhaus and James Tobin (1972), taking GDP as a starting point, but adjusting it for the value of leisure time, unpaid work, and environmental damage – making it a forerunner for later models.
- The Japanese Net National Welfare indicator (NNW) was developed in 1973, mainly as an extension and adaptation of the MEW index (Shingikai, 1974; Uno, 1989).
- The Economic Aspects of Welfare index (EAW) was introduced by the Greek economist Xenophōn Euthymiou Zolotas (1981). Applied to the United States for the 1950–1977 period, the index showed diminishing social welfare returns as a function of economic growth in affluent societies, pointing towards some point in time when further increase in the GDP would lead to a decline in economic welfare.
- John B. Cobb (theologian) and Herman Daly (economist) launched the Index of Sustainable Economic Welfare - ISEW (1989). The ISEW summarizes economic welfare in a single figure, adjusting the GDP for income distribution, cost of pollution and harmful consequences of economic growth. Applied to the United States for the 1950-1986 period it was seen that average welfare stabilized after the 1970s, in spite of further economic growth, mainly because of environmental deterioration and increasing economic inequality.
- The UN's Human Development Index (HDI) was introduced in 1990, developed by Pakistani economist Mahbub ul-Hag and based on Amartya Sen's studies of human capabilities (Stanton, 2007; Suri and Chapman, 1998). The index includes measures of longevity, education, and economy. It was further developed to adjust for inequality, called the IHDI, in 2010.
- The Better Life Index was developed by the Organization for Economic Cooperation and Development's (OECD's) Better Life Initiative in 2011. It compares people's well-being on 11 "topics", including material living conditions and quality of life indicators, for 41 countries. Users are free to pick relevant indicators for comparison.

- The Genuine Progress Indicator (GPI) was developed in 1995 to account for a number of trade-offs of economic growth, including a broad specter of economic, environmental and social factors (Kubiszewski et al., 2013). The index was applied to US data for the 1950–1994 period as a modified version of the Sustainable Economic Welfare model (Cobb et al., 1995). The main result was here also, that in spite of substantial GDP growth, genuine progress and peoples' welfare did not improve. As we shall see, the same tendency has been observed in a number of other nations.

Added to these indicators are a great number of more specialized compilations for comparison of economic performance and social development. The Kingdom of Bhutan is using a Gross National Happiness index (GNH), favoring happiness over economic growth; Ecological Footprint measurements have been used for quantification of natural resources depletion, carbon emission, fresh water usage, and habitat destruction; Wellbeing-adjusted Life Years (WELLBY) combines longevity with measures of subjective well-being, health, happiness, and life satisfaction; The Thriving Places Index (TPI), developed in the UK by the Centre for Thriving Places, measures factors relevant to community development, such as green infrastructure, health, education and work life; The Green GDP is a measure of GDP adjusted for pollution and resource depletion; The Inclusive Wealth Index (IWI) has been developed by the UN to assess a country's capital assets, including manufactured, human and natural capital; The Genuine Savings Indicator (GSI), developed by the World Bank, is used to analyze the real savings, adjusting for natural resource depletion and social costs of pollution; The Happy Planet Index (HPI) has been developed by the New Economic Foundation in the UK, combining life expectancy, wellbeing and ecological footprint for the aim of exposing nations' ability to convert environmental resources exploitation into the betterment of lives; and finally The World Happiness Report, a partnership between Gallup, the Oxford Wellbeing Research Centre, the UN Sustainable Development Solutions Network, and the WHR's Editorial Board, seek to measure happiness and well-being as criteria for government policy.

These indicators, and many more, may all contribute to an increased understanding of the consequences of growth, and varieties of genuine progress. In June, 2022, on the 50th World Environment Day, Secretary-General Antonio Guterres reminded us that our traditional GDP measure of economic growth includes the rewarding of pollution and waste. Destruction of forests adds to GDP growth, as does a long list of basically negative environmental factors.

The Genuine Progress Indicator (GPI)

Among the alternatives to GDP, the GPI appears to be the most promising for assessing overall economic, social, and environmental progress. It takes GDP as a starting point, and it makes adjustments for a number of shortcomings. Among the environmental corrections, the costs of water pollution, air pollution, noise pollution, loss of wetlands, habitats and arable land, loss of forests, biodiversity, carbon storage capacity and recreational opportunities, the costs of CO_2 emissions, the weakening of the ozone layer and consumption of non-renewable resources are added to the cost side of the accounts. Also on the cost side are deductions associated with crime, loss of leisure time, weakening of family ties and upbringing, costs of commuting and car accidents, and generally economic inequality. Added to benefits are correction for social development including the value of housework and parenting, voluntary efforts in neighborhoods and communities, the value of good schools, higher education, and infrastructure.

Of particular interest are comparisons between growth in GDP and growth in GPI in different countries. In 2013, an international research group reviewed reports on GPI measurements in different countries, held up against GDP measurements (Kubiszewski et al., 2013). The authors looked at figures from 17 nations that together represent 53 percent of the world's population and 59 percent of the world's GDP. The countries' GDP growth was positive throughout the period, and the level had, on average more than tripled since 1950. In comparison, the average GPI level stagnated or had been in decline since around 1978. The countries and their populations were actually worse off in terms of genuine progress, while the official GDP reports showed significant progress. It is about fooling ourselves.

The researchers also found that there was a strong correlation between life satisfaction and GPI, but no such correlation between satisfaction and GDP growth. People are happier when they experience genuine progress, as measured through the GPI, and a major factor is the distribution of income and wealth. It is possible to get a high GDP average by everyone earning roughly the same, but also to get the same average figure by a highly skewed economic distribution. The researchers concluded that if the world's total incomes were distributed approximately equally, 9.6 billion people would be able to live well on the average income, and total incomes above this would do more harm than good.

Narrowly defined economic growth is also a cause of social anomie, fragmentation, and *"ressentiment"*, what Tim Jackson calls "the paradox of materialism" (2017). He refers to "Bowling alone", by sociologist Robert Putnam (2000), for a description of the collapse of community and belongingness in modern, wealthy societies, and to a BBC's "loneliness index", showing that with a doubling of GDP, loneliness has increased in every UK region. The weakest communities in terms of belongingness and companionship in 1971, were stronger than the best of communities 30 years later. Jackson also refers to the classical economics school. Adam Smith shared with Amartia Sen (Nobel prize winner in economics) a concern about shame as a driving force of growth, making the purchase of a linen shirt a step up for pride and inclusion, at an ultimate level causing the "paradox of materialism" – with no end in sight. John Stuart Mill, likely the most brilliant of the classicists, railed against material advantage as a standard of success. "Trampling, crushing, elbowing, and treading on each other's heels" could never be the most desirable state for humankind. A best state would be one of no poverty and no one desiring to be richer, and also of no one fearing to be thrusted back by others seeking advancement for themselves.

Growth content and localization

There are two main concerns when it comes to economic growth as related to environmental threats. There is a content factor – the planet cannot tolerate growth of production and consumption of fossil fuels and material products that add to the environmental burden. The content factor calls for a decoupling of environmentally burdensome production from the total of economic activity. The total economy include areas of non-threatening growth, such as in the social, cultural, educational, and health fields. More people may write poetry without threatening the planet or impacting the global overshoot date.

Besides content, there is also a placement factor for growth. For the sake of justice, growth is needed to fulfill the basic needs of poor people generally, and of poor people in poor nations in particular. Today's global economy is geared to sin on both premises. In order to satisfy basic needs of people in the poor countries, luxury products for the wealthy are produced and exported to wealthy people in mature economies. I think of this as a "detour to development": In order for poor people to access what they need for basic life satisfaction they have to produce thing that rich people don't need. The placement issue calls for a second type of decoupling, between prosperity addition to the global top incomes and the basic needs satisfaction of the global bottom incomes. The overall message is that if we are to avoid a planetary crisis, limits must be set for environmentally burdensome material growth, and increased material consumption must be reserved for the fulfillment of basic needs for the world's poor by means other than the export of luxury products. The "detour" version of development, so much loved by BAU proponents, is simply a luxury that the planet does not allow.

Not everything that is part of BAU-GDP growth is good for the climate, for the natural environment, or for social sustainability. We must identify areas in the economy where growth leads to damage and areas where there is still room for growth, especially where growth is strongly desirable, as for the satisfaction of basic needs for poor people.

Relative and absolute decoupling

We have seen that GDP-type growth at some point contradicts the ambition to improve genuine progress, as measured by GPI. Using GDP figures to indicate progress is often simply fooling ourselves – saying that things are going better when in fact they are getting worse.

There is also a clear historical connection between GDP-type economic growth, nature destruction, pollution, and climate change. We may refer to this as "coupled growth". "Decoupling" of economic growth is about the reversal of the historical correlation. Without decoupling, for every percentage point of economic growth, there is an addition to the environmental burden, even without producing genuine progress for people. This needs to be corrected, and it will take some economic sacrifice.

Economic growth so far has come with a strong coupling to fossil fuel, pollution and exploitation of natural resources. Decoupling is about economic growth without such binding. In the discussion and assessment of economic growth, the potential of decoupling is a number one issue. A secondary issue of equal importance is a fundamental distinction between relative and absolute decoupling (Ward et al., 2016). Relative decoupling means that the environmentally taxing part of the growth declines over time, relative to overall growth, making the growing economy relatively cleaner, even as the overall pollution and destruction continues to rise. It means that the total growth in the economy is greater than the growth of environmentally taxing parts, but growth is still directly adding to the environmental burden. For a concrete example. Electric cars represent a step for relative decoupling. But still, every electric car produced and used represents an addition to the overall environmental burden, undermining the ambition of absolute decoupling.

Absolute decoupling means that even as the economy continues to grow, there is no increase in environmentally taxing production and consumption, mainly materials, energy use, and pollution. With absolute decoupling total growth in the economy may continue, but not necessarily so. If the halt of environmentally taxing production is a number one priority, overall growth may either come to a halt or continue, depending on the growth potential of non-taxing production.

It is debatable whether any country has been able to implement an absolute decoupling – except during periods of pandemic or economic crisis. It depends on definitions and methods of calculation. According to a report from the International Labor Organization (ILO, 2018) 23 countries had taken steps to some form of decoupling, most of them European. Among measures taken were investments in renewable energy, carbon pricing, and the subsidizing of green products and green jobs – thereby reducing CO_2 emissions. Still, no country has obtained an absolute decoupling across environmental sectors (Ward et al., 2016). Also, not reported in the ILO and similar data is the burden of the "transfer phenomenon" linked to the growth of imports that imply "export" of emissions and environment-taxing production to other countries. It is a problem of globalization that footprints and polluting production are simply "exported", while finished consumer goods are imported, not adding to CO_2 and pollution statistics in the importing nations.

A meta-study of 179 reports on decoupling in the 1990–2019 period (Vadén et al., 2020) found cases of national level CO_2 decoupling, also of land use and blue water use, but no evidence of economy-wide, national/international absolute resource decoupling, and no evidence of decoupling for overall ecological sustainability. Also, a study by the European Environment Bureau – EEB (Parrique et al., 2019) is quite explicit that while "green growth" policy thinking has dominated policy making both at the European and United Nations levels, and in numerous countries around the world, the conclusion is overwhelmingly clear and sobering: The authors find no evidence of decoupling of economic growth at a scale needed to deal with the environmental burden, and they even state that such decoupling appears unlikely for the future. While most governments have pursued "green growth" policies, expecting that decoupling of environmental pressures from GDP would allow for economic growth "without end", the report concludes that high-consumption countries will need to depart from economic growth policies, with direct downscaling of production in many sectors, in accordance with the EU's "precautionary principle" of reducing consumption to a level that will support good life within the ecological limits of the planet.

Instead of overly optimistic "green growth" strategies, nations need to realize the main challenges behind decoupling and downscaling at levels that are global or local, absolute or relative, based on territories or footprints, for short-run and long-term consequences and for all dimensions of the environment and ecosystems. Decoupling so far has been mainly relative, and for absolute decoupling only for short periods of time, for local impact, for certain resources and impact areas only, and for very small rates of mitigation. The authors (Parrique et al., 2019) suggest seven areas making absolute decoupling difficult in the future, including much of what I in this book have referred to as stickiness.

- ✓ For the extraction of energy and other resources, cheaper options tend to be used first.
- ✓ The rebound effect (Jevons paradox) of fuel efficiency and other potential material saving opportunities tend to drive up demand.
- ✓ New solutions to old problems tend to come with new problems (such as EVs adding to resource extraction pollution, or nuclear energy coming with nuclear risks and waste disposal problems).
- ✓ Shifting from material consumption to services comes with underestimated footprints.
- ✓ Recycling requires energy use and has a limited capacity for handling sustained growth.
- ✓ Much of technological progress is not targeted to decoupling and often fails to displace undesirable technologies.
- ✓ Much reported (relative) decoupling so far is due to a re-location of polluting industries to emerging markets ("the transfer phenomenon").

Unless absolute decoupling of the overall economy and in particular of economic growth is achieved, direct downscaling of consumption in the wealthiest countries is the number one option for reducing the environmental burden.

Overall, only long-term absolute decoupling is in accordance with the stabilization of climate and ecosystems. This can be achieved either by reducing the production and consumption of goods and services that add to the environmental burden, or by inventing products and processes that are less burdensome. An important argument in favor of the market economy is that the profit motive drives producers to minimize the use of energy and natural resources, making production less energy intensive and increasing energy efficiency. Historically, on the contrary, the trend has been more to save labor costs through increased use of fossil energy.

If we look at the situation for oil in our time, we note that higher oil prices since the 1970s have led to reduced energy use in production. Energy intensity in the economy is now globally around 25 per cent lower than it was in 1980 (Jackson, 2017, p. 88), with the strongest improvement in the richest countries, implying a relative decoupling from fossil fuels caused by improved production efficiency and the impact of problem "export", moving dirty production and dirty products to developing countries, contributing to the "transfer" issue. Since the year 2000, energy efficiency improvement has largely stagnated, however. Since there has been global economic growth at the same time, total energy consumption and total carbon emissions have continued to increase. Emissions of CO_2 have more than tripled since 1965, and emission reductions in the high-income countries have been more than diluted by increases in the newly industrialized countries.

Overall, the failure to succeed with absolute decoupling, and the meager results from relative decoupling has led to some resignation and dismissal of the decoupling strategy as such, going instead for "de-growth" or "negative growth". Instead of dismissal, however, we may rather admit that absolute decoupling is a strategy that has not really been tested out, and it is a strategy that in some form or other will have to be implemented in order to save the Earth. There is no reason to stop people from being more social, culturally more productive, better caretakers and even more inventive of non-taxing products. Some forms of growth will be positive for people and ecosystems and will have to be encouraged. Other forms will cause deterioration and will have to be phased out.

Rather than seeing the principle of decoupling as being bankrupt, we must admit that attempts at relative decoupling have been unambitious, and absolute decoupling is yet to be attempted!

Alternative paths to decoupling

In a small but informative book, "Farewell to growth" (2009), Serge Latouch wrote about "degrowth" as an alternative to "negative growth". Negative growth, he says, is a recipe for unemployment and economic chaos. When we think of negative growth, we can imagine a run on the banks, business bankruptcies, unemployment, tax revenues stagnation, and the loss of public services. He claims that the worst thing you can have is a growth-based society without economic growth.

"Degrowth" or "counter-growth", on the other hand, represents a quest for an economic system that protects the well-being of people and ecosystems, while setting limits for the material exploitation of nature. It means to allow for satisfaction of basic material needs, social needs, and health needs, supporting a variety of local initiatives in line with circular economic principles, and developing alternative business organizations and new means for environmental protection.

Many alternatives

There is currently no unifying de-growth, counter-growth, or post-growth agenda. On the other hand, there is a growing counter-movement for societal change, away from "more and faster", towards slow time and more quality of life, away from an extractive economy towards a circular economy, away from colonization of nature toward harmony with nature, and away from a "useless class" development in favor of an emerging voluntary sector economy. Latouche refers to it as a utopia – a source of hope and dreams, and he suggests eight guiding "R-words" on the way to a "concrete utopia":

- ✓ *Re-evaluating* values that have been suppressed in the mass consumer society, including altruism, solidarity, and truth, time for leisure, play and socializing, local production and cultivation of crafts and arts.
- ✓ *Re-conceptualizing* perspectives on wealth and poverty, scarcity and abundance.
- ✓ *Restructuring* to develop new and more suitable forms of business organizations, with greater responsiveness to basic individual and societal values and less responsiveness to ownership and the financial world.
- ✓ *Redistribution* of income and wealth between north and south and within each country, between classes, generations and individuals.
- ✓ *Relocation* in the direction of local production and localized activity, including local repair shops, local savings banks and funding sources, and local business.
- ✓ *Reduction* of all production and consumption that threatens climate and the biological world.
- ✓ *Reduction* of the working week for most people and of health risks for the most vulnerable.
- ✓ *Recycling* and *reuse* are core ingredients of a more circular economy.

The hardest part may be to breach the bonding to the consumerist society. Many have pointed to consumerism as the dominating religion of our time, "going shopping" as a chosen leisure activity, and "conspicuous consumption" as a dominating status marker. Within the field of economics, a number of specialized research areas have emerged, such as ecological economics, green economics, doughnut economics, feminist economics, circular economics, or bio-economics. Common to these are emphasis on ecosystem supremacy over economic systems, the preservation of nature (or "nature capital"), sometimes for its own sake, on social responsibility and the problems of inequality, and responsibility for later generations.

Many of the themes covered have been voiced by classical economists, such as Thomas Malthus on population growth and John Stuart Mill on "stationary economics". Kenneth Boulding is generally seen as the main 20th century thinker behind circular economics, and from 1966 the architect behind the "spaceship earth" metaphor (2013). Ernst Friedrich Schumacher has since 1973 been the main reference for the "Small Is Beautiful" type of local, people-centered bottom-up development (2011 [1973]). Herman Daly is, on similar grounds, a main source on steady-state economics (1991).

In their footsteps are hundreds of prominent economists contributing analyses and working out the finer differences among schools of thought. Clive Spash has pointed out the main differences between different schools (1999), and he has pointed out three main orientations (Spash, 2009, 2011, 2017), referred to as "mainstream new resource economists", "the new environmental pragmatists", and "the more radical social ecological economists". But his attempts have also been criticized for being overly normative (making for a "good", "bad" and "ugly" type of hierarchy), with no clear criterion, and overlooking some of the pluralism in the field (Levrel and Martinet, 2021). It is beyond the scope of this book to report on the many varieties of schools, orientations and perspectives within the field of economics, or to seek alternative classifications. An important clarification, however, is useful in the case of de-growth.

De-growth economics

Until now, there has been an unequivocal connection between GDP growth, increased materials and energy consumption, ecosystems destruction and emissions of greenhouse gases. On the other hand, there is also beyond a certain threshold no longer a match between "genuine progress" (as measured by the GPI) and GDP growth. This also means that instead of just decoupling within the confines of BAU and GDP growth, decoupling has to be based on an alternative footing, that allows for an equitable standard of well-being within the confines of planetary material resources and bio-diverse flourishing

As a young business student in Oregon in the late 1960s, and inspired by Kenneth Boulding's metaphor "spaceship earth" (first presented in 1966), I started doing some non-business calculations on my own. In a "spaceship" economy of limited resources there is no ethically legitimate reason for the most wealthy to continue consumption growth while others do not have enough for a decent living. I looked at data for the overall materials and energy availability in the US and Europe to see how much global production would have to be increased in order for all to consume at a level similar to that of the US and Europe. The US in 1970 represented 5.5 percent of the global population, while consuming some 31 percent of resources as measured by share of global GDP, meaning that for all to consume at the same level as the US, global production would have to be more than five-doubled.

A more reasonable target (than for an immediate catch-up) would be for the rest of the world to catch up over a prolonged growth period of say 30 years. At the time (1970), the expected growth rate in America towards the year 2000 was about 3 percent. With higher population growth rates for the world than for the wealthiest countries, the US share of world population was expected to shrink to below 5 percent meaning that global production would have to be around forty-doubled (depending on a number of prerequisites) in order for all to attain the US or European level of consumption by the year 2000.

Clearly, no such result was anticipated, nor was it obtained. Still, remarkable growth was achieved in many countries, especially in China and the group of NICs (Newly Industrialized Countries). Even so, the United states with (2024) slightly above 4 percent of the global population by IMF accounts represents some 25 percent of resource consumption, meaning that for the rest of the world to reach the same level of affluence, global output would have to be more than six-doubled, making the step up for the rest of the world even more quixotic. The numbers are absurd, and they get even more absurd as we update the data and remind ourselves of approaching tipping points. Also to be remembered is that we already reach overshoot consumption globally by the end of July, and for the wealthiest countries sometime in March every year, meaning that we already are consuming at a level far beyond planetary carrying capacity.

An Australian champion of de-growth, Ted Trainer (2012) has provided relevant numbers. The present world population has passed eight billion, and Trainer foresees continued population growth to a plateau of 10 billion. If all of these global inhabitants were to reach a level of resource use at the per capita rate of the rich countries, Trainer found that the annual global production of material output would have to increase by a factor of eight in a time span of 40 years from 2012. Trainer cite fresh-water resources, land areas and wilderness, fish catches, meat consumptions, and a number of minerals among resources that are already scarce. Also, dates for peak gas and peak oil are coming close, and for peak coal sometimes in the second part of our century. He also figures that it takes 8 hectares of productive land to provide fresh water, energy, settlements and food per person in Australia. If 10 billion people were to attain an Australian level of material consumption and space, it would take 80 billion ha of productive land – about 10 times what is available on the planet. Similar "impossible" numbers are also present in the case of greenhouse gas emission and sequestration, green energy transition, and near all elements associated with material growth.

But, as I concluded already in 1967, this is just the beginning of the problem. Anticipating an economic GDP-growth of 3 percent per year in Australia, if all 10 billion people were to reach an Australian level of output by 2050, the global economy would have to 20-double, producing 20 times as much in 2050 as today. Furthermore, if instead of using rich countries averages (such as Australia) we were to use the wealthiest 10 percent or 1.0 percent, or even 0.1 percent of the same countries, as a basis for setting growth and equalization targets for the earth by 2050, global production would have to grow by a multiple of the 20-doubling indicated by Trainer.

Given that the present level of production and consumption, already today is based on a global overshoot the double of sustainable carrying capacity, the figures imply that the living standards and the material consumption models of the wealthiest people in the world, and also the average consumption standards of the wealthiest countries, are only attainable as long as we are absolutely sure that the poorer people and the poorer nations do not achieve their GDP-type growth targets. We therefore need to look for alternative societal solutions, within the boundaries of global carrying capacity and above the lower limits on social sustainability, and we need to base our conclusions on ethics beyond the GDP-mantra.

An ethical categorical imperative

The present political-economic growth model, and the political order behind, is legitimized on the premise that it will also contribute to the rest of the world achieving similar prosperity in the near future. The UN sustainability goals are targeted for 2030, set to offer satisfaction of basic needs for the world population. There is no target for prosperity above that, on par with incomes in the richest countries. But the basic assumption is that everyone at least should have a chance to reach the level of consumption in the wealthiest countries – at least there should be no barriers to such achievement. Our present societal model is considered legitimate as long as the same model also can be employed by the rest of people in the world. In a "Spaceship earth" economy, however, if passengers on business class eats unproportionally much of the salmon and steak sandwiches, there will not be much left for the economy travelers.

Not all global citizens may want the salmon and steak sandwiches. Some may find much of modernity meaningless, but at least we cannot build our moral reasoning on a premise of hindering other people in achieving similar living standards. The argument can be derived from Immanuel Kant's categorical moral imperative "Act only according to that maxim whereby you can at the same time will that it should become a universal law" – from his "Groundwork of the Metaphysics of Morals". Kant felt that his categorical imperative was the solution to his search for a reason-base for all morality, for duties, obligations and moral advice – forming a new core of deontological ethics. Living by the categorical imperative was for Kant more than just a norm among many norms, it was to be a superior guiding principle for all human endeavor.

In a world of seemingly unlimited natural resources, and endless opportunity for growth and development, one might contend that "the white man's burden" was a worthy case of development – as argued by Kipling, setting a worthy example for other people to follow, with economic growth, democracy and freedom, and social sustainability as added benefits. In a world of limited resources and carrying capacity, on the other hand, the ethical norm has to be one of sharing. As noted by Trainer (2012), even the "Factor Four" improvement in resource use efficiency (Lovins et al., 1998; von Weizsacker, 2013) falls far short of legitimizing the present BAU model. We noted in chapter 5 the enormous challenges associated with climate gas sequestering and green energy production. Doing this at a 20-fold level in a "Spaceship Earth" world is very far beyond reach.

Adapting the world economy to a de-growth decoupling alternative, is not impossible, but will require significant adaptation. Heroes in the system will no more be the growth-promoting entrepreneurial talents and the finance sector behind. Markers of success, such as huge properties, private jet fleets, car-fleets and superyachts, will turn into markers of pathetic and failed living. Reinventors of subsistence living and producers of non-material values, solidarity, and care, as suggested among others by pope Francis, will be the new heroes. Artists and artisans, sport stars and cultural entrepreneurs may continue to enjoy respect. Changes such as these are not just about a green transition – rather they involve a qualitative transformation of society, with a new mentality of austerity and sharing, but also of creative and transformative living. It is about a transition task that is "far more daunting than most green and left people imagine" (Trainer, 2012, part 4.3).

The capitalist political-economic system is based on accumulation of profits to be turned into further accumulation and wealth, deemed to be impossible or meaningless in a de-growth society. Such transformation implies a great overhaul of values and changes to most elements of the economy, but not the elimination of every element of the system. Some elements may be scrapped, some will be adapted, and some will largely be retained. Most likely, market- based transactions, private investments, and consumer choice will still be at the center of the "possibilist" political economy. And while the materialist mass-consumption society may give way to limits and scarcity, The Simpler Way project (Trainer, 2010) points towards liberation to enjoy a higher quality of life, with a micro-economic focus on small-size, local, and neighborhood life, and on local bonds and networks for exchange and support, on acceptance of equity and sustainability as norms for distribution and consumption of material goods and beyond that a widening of non-material life satisfactions.

Many authors seem to take for granted that capitalism without growth is impossible. The impossibility may not lie with the market system as such, or with ownership, or the profitability elements of the system, but with legitimacy. Economic growth has since the industrial revolution added legitimacy to the system. With growth, some level of sharing has been possible, mainly between owners and employees, and not necessarily in a fair manner, but still somehow shared. Without growth, owners (and top executives) can only grow richer by making someone else (employees, customers, other stakeholders) poorer. That would undermine the legitimacy of the system.

Doughnut economics

An interesting attempt at capturing the essence of a truly green transition and sustainable development within social and planetary boundaries is captured in the metaphor, and the book and writings about "donut economics". Oxford economist, Kate Raworth, presented an outline about the "space for humanity" idea (2012), and followed up in her book on "Doughnut Economics: Seven Ways to Think Like a 21st-Century Economist" (2017). The "doughnut" metaphor has three main elements – (1) an inner "social foundation", a space of what humans need for a good life – including 12 main elements for health, education, income and equality, (2) the doughnut itself defined as "the safe and just space for humanity", and (3) the outer rim or "ecological ceiling" – including nine main elements for climate change, biodiversity loss and land conversion, and synthetic fertilizer and chemical pollution. A planetary overshoot occurs when the ecological ceiling is trespassed. A human catastrophe occurs when the social foundation is shaken. A truly prosperous economy exists when all 12 social elements are satisfied within the boundaries of the nine ceiling elements.

The book is also a critique of mainstream, or neoclassical economics, that have tended to ignore environmental constraints and non-commercial values. As long as the ecological ceiling was not overshot, this was of less concern. In our world, with an Earth Overshoot Day on August 2nd and an economic output adapted to the capacity of near two earths, with the consumption patterns of the wealthiest countries adapted to the carrying capacity of three or four earths, the outer rim of an ecological ceiling should be of main concern to economists. Also, since the inner social foundation is failing especially when it comes to economic distribution and power distribution, the carrying capacity of the Earth is shivering, while a majority of people do not even have a fair share in the spoils. In a doughnut economy, the needs of all are met within the limits of the planet – a criterium that has been tested in a number of locations in Scotland and Wales, South Africa, China, and India, and more are being added.

The stickiness of economic growth

The sum of challenges, or better – the product of challenges, and the complexity of the series of interwoven challenges, presents the world community with highly complex change processes. Some changes will occur by necessity, sooner or later, because of absolute environmental and resource limits (such as "peak oil", "peak gas"), some will be self-imposed for mitigation purposes, and some will come from causes not directly related to the environment, such as AI-development and everything related to social sustainability. The main question of this chapter is to what extent the problems can be solved within our current, BAU capitalist political-economic system and our mass-consumption economic growth-culture.

Again, we have to consider stickiness factors – stuff that add to the hurdles of decoupling and de-growth.

Conspicuous consumption

Thorstein Veblen wrote "The Theory of the Leisure Class: An Economic Study in the Evolution of Institutions" in 1899 (2017). He described conspicuous consumption as a tendency of consumers, in particular the newly rich, to buy and use products of higher price and quality, or in greater quantities than needed, essentially for the enhancement of status and prestige through the display of luxury. It is about consumption for the sake of consumption and status.

Other motivations can be to create the envy of other people, gain political power, or whatever it takes to impress a reference group. Also, other means than regular consumption can be used for the same purposes, such as conspicuous compassion and charity (West, 2004), conspicuous leisure expenditure, luxury tourism, oversized houses and estates, luxury watches, generous hospitality, and even conspicuous frugality (Stanley et al., 1996).

While Veblen was mainly concerned with the rich people of his time, later research, reflecting the growing prosperity of the 20th century, shows that conspicuous consumption has increasingly become habitual among the middle class and even the working class, perhaps more for the demonstration of belongingness than for status or envy. In emerging economies, conspicuous consumption may signal that the buyer has come out of poverty and made a successful transition to modern lifestyle (Postrel, 2008).

Related to our theme of environmental threats, growth and de-growth, conspicuous consumption clearly adds to the environmental burden of growth. It is also about what most people come to consider necessary for a good life in a modern society, not just for prestige and status, but driven by advertising and the fashion industry, also a force of belongingness and responsible provision for children, even for personal identity (Hebdige, 1994). It means that taking away that piece of luxury from people will often hurt. Conspicuous consumption may also be a form of revenge by people who feel "*ressentiment*" and alienated, as noted in Chapter 9.

In terms of absolute impact, the annual revenue of the luxury fashion industry was estimated at €1.64 trillion in 2019 (Cabigiosu and Cabigiosu, 2020). But that is only a portion of the total, since consumerism and the materialism behind, together with advertising, "logomania", social media "influencers", and financialization drive the overall spending in society towards an unsustainable level. It has also been shown that conspicuous consumption drives people to more borrowing and debt, more so in societies of high economic inequality and easy access to credit (Bofinger and Scheuermeyer, 2019). Poor people borrow for conspicuous consumption in the hope of signaling belongingness and equal status. It produces a vicious circle of getting poorer in the process while adding to national economic inequality, a weakening of saving levels, and adding to macroeconomic volatility and crisis vulnerability at the national level (Banuri and Nguyen, 2023).

Besides the environmentally based criticism of economic growth, we also owe to include the more general criticism of consumerism – our new world religion. Before the environmental movements of the 20th century, there is a long history of anti-materialist critique, anchored in religion, historical tradition, indigenous cultures, art and poetry, and presently in local communities and neighborhoods, in circles of friends and family, in voluntary work and social care. And they may all come in useful in a "De-growth" or "Post Growth" society, sometimes offering surprising opportunities for decoupled growth in an otherwise growth-driven society!

Unlimited prosperity

For environmental damage mitigation, conspicuous consumption needs to be limited, starting at the top, for which the main instrument is to increase taxation. John Stuart Mill, in his Principles of Political Economy (Mill, 1871) made a strict differentiation between consumption based on enjoyment of the "thing in itself", and on the other hand, the great portion of expenses incurred for enjoyment of the "regard to opinion" and for "appendage". He looked upon the latter as "a most desirable subject of taxation". If taxation came to discourage such spending, "some good is done", and if not, "no harm" – since nobody will be worse off (book 5, Chapter 6). Later research has confirmed that such (luxury) taxation is less distortive than regular labor and capital income taxes (Yew-Kwang, 1987), and also contribute to raising the optimal rate of inheritance taxation (Cremer and Pestieau, 2011). A radical alternative solution has been suggested by John Tierney in a New York Times article (1998), suggesting a progressive tax upon annual expenditure on conspicuous consumption of goods and services to replace the personal income tax. At a more general level, conspicuous consumption can be reduced by ordinary means for income and wealth equalization, and by the provision of public and social goods and services, to be explored in later chapters.

Preceding by 2200 years John Stuart Mill's categorization of "thing in itself" versus "appendage" elements of the economy, Aristotle (in Politics, book I, Chapters vii-x) observed that there were two quite different modes of seeking property or wealth – *chrematistike*. "Natural wealth" included everything necessary for self-sufficiency and a decent life – such as shelter, food, clothing. The seeking of such wealth was at the heart of household management (*oikos nomos*) – at the heart of the economy, including responsibility for storage and sharing. Since there were definite limits to how much "natural wealth" a person or a family would need, there would also be a limit to how much money people would strive to earn.

A quite different mode of "wealth getting", according to Aristotle, was where money acquisition was an end in itself, rather than a means for the acquisition of necessities. Means for such non-natural "wealth getting" were trade and the lending of money, and it was facilitated by the coining of money. A distinguishing factor of "money-getting" was that there was no limit to how much was enough – a theme to which we shall return in the final chapter.

As a name for the money-seeking mode, Sarah Chayes (2020) refers to the "Midas disease". According to Greek legend, King Midas of Phrygia was offered a reward from the god Dionysus, and he chose the gift of having everything he touched being turned to gold. Happily, Midas touched a twig, and it turned to gold. Then a clod, and it also turned to gold. But then the king was ready to celebrate with a big meal, and at the moment he touched a loaf of bread, it turned to gold. And when finally, his dear daughter came to sit on his lap, she also turned to gold. The gift came to be a curse.

The general economist assumption of diminishing marginal utility of acquisition does not seem to work for money-seeking and the Midas disease. Marginal utility ideas have been traced to Aristotle who noted that otherwise useful things can be of little use, even of harm, if there is too much of them (anticipating the question of why we to-day need have "mountains" of little-used, even not-used, scrapped textiles). The modern term "marginal utility" – "Grenznutzen", was developed by the Austrian economist Friedrich von Wieser (Marshall, 2009 [1890]), while a number of economics classics witnessed the phenomenon. Adam Smith speculated on the absurdity that diamonds are more worth than fresh water!

The "law of diminishing return" states that if other factors are held equal, the addition of a unit of some product will add gradually less to people's utility. If you own only one shirt, the addition of one more will add great utility. But if you already own a hundred shirts, one more will add little to satisfaction. But for money-seeking, a near opposite seems to hold true. If you own one million dollars, you will go for more than an extra million. If a colleague in investment banking has made a billion, and you have caught the "Midas disease", you will want the triple. An underlying contradiction in both modern economies and global distribution is the conflict between people striving to make a living based on "natural wealth" on the one hand, and the money-seekers or "wealth getters" on the other hand – those suffering from the "Midas-disease".

Ideally, at a time of approaching climate tipping points, a first priority in economic activity should be to satisfy the most basic needs for all, and in particular for those who need it most – with the smallest possible "footprint". Satisfaction of higher-up material "needs", and in particular the needs for conspicuous consumption and luxury, should take second place.

Much of globalization can be linked to the kind of "detour to development" explained above. If we think of the Earth's carrying capacity as a scarce resource, we will have to ration the productive capacity and utilize whatever room there may be for growth to satisfy the most basic needs of those who need it most – which is most people in poor countries and poor people in rich countries. And we shall need to do it by other means than producing more and more luxury for people hit by the "Midas disease". In order to manage production development in such a direction, we will need both tax corrections and income redistribution – a very demanding change to be dealt with in the next chapter.

Genes for Growth

Humans seem to have a biologically conditioned, gene-engrained preference for adding rather than subtracting. The same seems to be manifest at the cultural level. Leidy Klotz has shown in the book "Subtract - The Untapped Science of Less" (2021) that the vast majority of people in experimental situations choose solutions that represent "add", "something more", "growth", etc., rather than the opposite. If you build with Lego bricks and the building is slightly crooked, you easily choose to build on rather than take something away, even when the latter solution would be just as easy. The same happens in town planning offices, government offices, product development departments in business, and in families and neighborhoods. 200,000 years of pre-programming seen to have taught humans to gather stuff – such as firewood for the winter and surplus stores of food when the crops are good. The modern, high-income version of this has come to add stuff that threatens the Earth's sustainability and gives us an "overshoot day" in wealthy societies in early spring.

In "Laudate Si'", Pope Francis has given us a recipe for saving the Earth based on austerity. It can be about moderation of our private choices, but also about creating a culture of collective thrift – the opposite of collective consumerism. For absolute decoupling, a great many of us need to become better at subtracting than adding, overcoming an urge for conspicuous consumption or for spending holidays on shopping. We must also be willing to use the political-economic brakes on over-consumption, and allow for slow-time, rest, and the ability to be grateful for what we have more than being unsatisfied for what we don't have.

The premise of growth, and the urge for more, may have an anchoring in Darwinian evolution, but it is also embedded in culture and organizational systems. Consumption is linked to perceived freedom, personal success and identity, popularity and security, and not least employment and economic growth. People don't get rid of old things just because they're worn out. The circular economy is about more than recycling and repairs. Having something new is for many people of special value. For some of us it gives greater joy to see something old being restored.

Growth and organizational rationality

At the level of organizations, there is similarly an imperative of growth and "more" built into the fabric of most institutions – not only business firms, but also in government, healthcare, education and communication, and in non-governmental organizations, philanthropy, and sports. The article "Theoretical perspectives on organizations and organizing in a post-growth era" (Banerjee et al., 2021) sees economic growth as the "ruling idea" of our age, entrenched in everyday life, but also in organization-theoretical thinking and research traditions. The imperative of growth seems to apply to all disciplines and specialized study fields within the area of management, including organization theory, strategic management, leadership styles, and personnel policies and recruitment. Methods and approaches that add to efficiency and growth are generally embraced, and growth is the main driving force behind capital accumulation, wealth, and employment – seemingly the answer to just about everything..... Dissenting views are at best seen as something odd, and more likely as something irrational. Also, apart from regular taxation and a few regulations, there are presently no public policies in place for the limitation of overall consumption and production, as these would tend to undermine not only economic growth, but also the prospects for political popularity and reelection.

The problem is that "the ruling idea" of growth, without absolute decoupling, adds to the stickiness of the environmental burden, making growth itself the main problem and not the solution. The Dow Jones Sustainability Index, was established in 1999 to provide sustainability guidance for firms, but it still bases assessments and advise on long-term shareholder value growth based on the integration of economic, environmental and social growth. The same holds for mission statements and policy formulations at the company level. Banerjee and colleagues (2021) mention several cases of companies with green and social sustainability profiles, that still define their mission in growth terms, including BlackRock (the world's largest asset manager) and Unilever. They also point out that enterprises with alternative forms of ownership, or more democratic organizational structures, generally also share the growth imperative, such as the Mondragon (a Spanish worker-owned cooperative, with a management philosophy of participation and empowerment, labor solidarity, and social transformation), and Patagonia (a Californian textiles company with a green profile, democratic governance, and a slow-growth ambition).

While small and medium size enterprises (SMEs) tend to have slower growth than the larger nationals and multinationals, they are still part of the growth economy. At least for some purposes, they tend to be judged more by survivability results than by growth, however. As a class of companies, SMEs may point in the direction of organizational solutions to a post-growth or de-growth society, providing for an enormous diversity of solutions to an enormous variety of challenges.

Generally, to the extent that enterprise growth is environmentally decoupled, it is welcome, but without decoupling, all elements of growth (in whatever organization) just add to the environmental burden, increasing the level of threats while also increasing the level of ethical tension (Trainer, 2012). Coupled growth serves to establish a barrier against development, growth, and hope for poor countries.

As main premises for a concept of organizational and political-economic development according to de-growth principles, Banarjee and colleagues (2021) mention four main transitions: From accumulation and growth to redistribution and sharing; from extraction to recirculation; from competition to cooperation; and from consumerism to sufficiency and the satisfaction of needs. These are issues to which we shall return in subsequent chapters.

Chapter 12

From Inequality to Redistribution

There is a gap between the clarity and seriousness of reports about environmental deterioration on the one hand, and the confusion and delays of mitigative political action on the other. To the question of why, several factors have already been mentioned, such as the weaker status of latent damage as compared to manifest damage, the influence of deniers and the spread of doubt and confusion, the predominance of a growth orientation in societies, at the personal level and at the level of organizations and institutions, and behind it all tendencies for disintegration following immiserations and elite rivalry.

In the present chapter, the effects of inequality of wealth and political power will be examined. Economic inequality, both within and among nations, is a main cause of social unsustainability, which is again behind the failure to deal adequately with the environmental burden. Making up for inequality is therefore both an ethical issue of doing good for poor people, and a matter of strategic importance for societal cohesion and agency, as well as a necessary step towards concerted action for the environment.

Asymmetries behind inertia

In an article about climate change and "the mismatch between scientific urgency and political inertia", Jonathan Boston and Frieder Lempp (2011) focused on four main "asymmetries" behind the inertia, to be grouped in time and space. First, the voting asymmetry is about political decision-making referring to the obvious fact that children at present, and future generations, both expected to suffer most from climate change, have no vote in the present political system. Of course, politicians are free to take their interests into account. There is, however, a tendency to overestimate the present costs of mitigation relative to the mitigation-benefits to later generations. There is also the "political myopia" factor – general shortsightedness and a preference for staying in power by pursuing short-sighted policies (Congleton, 1992).

Secondly, also about timing, the cost-benefit asymmetry is about the fact that the costs and benefits of mitigation are quite different in terms of visibility and tangibility, reinforcing the voting asymmetry. Mitigation incurs immediate costs, experienced by present voters (including lower growth, carbon taxes, and higher prices), whereas the benefits of a stable climate and preservation of ecosystems accrue over centuries. It is a case of political "intergenerational buck-passing" (Gardiner, 2009), and of the time-value of money, or social discounting, as exemplified in the controversy between the Nicholas Stern (2007) and (Nobel laureate) William Nordhaus (1997). The asymmetry is also present at the special relationship between mitigating nations and "free riding" nations in the sharing of mitigation costs.

Thirdly, the interest group asymmetry refers to an asymmetry in space, in that the main beneficiaries of mitigation (such as poor people in less developed countries) tend to be highly

DOI: 10.1201/9781003518815-14

dispersed and not highly organized, whereas privileged groups and "free riders" tend to be more concentrated, better organized and better equipped to resist paying. In the case of future generations of beneficiaries, these are entirely excluded from organizing and involvement, whereas strong groups and networks with money to invest are well organized with a strong case for not paying. They also have strong lobbying powers, enjoying home-field advantages even at the annual United Nations Climate Change Conferences (like the COP28 UAE in Dubai). Corruption and bribery are also part of the game, with outcomes biased in favor of special interests at the expense of the general public and future generations. The result tend to be economically inefficient, where the gains to special interests are generally less than the losses to the majority of people (Lohmann, 1998).

Finally, Boston and Lempp's accounting asymmetry is about the fact that neither at the firm level nor at the level of national GDP are the effects of climate-change and nature-depletion effectively recorded, adding to the space assymetry. Since 2011, when the article was written, more firms are today doing "Triple Bottom Line Reporting" in Annual Reports, including financial, social, and environmental results, but only the financial numbers enter into the national statistics, and in the national GDP-accounts. And there is no subtraction line for depletion of national resources, ecosystems and habitats. Also, as noted in Chapter 8, even the Triple Bottom Line Reporting allows for "green-washing", omission of problems and exaggeration of green efforts.

In the following we shall return to some of the asymmetry problems, and the main structure of the chapter is along the lines of inequality and economic and political power distribution – in time and space, targeting adjustments from the top and from the bottom of income and wealth. Distribution in time is about sacrifices of the present generation for the security of future generations. Distribution in space is about the skewed distribution of costs and benefits of mitigation among rich and poor, about social and political costs of inequality in the present, and about distribution of income and wealth between the haves and the have nots – with consequences for social sustainability and environmental resolve.

Distribution in time

Questions about economic distribution and redistribution in time are about how we assess the costs of mitigation now compared to the costs of adaptation for our descendants. It is also about the reduction of consumption for the world's richest now, relative to the benefits of economic security and consumption for later generations. In economics, the tradeoffs are narrowed down to a choice of a social discount rate. In politics it is about the power behind mitigation versus delays and later adaptation.

The question is basically about how much our generation and the wealthiest of the present generation are willing to invest voluntarily now in mitigation for environmental and ethical reasons, versus waiting to be forced to mitigate by resource shortages or other environmental barriers.

The discounting dilemma

How much should we invest now to save the Earth for our descendants? In the last chapter we considered questions about material economic growth, and about the limits of what should be produced and for whom. Here we ask the question "for whom" with respect to our generation versus future generations.

The discounting issue is at the heart of all investment decisions. It is about how much we are willing to pay today to obtain certain benefits in the future. When an entrepreneur considers borrowing money in the bank for an investment in new equipment, the interest rate and the level of expected payoff will decide the outcome. Such an investment normally implies a considerable financial outlay the first year (the cost side), and then several years of added income to pay for the outlay (the benefit side). But if the interest rate paid for the loan is very high, it will more than swallow the added income, and the investment will not be profitable. The expected payoff will not be sufficient to cover the interest. So also for mitigation investments.

In investment models, the calculation will be stated in terms of present value. The present value of the investment is at start a negative one, equal to the price of the equipment, to which is added the positive value of future profits adjusted for the interest rate. If the interest rate is near zero, the proceeds will be counted at near full benefit value, tipping the choice in favor of making the investment. If the interest rate is high, the proceeds will be counted as if they are gradually less and less worth. And if the interest rate is set very high, very few investment proposals will pass the test of the investment calculation. The calculation does not only apply to bank loans and bank interest rate. The entrepreneur may use an internal rate, based on past performance or future opportunities, or a mix of return on owner's equity and bank loans, normally implying a tougher screening of investment proposals than when using bank rates alone.

This is also the case when decisions are made about mitigation versus adaptation type of investments for climate and nature. Mitigative regulations, bans, taxes and degrowth policies are bound to make many things more expensive now, adding to the direct investments outlays. But the positive payback in terms of future economic, social and environmental benefits may more than compensate for the initial investment. That all depends on the interest rate, or discount rate.

At an interest rate of seven percent, a bank deposit would double in 10 years. But that also means that a benefit 10 years hence would only be counted as worth 50 percent of face value in an investment calculation. In 30 years, the deposit would increase by a factor of eight, meaning that the future benefit to our grandchildren would only count as 1/8 of face value, again meaning that whatever the mitigation-benefit to our grandchildren, it would hardly pay for the present generation to invest for climate and environmental purposes. Even if the future costs resulting from damage to nature and more and more extreme weather should become astronomical in the future, the (negative) present value of these damages will be low if we operate with a high discount factor. If on the other hand, we decided on an interest rate of one or two percent, we would definitely go for mitigation, we would count the suffering of our grandchildren on par with similar suffering for ourselves. If the interest rate were zero, we would be willing to invest a billion today for every billion of future benefit. And if the interest rate was set at a negative, it would mean to estimate the suffering of descendants as worse than corresponding suffering for us, wanting a better life for our descendants than for ourselves.

Values for posterity

Using an unduly high discount factor for climate and environmental mitigation can be seen as a way of trivializing the problem – another form of denial tactics. Environmentalists insist that a low discount factor has to be employed in order to save the environment for the future! The opposite position is held by critics who see the use of a low discount factor as a form of self-harm, a way of destroying the good life for our own generation. The combined effects of economic growth and technological invention may render mitigation unnecessary, suggesting a high discount factor! No wonder there is tension linked to the interest rate in matters of mitigation, causing conflicts over interventions such as carbon taxes, the phasing out of fossil energy, and subsidies for green energy projects. If you imagine a flourishing, unrestrained growth economy in the future, where environmental problems are solved by AI and advanced technology, and where our descendants can easily pay for the adaptation costs of whatever problems arise, it may seem pointless to set aside mitigation money from our strained budgets today.

Nobel laureate William Dabney Nordhaus is probably the economists who has done most work specifically on questions of discount rates in connection with mitigation and adaptation, also developing highly advanced models for policy advice. Together with James Tobin, he introduced the Measure of Economic Welfare (MEW), and the Index of Sustainable Economic Welfare as a model of environmental accounting (Nordhaus and Tobin, 1972). And as a long-time supporter of (late) adaptation over (immediate) mitigation, in the 1990s he was the main developer of the DICE and RICE models to study the effect of energy transition and climate change (Nordhaus, 1993). By many, however, he is mainly remembered for his controversy with the British economist, Nicholas Stern, formerly the chief economist of the World Bank, the second permanent secretary of Her Majety's Treasury, head of the British Government Economic Service, and the main author of "The Stern Report" (Stern, 2007). Nordhaus argued for a discount rate of 4.3 percent, while The Stern Review suggested a rate of 1.4 percent, emphasizing worst-case scenarios and environmental and social risks. Many of the critics of both argued for rates in between (Ackeman, 2007; Nordhaus, 2007). Generally, as manifest climate damage and dramatic tipping points are coming closer, economists tend to argue for lower discount rates. Most advice now seem to encourage a rate of about two percent (Groom et al., 2022).

Below zero and beyond GDP

Alternatively, it can also be argued that in cases of abrupt damage to our descendants, a rate of zero should be used: there is no ethical cause for estimating damage and pain to our grandchildren and later descendants any less than damage to ourselves. And in the case of possible "threats to humanity's long-term potential" or extinction (Ord, 2020), discount rates are beyond relevance. If there is a "Russian roulette" chance of irrevocable collapse, only the utmost efforts for survival are legitimate. Even investments in mitigation that later turn out to have been unnecessary are defendable.

A general observation so far is that as the cost estimates for future damage have been upgraded, and the discount rate for calculating returns on mitigation has been downgraded. A conclusion is that there are both ethical, environmental and economic reasons for talking about an almost zero interest rate on mitigation. Arguments for a high discount rate, in line with some level of "climate denial" and doubt, generally assumes that BAU economic growth will continue, and that our descendants will have a greater chest to pay from for damage. Observations so far (in Chapter 11) indicate that either there will have to be a voluntary restriction on growth, implying some form of decoupling or degrowth, or there will be a period of prolonged economic growth followed by future collapse – actually making it harder for our descendants to pay for damage and adaptation, suggesting a very low discount rate on mitigation. Generally, mitigation has so far been shown to be both costlier and more urgent than expected, as shown by rising insurance costs and also by the problems of stickiness, cannibalism and descent (in Chapters 4 and 5).

It should be noted that most estimates of appropriate social discount rates operate with a "payoff" factor defined in terms of GDP or a similar economic growth measures. The worth of present mitigation investments is estimated in terms of impact on economic growth in the future. Future GDP is the payoff. But the fact that economic growth is also a net contributor to the environmental burden, and beyond a certain level a poor indicator of well-being, tends to be inadequately considered in debates over social discount rates related to mitigation. Under BAU assumptions, GDP-style economic growth tend to be considered an appropriate general measure of future well-being. Our reasoning around the environmental burden and the need for absolute decoupling, and the alternative use of a genuine progress indicator (GPI) of growth, tells us that much of the social discount discourse rests upon a dramatic fallacy of premises. GDP is simply not a valid indicator of future welfare.

The reality of lifetime exposure

Swedish climate activist Greta Thunberg at 16 founded the children's climate movement Fridays for Future, calling for urgent action to halt global heating. Her calls for action were followed by hundreds of thousands of young people, reminding politicians and business leaders of who were to bear the consequences of present generation neglect.

The extent of the generational impact of the greenhouse effect and extreme weather exposure has now been mapped out (Grant et al., 2025). Linking exposure to temperature rises (of 1.5°C, 2.7°C, and 3.5°C), to extreme weather frequences (heatwaves, crop failures, river floods, droughts, wildfires and tropical cyclones), and to birthdate cohorts (1960, 1990, 2020). For example, people (in Brussels) born in 1960 should expect to experience three heatwaves during their lifetimes. For comparison, people born in 2020 in the same area should expect to face 11 heatwaves during their lifetime if the temperature rise can be controlled at the 1.5°C target level. If the temperature rise instead reaches 2.7°C, the same 2020 cohort will face 18 heatwaves, and with a 3.5°C rise, they will face 26 heatwaves. Reminding ourselves that heatwaves and high temperatures are already taking thousands of lives annually even in the wealthy Europe and America, and that heat and pollution in combination is taking millions of lives globally, also that heat-strokes represent extremely painful suffering and death, the generational issue should be taken with maximal seriousness. Cohort exposure to heatwaves is also paralleled for all kinds of extreme weather events, crop failures, etc. As we shall in the following, the burden of it all is also disproportionally carried by the economically and socially most disadvantages in the global society.

Distribution in space

Economic inequality seems to pop up in all areas of relevance to the present threats. Inequality undermines social sustainability, which again undermines prospects for peace, cooperation and effective, global climate and wilderness mitigation. Inequality reduction is also a prime target according to the UN Sustainable Development Goals. Goal 10 is to reduce inequality within and among countries, but inequality is also behind most of the other social goals, such as 1 – ending poverty; 2 – ending hunger; 3 – supporting health and well-being; 4 – access to education; 5 – gender equality; 6 – access to clean water and sanitation; 7 – access to energy; 11 –city and community development; and 16 – peace and justice. As we shall see, luxurious and conspicuous consumption by the wealthiest is also a main driver of climate change – targeting Development Goal 13 – on climate action. In an indirect way economic inequality is related to all the goals.

Economic inequality, in addition to making life miserable for 60 percent of the world's population, is also about inequality of political power and state disintegration, about corruption, and the undermining of democracy, justice, and social cohesion – severely undermining political, concerted action to deal with planetary threats. Before attending to social and political issues of inequality, we shall first look at alternative ways of measuring inequality.

Inequality measurements

There are different ways to measure economic differences. The most common has been the so-called "Gini coefficient", named after the Italian sociologist and statistician Corrado Gini. Each country is here assigned a numerical value between 0 and 1, where 0 stands for total equality and 1 stands for total inequality (where one person earns and owns everything). The figures can also be expressed in percentages. The Scandinavian countries and other countries in Europe traditionally have the best distribution, with the lowest index number indicting the lowest inequality. America (North and South), the Middle East and Asia have generally greater differences, and the poorest countries in Africa consistently score the largest differences.

The Gini index comes with a number of problems, however. The figures are in many cases too old, and updating them is difficult. Moreover, the index glosses over important demographic differences of age that have an effect on the real distribution of living standards. It also tends to exaggerate the differences in the average zone of income and underestimate the differences between the very richest and poorest in a society.

A simpler measure is the so-called Palma index, named after Chilean economist Gabriel Palma. Palma observed that in the vast majority of societies, the middle-class accounts for around 50 per cent of the income. The remaining 50 per cent tends to be distributed between the richest 10 per cent and the poorest 40 per cent. He also argued that it is the distribution between these two extreme groups that reveals the most about the economic differences in a country and about the differences between countries. It also explains why the position of the middle-class in relation to inequality tends to be decisive for national equalization policy.

Generally, the results from the Palma index tend to agree with the Gini index, but the Palma index gives a more striking presentation. The Scandinavian countries are at 1.0 on the Palma index, which means that the richest 10 percent have the same income as the poorest 40 percent. The UK is at 2.0, the USA at 3.0 – and South Africa at 7.0.

A link has also been made between the Palma index and figures for carbon emissions (Pan et al., 2019). Brazil and the countries in southern Africa, as well as most developing countries come out worst, with over-proportional, extremely high figures for emissions from the richest 10 per cent. In general, countries with the smallest economic differences fare better when it comes to emissions of greenhouse gases, but since there are so many more rich people in the richest countries, it is the group of the 10 percent richest in these countries that contributes the most to the climate problem.

Increased inequality, increased expectation gaps, and the spread of radical and undemocratic ideologies often operate together. They also tend to threaten the maintenance of ordinary prosperity, which may further expand the inequality in a negative spiral.

Social consequences of differences

Large economic differences contribute to unhappiness for poor people and to major social problems. The level of inequality is more important for citizens' well-being and experience of happiness than the level of average GNP-type national income. This applies to almost all indicators of well-being: life expectancy, children's well-being, education, health, crime, trust and belongingness. It also applies to drug and alcohol abuse, mental health problems, social mobility, and income mobility between generations (Jackson, 2017; Raworth, 2017; Wilkinson and Pickett, 2009).

The connection between inequality, impaired health and social problems is high and linear. Societies characterized by great inequality are also characterized by more status competition, status anxiety, "consumerism", "conspicuous consumption", and depression! Workers in such communities tend to have longer work hours, be forced to hold more than one job, and end up with more debt. Growing differences between rich and poor lead to social decay. In their book "Deaths of Despair and the Future of Capitalism" (2020), Anne Case and Angus Deaton have shown how increased death rates and reduced life expectancy in the USA due to suicide, overdoses and alcoholism, is linked to poverty and lack of access to higher education. Also, public health systems are in several developing countries markedly better than for poor people in the United States – a nation with the world's largest health costs, but only ranked in about 30th place in public health.

The large income differences also affect the entire culture and "psyche" of a society. It is worse to live on a minimal income in a country where you compare yourself to a wealthy upper class, than to live with a similar income in a society of equality. Large economic differences contribute to alienation and radicalization, which undermines cohesion and democratic action and at the same time leads to an increased risk of conflict, terrorism and acts of sabotage.

Political gains from wealth

Wealth leads to political power for a rich elite and to increased opportunities for influence on political processes. There is a rich literature covering the processes, the networks, the lobbying mechanisms and corrupt practices associated with power-grabbing. Through combinations of lobbying and corruption, promises of investment and employment, threats to relocate headquarters and threats to close down local mills, or outsource production, the richest can influence tax systems and seek out loopholes in regulations to their advantage. They can also influence politics by purchasing media control, and by hiring the best and most expensive lawyers and lobbyists, gaining a disproportionately large stake in the planet's destruction.

Rich people and large companies have immense funds for lobbying and political influence. Time and again they win court settlements and political victories over opponents with lesser means and less access to influential lawyers and lobbyists. This also, among other things, leads to deficiencies in the political-economic system not being corrected, with great scope for "rent seekers" to prevail, and for inequality to persist. "Rent-seeking" is about exploiting loopholes in market regulations and tax systems, and opening up for corruption in order to obtain financial advantages.

The purchase of political influence is also extremely important for availability or non-availability of welfare provisions for poor people, often in the form of public, social services and health benefits. For the wealthy, public services mean higher taxes, and in many countries, the wealthy use their power to reduce taxes and the production of public goods, normally provided for free through taxation.

Related to global environmental threats, we also note that the direct contribution of the richest to planet destruction is massive. The relationship is positive and almost linear: The richer the person, the country, or the company, the greater the damage to nature and the environment, to climate and sustainability. We noticed in Chapter 1 that two thirds of global warming is attributable to the wealthiest 10 percent of the global population – 6.5 times the average, and one fifth of warming is attributable to the wealthiest one percent – 20 times the average (Schöngart et al., 2025). The sad counterpart to these figures is the fact that the poorest people generally are the ones worst exposed to extreme weather of all kinds. The poorest account for only 7 percent of emissions, but they are generally the ones who will have to endure the worst consequences of planetary destruction. On top of this, we have the catastrophic influences from the very super-richest 0.1 percent (Kenner, 2019; Paddison, 2021). In sum, economic inequality undermines democracy, general well-being, and the power to act concertedly for an endangered planet.

Undetonated bombs

In the context of dis-integration and failure to mobilize collectively for a green transition, the colonial-type cash streams, still flowing overwhelmingly from South to North, work like undetonated bombs. Piketti (2022) traces the cruelty of global inequality to mainly colonialism and slavery, historically justified by implied racism, and he mentions cases where under norms of equity and righteousness, only direct economic compensation can be justified (such as for a settlement between France and Haiti). In other cases, improvements can be installed by financing equal rights to education, jobs, and property in former colonies and slave societies. The debt crisis, on the other hand, works to increase deprivation and reduce the opportunity for decent living, spreading anger, "*ressentiment*", and violence in the process (Mishra, 2017).

Several factors have contributed to the present debt situation. Many among the poorest nations never recovered from the debt crisis of the 1980s and 1990s. Still the first two decades of the 2000s were seen as a "low everything" period (including low interest rates, inflation rates, and debt rates). In expectation of continued "low everything" conditions, even until 2021, interest rates were kept at a minimum. Shocks occurred around the world when dramatic increases were to follow.

A slowing world economy, with negative growth of three percent during the Covid-19 slowdown of 2020–2021, and increasing costs of fuel and food staples following Russia's war in Ukraine from 2022, caused leading economies, led by the US, to step up anti-inflationary policies with increasing interest rates, thereby strengthening their currencies. This caused debtor nations in the South to pay for higher interest rates, and for uphold of repayments, with weakened currencies. By 2022 at least 60 countries had come to experience debt distress – double the number from 2015. The Institute of International Finance (IIF) estimated total debt by developing nations at US $ 98 trillion by the end of 2022. In 2020 alone, the total debt burden of low and middle-income nations increased by 9 percent (Wade, 2023). According to data from the 2023 World Bank report on the debts of developing countries, published December 13, 2023, LDSs as a whole spent a record US$ 443.5 billion to pay for their external public debt. Increasingly, poor countries also have had to add new debt in order to service their old loans, generally at worse conditions following lower credit ratings, weakened currencies, worse outlook for growth, and generally higher risks.

The tragedy of all this is that beyond the financial implications of rescheduling, the net effect is for poor countries to limit public expenditure for social welfare, education, food, public services, infrastructure and domestic investment (Maihali, 2023; Stiglitz, 2017). For households it means austerity, sometimes at extreme levels. And for the world situation it means that more people again have strong reasons to hate people, financial institutions, and governments in the wealthy North. The lending nations and institutions of the North, often instead of assuming the consequences of the risks that justified high interest rates in the first place, seek governmental support for imposing austerity measures to force poor people to prioritize debt servicing at the expense of life support – also at the expense of climate and wildlife mitigation. The anger and tensions associated with high debt in poor societies is like an undetonated bomb, ready to explode at some point.

We may sometimes think of critically high public debt by poor countries as a result of frivolous borrowing by over-optimistic enterprise and national leaders. What is at the same time overlooked, however, is the deliberate policies of Western banks to grant such loans in spite of high risks, speculating that home or host country governments will step in and pay. Unfortunately, the end result is that instead of the banks taking their calculated losses, and generally without the home country governments stepping in, the burden of payment is passed on to the population, often through austerity programs mandated by The World Bank and The International Monetary Fund (Maihali, 2023). To the misery of the debt trap, there is a tendency for corruption and kleptocracy to be part of the size and the stickiness of the problem, undermining the effectiveness of development programs, democratization and competence building.

The debt crisis and default of Russia in 1998, marked an end to democratization and the beginning of Putin-fascism, detonating far-reaching geo-political and military consequences, including war with Ukraine 24 years later. Around the world, economic break-downs following debt-crises have caused endless problems, especially in poor countries, but also including Argentina, never recovering from the Latin-American debt crises, with a present political leadership heading the nation towards democratic collapse.

As brilliantly told by Graeber (2011), the institutionalization of borrowing and debt has served peaceful purposes of trade, settlements, and development for thousands of years. Entering into debt as a free choice, ideally serving both the lender and the borrower, is an arrangement sought by both parties for very good reasons (such as seeking education, buying a house, or starting a business). But debt may also result from repression, humiliation, and surrender, and it may detonate to create the worst of consequences.

Adding inequality from monopolization

Economic inequality implies a huge gap in consumption levels. Extreme inequality implies that extremely rich people consume at a level many orders of magnitude higher than the median, and still more orders of magnitude higher than the poorest. Since there is a direct, near linear correlation between wealth, consumption, and the environmental burden, extreme inequality also means that the wealthiest people, based on mere consumption, contribute over-proportionally to environmental threats (such as cruise-ship-size private yachts, castle-size housing, fleets of luxury cars and private jets). It also means that above median incomes in wealthy countries, due to norms of consumerism and conspicuous consumption contribute disproportionally to waste and nature depletion.

As noted, extreme wealth also implies an undemocratic and disproportional impact on politics, with tax advantages and market regulations for themselves, and the erosion of public goods and social benefits for other citizens. To this must be added the capacity of tech-billionaires to pump enormous sums of tax money into their own pockets for governmental finance of both military and civilian projects.

Beyond the personal wealth of individuals, extreme wealth and extreme incomes are also interwoven with large-scale business operations and monopolization. And with monopolization follows still more lobbying and rigging of regulations, more corruption, taxation laws in favor of large companies, and extra rents from a large market share and market power.

The market mechanism is regularly premised as the core of a capitalist economy. What legitimizes both the market mechanism and capitalism in economic theory, is that competition between many producers on the supply side and many consumers on the demand side leads to efficient distribution and use of resources. To promote market efficiency, and a shrinking of the public sector, the United States in the 1980s, under "reaganomics" and "supply side" economics removed a number of business regulations, expanded privatization of public work, lowered business taxes, and reduced public spending and social benefits for the poor – all legitimized by reference to the efficiency of markets and competition. The European Union in 1986 introduced rules for a freer and more open internal market in order to stimulate Pan-European competition. And globally, industrialization and development policies were pursued in country after country, both rich and poor, by removal of trade barriers, market regulations, public monopolies, and social spending – often at the urging of the United States, the International Monetary Fund, and the World Bank. The opening of markets accelerated globalization, and many societies achieved rapid industrialization and economic growth. Multinational companies thrived, while some nations were left with little growth and the loss of social welfare, downsizing of vital public services, and increased economic inequality.

Liberalization abroad - concentration at home

The reforms were legitimized by reference to freer and more equal market competition, much of it initiated by President Ronald Reagan in the US and Prime Minister Margareth Thatcher in the UK. But while hard competition and privatization measures were introduced in Europe and elsewhere abroad, it is interesting to observe the long-range trend in the US. It is not only the contradiction between the public and the private sectors that are at the core of economic controversy, as in the 1980s, but also the contradiction between monopoly business interests and small firms in open markets. In so-called "atomistic markets", numerous small firms acting as "price takers" without any of them being able to determine price levels, bring about competition and no opportunity for monopolization or rent-seeking. It is therefore of interest to learn whether the US economy is closer to this ideal to-day than thirty years ago, before deregulation.

A study by Gustavo Grullon and colleagues (2019) asked whether Americans have taken the necessary steps for more open competition at home – as they have pushed in other countries. The result of the study is a quite solid documentation that 75 percent of American business, since the late 1990s has experienced increased concentration and monopolization on home ground. The data also shows that the companies in the most concentrated industries simultaneously experienced the highest profit levels and got the most (profit) out of mergers and acquisitions. At the same time, no clear increase in efficiency could be demonstrated. The profit margin increased by 142 percent, while efficiency only increased by six percent. In other words, the profit increase was attributed to the increased market power as a result of concentration, mainly based on monopolistic pricing, not on productivity gains. Concentration took place in most sectors, measured as the market share of the four largest companies in each industry. During the period, the leading companies tripled in size and saw an enormous increase in income. The profits from monopolization went to owners and top managers – not to consumers and employees.

In 2016, "The Economist" magazine ran several articles warning about the consequences of monopolization, with special reference to the US, and with provocative headlines: "The problem with profits: Big firms in the United States have never had it so good. Time for more competition"; "Winners take all: Why high profits are a problem for America"; and "Too much of a good thing: Profits are too high. America needs a giant dose of competition". Noting that the level of profits in the 2000s has risen in most rich countries, but most of all in the US and most directly in connection with monopolization and concentration of ownership, the magazine also noticed that with increasing returns on capital, and with profit's taking an increasing share of the national product, incredible values had passed into the pockets of owners and top managers of the largest companies, not to consumers and workers. In many industries, such as banking and finance, but also in technology and pharmaceutical companies, a sizeable share of the profit was also distributed to an elite of top employees and experts. It has also been shown that the average markups in US business have risen from around 1.2 in 1980 to 1.6 in 2016 – signaling a massive decline in competition.

In line with John Kenneth Galbraight's predictions in "The new industrial state", the majority of smaller firms saw no corresponding markup increase (De Loecker et al., 2020; Eeckhout, 2022). It is open to debate whether or not average markups is a good indicator of monopolized profits (Grüne-Yanoff and Verreault-Julien, 2021), but the trend towards monopolization can be verified by going directly to companies' market share, and the macroeconomic implications (Eeckhout, 2022). The resulting monopolization trend may also serve as a warning of what the "revised sequence" of "surveillance capitalism" may bring in the future.

According to data from the Open Markets Institute (2023) a large majority of Americans, of both political parties, now hold that the US has a monopoly problem, with wealth, power and control increasingly concentrated in a few hands. More than 60 percent of the population say that the distribution of wealth and power is unfair. The fruits of economic growth are being seized by corporate giants, and with the monopolization of markets also comes increasing inequality of wealth and income, increasing geographical differences, and generally increasing barriers to entrepreneurship through exclusionary business practices.

Many people have realized that Microsoft has a monopolistic position in the market for operating systems. For Intel-compatible computers (which are the vast majority), Microsoft accounts for over 90 percent market share. Many have also noticed that Walmart has a dominant position among American supermarkets. The company now controls 72 percent of the American market. Also well-known is that Amazon dominates e-commerce in books, with sales of 74 percent of all e-books and 64 percent of all paper editions. What is less well known is that the company is gradually increasing its e-market shares in more and more product areas, while at the same time tailing Google and Facebook in the market for internet advertising.

The range of monopolized industries includes a number of examples of what one might expect, such as car rental, oil equipment and production, and the production of electronic equipment. But we also find a series of unexpected entries, such as cowboy boots, laboratory equipment, catering and food service, champagne (controlled by a French multi-billionaire, Bernard Arnault), home improvement, chocolate and sweets, or mattress manufacturing. Or take Luxxotica, which dominates the market for glasses.

Trend-changing monopolization

A very special development has taken place within the agricultural sector and food supply. A generation ago, the sector was dominated by traditional farmers, with full competition and independence. Today, the production of cattle, pigs and chicken meat at the national level is dominated by four large companies, with near full monopoly at local and state levels. Two of the companies, Dean Foods and Dairy Farmers of America control 80–90 percent of the milk supply in several states. The same pattern repeats itself in the production of eggs, grains and other foodstuffs, and in the industrial processes that produce unhealthy ultra-processed foods.

Even more conspicuous is the private monopolization of hospitals, health insurance, pharmaceutical production, and the production of medical equipment. Three hospital chains account for 77 percent of hospital admissions. At the same time, these chains have also acquired more and more medical centers and now own around 50 per cent of these. Two chains control the pharmacy business, CVS with almost 60 percent and Walgreen with just over 30 percent. Three private insurance giants control payment for health care, and five of the eight largest pharmaceutical companies in the world ("Big Pharma") are American-owned, with Pfizer at the forefront (Pfizer earned $35 billion from covid vaccine in 2021/2022, while BioNTech earned $20 billion and Moderna $20 billion – most of the income reported through tax havens such as Switzerland and Delaware, USA). The pharmaceutical industry is also the one that, over a long period of time, has most often been sentenced with multi-billion dollar fines for cheating with research and marketing.

While the pharmaceutical industry is special, there is a similar trend towards concentration in an increasing number of industries, such as white goods (Whirlpool), trainers and sports shoes (Nike), spirits, and office equipment. A somewhat special pattern can be found in the automobile industry. There is still the entire range of brands, with a good selection and strong competition. Thanks to imports, there is more competition on the American car market today than it was in 1970. Behind the finished cars, however, are a limited number of manufacturers of parts and equipment. The ten largest subcontractors account for 60 percent of the turnover.

Still more politically sensitive is the concentration of the armaments industry. While 107 companies competed for military assignments in 1993, according to the Open Markets Institute in 2020 the number was down to five. This is the industry that Dwight Eisenhower referred to in his farewell speech as president as "the military-industrial complex". The "complex" has been reduced in terms of the number of players, while turnover and earnings have multiplied. And within the "complex", much power has evidently shifted from the government and the military to the monopolized industrial sector, also facilitated by flexible career patterns including public service and private for-profit promotion (stimulating the "wealth pump").

While monopolistic dominance of markets is taken for granted in communist and authoritarian economies, such as the Chinese and the Russian, and less than open competition is expected in developing countries, recent trends in the US have given rise to a voluminous, critical economics literature. Recent trends contradict the premises of the capitalist system and market economics. Also, similar trends have been observed in Europe. The European "internal market", with all its freedoms, but also with a complex set of regulations (justified to ensure free and open competition!) often serves as a barrier against small national enterprises and young entrepreneurship. And while inflation in Europe is recently explained as a result of the war in Ukraine, there are strong indications that European companies themselves have driven up prices in order to increase profits – at the expense of consumers and workers (Hansen et al., 2023). The US is at the forefront, whereas Europe and other liberal economies follow a few years behind.

Small firms are stagnating, while ownership of the largest is becoming more concentrated. And the outlook for workers and low-end consumers is worse than for their parents a generation ago. Lobbying expenses have doubled, and appeals to legal rights has become exceedingly expensive both for small firms, local communities, and individuals who are the losers. Large companies spend enormous resources both on influencing regulations and tax rules and on getting away from responsibility for "externalities". They are sitting on huge legal resources, while smaller companies lack the means to defend their rights. The level of start-ups in the US in 2016 was at its lowest since 1970. The removal of "red tape" has been suggested, to give small businesses and open competition a chance. Where the political-economic confrontations in the 1980s were about the public versus the private sector, our present consideration is about how regulations often favor a monopolizing class of large-firms at the expense of consumers, small firms, and the working class.

From Adam Smith to present-day neo-classical economics, the public interest in avoiding monopolization and defending atomistic competition has been an urgent concern. Countries have used antitrust laws to curb monopolization and the restraining of competition by dominating firms. And while attempts and conspiracies to monopolize are banned in most market economies, present trends clearly point in the direction of increasing concentration, not the least in technology and science-based sectors.

Alternative creative taxation

As an alternative and addition to antitrust laws, there is room for experimentation with wealth taxation. Large, monopolistic companies may be taxed based on size and market power, encouraging them to (voluntarily) break up their operations into smaller units – a kind of voluntary antitrust policy. Such taxes can also be made progressive, with higher taxation for larger enterprises. Progressive taxation can also be applied to protect wilderness and wildlife habitats, and for rights to resource extraction (such as when a global timber trader, Samling Timber Malaysia, cuts down huge areas of rainforest in Borneo, uproots local communities, causes greenhouse gas emissions and local climate change, and drives red-listed species to extinction).

Even carbon taxes can be made progressive and linked to size and wealth. A number of price hikes – whether from food price inflation, green taxes on gasoline and diesel, green energy prices and the costs of heating a home – make a much greater cut in a low-income family budget than in a wealthy family budget. A chance to eat a meal and heat a home is a basic necessity, but transportation can also be a necessity for work and functioning in a modern society.

This means that increasing environmental taxation may come to function as a regressive taxation, hurting low-income people. This was part of the "yellow vests" protests in France, and similar protests elsewhere. Environmental taxes and carbon taxes are needed, but the proceeds should be ploughed back into the pockets of low-income people. Such taxes will guide consumer demand in eco-friendly directions, while compensation to low-income groups, who are least to blame for environmental collapse, will guarantee that they will not be carrying the burden. Unfortunately, subsidies and support of green electrification, such as EVs and home photovoltaic installations, have so far mainly benefited the well-to-do who can afford a second car and have money to invest in the greening of a home.

Progressive taxation linked to size, and a plan for further increases over time, would be an excellent climate measure, while at the same time weakening the monopolization of fossil-intensive businesses. Compensation to low-income groups would be an effective means for social sustainability and support of the green transition.

Cuts at the top

While conservative governments around the world have consistently argued that tax cuts for the wealthy serve the nation's interest through job creation and economic growth, research in the field largely shows the exact opposite. If job creation is the goal, tax cuts for the 90 percent lower income groups is a better solution. Tax cuts for the top 10 percent tend to have a minimal impact on total employment (Hope and Limberg, 2022; Zidar, 2019).

Hope and Limberg examined developments in 18 OECD countries over the past 50 years, from 1965 to 2015. They found that income differences increased with tax cuts for the wealthy, without having a positive impact on economic (GDP) growth and employment. The 18 countries examined were the United States, Great Britain, Sweden, New Zealand, Italy, the Netherlands, Germany, Canada, Australia, Norway, Japan, Finland, Denmark, Belgium, Austria, Switzerland, France and Ireland.

"Trickle down" or "trickle up"

"Trickle down" economics was re-named "supply-side economics" by David Stockman, Director of the Office of Management and Budget under President Ronald Reagan. "Supply side economics" was designed to cut business and investors' taxes and reduce the size and regulating powers of government. In "The Triumph of Politics" (2013), Stockman later reported on the political maneuvering, intrigues and manipulations that eventually led to failure.

There is a funny story behind the term "trickle down economics". The person who introduced the term was William Penn Adair Rogers (1879–1935), America's greatest comedian of the time –

actor, filmmaker, political and social commentator – and also a full-blood Indian of the Cherokee Nation. In 1932, Rogers made a satirical comment about President Herbert Clark Hoover. Hoover, who was president during the Great Crash of 1929, was convinced that in the same way that water trickles down and sprinkles gardens and fields, the money of the rich will also sprinkle the markets and create growth and jobs. Rodger's point was that while water drips downwards, money tends to disappear upwards. His observation was that "Give the money to the people at the bottom, and the people at the top will still have it before night, anyway. But then at least it will have passed through the poor man's hands". Whatever the quality of the source, it is clear that Franklin Delano Roosevelt, through the New Deal program did exactly as Rodgers said - departing from Hoover's solution. "Trickle up" seems to have worked better than "trickle down".

Roosevelt was not the first president to depart from "trickling down" policies, however. Democratic presidential candidate William Jennings Bryan delivered the "cross of gold" speech to the Democrat national convention in 1896, presenting a "leakage theory" of making wealthy people even richer, in order for their wealth to "trickle down" to those below. Instead, he argued for the spread of economic means to the masses of people in order to allow their prosperity to "trickle up" (Wilson, 1992). Another term for "trickle down" was used by the economist John Kenneth Galbraith (1982). He called it the "horse-and-sparrow theory". If you feed the horses, some of it will fall on the road and be picked up by the sparrows. But if the point is to feed the sparrows, it is better to give it directly to them instead of letting it go through the horse first.

While "trickle down" thinking associated with "Reaganism" in the US and "Thatcherism" in the UK and Europe has spread to most parts of the world, stimulating tax breaks for the wealthiest, it is still a dubious theory, sometimes referred to as "voodoo economics". The recurring rationale is that tax cuts for the wealthy generate additional growth in the economy, which leads to an increase in employment and the tax base. Therefore, the total tax revenue will be at least as large as without the tax reduction. In terms of economic inequality, and beyond the politics of the Reagan and Thatcher, this is still a much-debated issue. Most research, however, points in the direction of the "voodoo" classification. Since "woodoo" economics has been embraced mainly by the Republican Party, it is worth noting that the unemployment rate has over time tended to rise under Republican presidents, while it has fallen under Democratic presidents, and also that budget deficits relative to the size of the economy have been lower under Democratic presidents (Blinder and Watson, 2016). Similar results are scored for GDP growth (greater under Democratic governance), stock market returns, and job creation (Blinder and Watson, 2016). Out of 11 recessions in the 1953–2020 period, 10 occurred under Republican presidents (Egan, 2020).

Trickling research

The French economist, Thomas Piketty has contributed immensely to our understanding of inequality dynamics and remedies, and to scholarly discourse of inequality subjects. His best-selling book "Capital in the Twenty-First Century" (2014) traced inequality patterns over the past 250 years, and also the effects of progressive taxation for a more equitable distribution. Further mapping of inequality in different countries was presented in "Capital and Ideology" (2020), and the potential for wealth redistribution was popularized in "A Brief History of Equality" (2022). In the latter book he also argues for substantial compensation by former colonial powers to post-colonial societies today.

Piketty's main recipe for the reduction of economic inequality includes cuts from the top through progressive income-, wealth-, and inheritance taxation, global cooperation and standard-setting to prevent tax evasion, tax havens, and unfair trading practices by the wealthy and by multinational enterprises. It also includes an overhaul of international trade agreements, and ideas for equalization from below.

Globalization has added to suppression of wage level for low-skilled workers in the wealthiest countries, to the advantage of owners and high-skilled workers. For the betterment of welfare for the economic bottom classes, Piketty's recipe includes public investment in education and skills for the poor, stricter enforcement of minimum wage standards, and further experimentation and development of guaranteed minimum income (GMI) schemes – financed in part by progressive taxation of high incomes, wealth and inheritance, and in part by the replacement of various welfare schemes to be phased out.

In a number of additional publications, Piketty and his colleagues have also dealt with a wide variety of inequality issues, such as the fact that almost none of the payoffs from recent productivity increases and GDP-growth in America have benefited the bottom half of the population (Huber et al., 2022), the special position of the top one percent group (Alvaredo et al., 2013; Volscho and Kelly, 2012), long-term inequality trends (Roine et al., 2009), wealth and taxation contrasts between continental European policies and the English-speaking world (Atkinson and Piketty, 2007), and the effects of lower top taxes on top wages and CEO renumeration.

The latter is an observation of great importance. When top taxation is low, top executives will benefit more from higher salaries and extra bonuses, driving additional inequality (Piketty et al., 2014). This may explain the explosion in executive incomes in the US, also when compared to Europe and other parts of the world with lower top salaries.

While the connection between tax cuts for top incomes and failure to produce overall employment and prosperity benefits, according to most studies seem convincing, it has been difficult to demonstrate a direct causal connection. David Hope and Julian Limberg of the London School of Economics developed a more advanced system for registering tax reforms for the wealthiest and a more advanced research method to demonstrate causality from time series data (Hope and Limberg, 2022). The result seems fairly clear, namely that tax cuts on higher incomes lead to greater differences in society, without any positive gain in terms of GDP growth and employment.

The consequence of the "trickle-down" policy and tax cuts for the rich is, in short, that the rich get richer, and that the rest of society will have to absorb the negative consequences of increased differences. Also, the argument that lower taxes for the rich should act as an incentive for people in higher income groups to work harder, and thereby create more jobs, appears to fall. If this had been true, it would have had a positive effect on GDP growth. On the contrary, the study suggests that the tax advantage for higher income groups acts more as an incentive for the rich to demand an even higher share of the "cake" without any beneficial effort.

Beyond the economics discourse on inequality, growth and employment, there has also been scrutinizing analyses in political science (such as Bartels, 2009; Emmenegger and Marx, 2019; Gilens, 2005; Hacker and Pierson, 2010). Further examination of the political arguments is beyond the reach of this chapter. Instead, we shall have a closer look at the taxation-CEO connection.

Managerial salaries and taxation

In the US, top management salaries have soared over the last decades. Previous calculations have estimated CEO incomes at just under 400 times the average salary for employees. An updated report indicates that the average US top executive pay in 300 of the largest firms in 2022 was $10.6 million, while the median employee pay was $23,968 – a gap of 670-to-1 (Anderson and Alperstein, 2023). In a third of the same firms, employee pay did not keep pace with inflation.

Often, executive incomes increases are not recorded as salary increase. Instead, they may be part of share buyback programs – which involve the company buying its own shares, which causes the share price to rise, which in turn means that managers with shareholdings in the companies may receive additional profits. The report from the Institute for Policy Studies shows that in two-thirds of companies with low pay levels, billions of dollars were spent on increasing top management salaries through such buyback programs. In one of the companies, Lowe's, 13 billion dollars were spent on buybacks – an amount that could alternatively raise the salary level for ordinary employees by 40,000 dollars a year. Instead, the salary level fell by 7.6 percent to $22,697. The report also refers to a survey of Americans, showing that 87 percent see growing inequality as a problem for the nation.

It is also a problem that many of the companies with the largest gaps benefit from large, lucrative contracts for the government and public works.

The situation in Europe is not quite as absurd (In 2022, the executive level in Norway was approximately 20 times that of the average employee pay, the same as in the USA in 1964). But here too, arithmetic gymnastics may be required to estimate the real executive income levels. Often, it takes several pages in the Annual reports, with highly technical jargon, just to explain CEO renumeration. The nominal basic top salary may well be worth double in terms of total real income.

For a long time there has been a debate in the USA, and also in Europe and several other countries, about how big the gap can be before it taps into the work motivation of employees, customer loyalty and even the "glue" in society (Hendriks et al., 2023; Kiatpongsan and Norton, 2014; Schulz and Flickinger, 2020). Ordinary employees and other stakeholders are often poorly informed about top management real incomes and only react when the figures are highlighted in the media. The same also applies to the reputation of monopolized companies and wealthy individuals in society.

Some research suggests, however, that excessive top management salaries are interpreted by employees negatively as an expression of greed, cynicism and bad morals, attitudes that can easily permeate large parts of the organization with demotivation effects. But there are also concerns about the effects of greed, not only for the business community, but for the wider society. Thus, head of the US central bank, Alan Greenspan, in his report to Congress (2002) claimed that "an infectious greed seemed to grip much of our business community". If the greed infection was well underway in 2002, in the 2020s it has already become a national epidemic with the potential of reaching pandemic proportion in years to come.

The absurdity of the wage gaps between top executives and average employees, or bottom-paid employees, can be easily regulated, however, by just specifying a maximum legal difference. Of course, it is also possible to just introduce a cap on wealth. In her book "Limitarianism", Ingrid Robeyns has suggested that the way to deal with the present US level of inequality (with the top 1 percent wealthiest holding as much wealth as the bottom 90 percent), is to set a limit of 10 million euros or dollars per person (Robeyns, 2024).

The payment gaps seem to have many causes, including low taxation levels on top incomes, national masculine cultural values (such as competitiveness, assertiveness, and material success, Hofstede, 1984), strong lobbyism and skewed access to the best legal services. In addition, there is the monopolizing trends in many sectors, with monopolistic profits, huge company sizes and market shares, and corresponding social prestige – allowing for conspicuous and climate- and nature-hostile consumption, and for deep splits in the population, with the spread of anger, conflict, and extreme political positions, undermining solidarity and concerted action at a time of increasing manifest environmental damage.

In order to reduce inequality, progressive income-, wealth-, and inheritance taxation has to be considered, in addition to other measures, such as introducing international standards for taxation and corrections of global trade agreements. But there is clearly a limit to how much inequality can be corrected by progressive taxation alone. In the 1950s, the marginal tax rate on incomes over $400,000 in the United States was 91 percent. That was not a viable choice. Milton Friedman (in Capitalism and Freedom, 1962) was not the only economist to notice that the real tax level (the percentage of income really paid by the wealthiest) never reached that level. People with high incomes could well afford to hire tax-lawyers to find loopholes and deductibles, bringing taxable income and taxation down to a fraction of nominal figures.

Since then, progressive income taxation has been steadily weakened until it today is turned into regressive taxation – where employees often pay a higher percentage rate than owners and top managers, and low-income groups pay a considerably higher price for inflation, carbon taxes and mitigation than high-income groups. Taxes on income, wealth and inheritance must be set at levels that are in accordance with a general sense of fairness in the population, and we have to look for still other means (beyond progressive taxation) for serving both the equity concerns of society and the market demands for fair, open, and equitable competition – facing up to monopolization.

Boost at the bottom

Recognizing the drive towards inequality in capitalist economies, and the negative consequences of inequality, governments in democratic societies have for a long time installed social reforms and public, social benefits for low-income groups. At a time when the proceeds from digitalization, automation, productivity gains, and monopoly pricing are increasingly accruing to the highest income groups, there is an urgent need for reconsideration of social reforms. In the next chapter, AI will be added to the list of trends that accentuates and radicalizes the need for such reforms.

Inequality is still, at a time of unprecedented total wealth, hurting low-income groups, families and environments, undermining GPI-type genuine progress, and increasing environmental degradation and loss of social sustainability, and with that also a loss of potential for concerted environmental action to save the planet. We have seen how the inequality gap can be closed from the top, with progressive taxation and limitation of monopoly benefits. The following are some means for closing the gap through reforms at the income bottom. Variety of approaches is apparently greatest in nations with social-democratic traditions and "mixed economies". But new ways forward will have to be sought in all nations facing equity challenges (which means nearly all countries on earth).

Systems choices

When it comes to general rules for social security, there are two main systems, on the one hand so-called universal schemes (with automatic payment to all citizens and age groups covered), and on the other hand schemes which require some form of individual needs-testing. When it comes to universal schemes, we again have a question whether benefits should be graded according to income, or whether they should apply equally to everyone. When it comes to needs testing, there is a differentiation between highly automatized, unambiguous payment systems, and systems that require in-depth interviewing and cross-examination of needs and suitability – opening up for stigmatization and humiliating experiences. Finally, there is also a main distinction between systems based mainly on cash payments, and systems tuned to the provision of social benefits (such as social housing, free preschool attendance, free transportation or food coupons).

Over time, it seems that universal arrangements, with minimal needs for individual testing, and standardized cash benefits (such as old-age pensions and universal child benefits) have shown to be the most robust. Important topics today are related to the troublesome functioning of many national social security systems, and to the concretization of much-demanded universal guaranteed minimum income (GMI) schemes.

Needs testing means that social benefits are made dependent on a discretionary, individual assessment of the applicant's needs, and such testing is frequently associated with social stigma. Comprehensive systems may include universal, but graded provisions related to child benefits, unemployment benefit, sickness benefits and old-age pension. But even in cases of unemployment and sickness, there may be a need for individual clarification and needs testing. When it comes to support for people with special needs, systems tend to be largely based on needs testing. In addition to social welfare recipients, needs testing may be applied to applicants for housing benefits, support for funerals, study grants, settlement of survivors' pensions, and handicapped children support.

The threat of stigma

As a foreign (Norwegian) student in the USA in 1967, I witnessed social stigmatization first-hand, and for the first time. I shopped at Safeway and quickly learned to recognize people "on welfare". They paid with coupons and picked foods that provided the most calories for the value. They were worse dressed than other customers, and they showed signs of physical, bodily wear and worry. Their faces revealed stigma and shame, and hopelessness was printed in their eyes. I remembered the words of our post World War II social-democrat prime minister and post-war "land-father", Einar Gerhardsen. When introducing a new welfare reform, he insisted that no recipient should have to ask for it "hat in hand". Humiliation or stigmatization was forbidden. I have since learnt that, even in supposedly well-functioning social democracies, humiliation is not uncommon, and bureaucracies may do wrong. This became visible in Europe following the economic crisis of 2008 and during the Corona shutdowns of 2020–2021. Searching for new solutions, befitting accelerating change and increasing economic insecurity, is of utmost importance.

Many researchers have in recent years addressed the phenomenon of stigma in connection with welfare schemes (Baumberg Geiger, 2016; Roelen, 2020). If we look to history, aid to the poor has very often been associated with shame and ostracism, also cynically designed to reduce the demand for benefits. Stigma makes others look down on the recipient, which again makes recipients look down on themselves. This was the logic behind so-called "workhouses" in England in the 19th century, where the unemployed could get a job at a miserable pay (Baumberg Geiger et al., 2012). An American political scientist and conservative thinktank commentator, Charles Murray, even said bluntly that "stigma makes generosity feasible" – without stigma the payouts would be too large. The underclass is deplorable and stigma is necessary (Murray and Phillips, 2001).

Designing a system for negative stigma can be seen as an evil act. Stigma has been referred to as "symbolic violence". The British sociologist, Robert Arthur Pinker, has said that while various forms of stigmatization have been with us for centuries, inflicting stigma has been the most widespread form of violence in democratic societies, comparable to psychological torture, breaking down victims psychologically and physically, while they outwardly appear to be untouched (Pinker, 2006).

According to Baumberg Geiger (2016) stigma appears to be a recurring problem in most welfare schemes. Welfare recipients are often seen as unwilling to work, or as unable to make sensible life choices, worse so in societies of high inequality (Offer and Pinker, 2017). They are also seen as breaking norms of reciprocity: People are expected to earn what they receive rather than base their lives on "gifts" (Rogers-Dillon, 1995).

While news about poor people "exploiting the system" are repeated quite regularly, there are also frequent utterances about harassment and humiliation, even in the most developed welfare states. Recipients have experienced disrespect, unnecessary waiting times, and suspicion of personal statements about life difficulties. During recent economic setbacks and inflation, such messages have become more persistent. And of course, we can assume that the proportion of recipients who write critical articles and letters to the editors belong to the most robust and resourceful of recipients. The poorest, least resourceful, and most humiliated are not likely to trumpet their criticism.

Variety of welfare schemes

In sum, humiliation and stigma are most relevant in connection with needs-tested benefits, and they offer solid grounds for change in the direction of universal and automatized welfare solutions. Poverty is often conceptualized in terms of financial inadequacy – lack of food, lack of money to pay for an apartment or for the heating of a home, lack of money to bring up children, with food and clothing and the means to keep up with normal social and sporting activities. But it is not just a material thing. With a lack of money also comes a series of psychological costs, such as a lack of security, inclusion and belongingness, lack of a chance to predict a life situation and plan for a decent living, and a lack of hope for oneself and for children and the next generation.

The welfare delivery system, beyond problems of humiliation and stigma, may add to problems and continued poverty because of technicalities in the system itself. In progressive welfare systems, recipients often qualify for several different provisions, also meaning that they may have to adapt planning and consumption to six or seven different payment dates. Also, some benefits may be truncated against each other, making it unclear how much money will be paid and at what dates. Lack of clarity may undermine family security and planning possibilities – contradicting a main purpose of the welfare state. In a local case in Norway, a family of parents and three children, with the father capable of part-time work and the mother on an "employment verification allowance" had the following income pattern: On the 9th of the month, the mother received "work clearance money". On the 12th, the father earned a minimal wage from his part-time job. On the 20th, they together collected a social housing benefit, and on the same day "supplementary social assistance". On the 23rd, the mother received work clearance money again (paid every 14 days), and on the 30th, the family received child benefits for all three children. The technicalities of the system added to the psychological costs of poverty, undermining security and capacity for planning and adaptation.

The psychological costs of poverty are often underrated. Social stigma, and stigma directly connected with the welfare system, add to the pain of being financially poor, and may hurt more than missing a meal. Still, the psychological costs of poverty are intimately tied to the financial burden. An overriding challenge then is to define and implement a system for financial alleviation with a minimum of psychological costs. A great number of local experiments with such systems have been carried out, but so far, no nation has been able or willing to implement a fully developed program, combining two strong demands from such systems – guaranteeing a basic income for all and securing a minimum incentive to work and perform.

Towards guaranteed minimum income - GMI

A recurrent problem with welfare systems is that politicians and administrators, in honest attempts to do good for the poor, end up designing schemes that include a variety of compensation items. Included may be free bus or train services for the elderly, social housing for the poor, other social services and acute economic assistance, special assistance for immigrants, special support to farmers and farm product consumers, minimum wage legislation. Some of these work quite well, at least under stable economic and demographic conditions. But they add to the complexity of welfare, and they sometimes compromise poor people's freedom of choice. Perhaps a retired person would be better off travelling by plane than by train, or a poor family would rather have cash in hand and choose their own solution to housing.

Rutger Bregman, the Dutch historian, wrote "Utopia for realists: How we can build the ideal world" (2018), starting with the story of a simple experiment from May 2009, including 13 homeless men, living on the streets of London City – some of them for 40 years, many of them classified as troublemakers. Expenditure on police, justice and social services was estimated at upwards of £400,000 a year. In addition, there was considerable voluntary activity to help the 13 and others in a similar situation – all to little avail. The 13 were considered hopeless cases.

In 2009, one of the aid organizations, Broadway, took a radical initiative. The 13 were selected among the most hopeless, and the experiment included setting an end to food stamps, soup kitchens, hospices and the series of targeted measures. Instead, the participants were to receive £3,000 each in cash – without any application form, no strings attached, and no controls of any kind. They were free to do whatever they wanted with the money. Free social counseling was available if they wanted. The main thing was that they had to figure out for themselves what to do with the money and with their lives.

So, how did they spend the money? After a year, they had spent an average of just £800. They had actually saved most of it. What else? A telephone, a dictionary, a hearing aid. One of them, who had used heroin for 20 years, had become sober during the year. He was now able to look after himself and eat sensibly. And besides, he had signed up for a course in gardening. "Things fell into place", he said. After a year and a half, 7 of the 13 had found roof over their heads, and two more had definite housing plans. All 13 had found a new start in life. Several had signed up for courses, mainly in useful practical things such as cooking. Several had renewed contact with family. The total cost of the experiment was £50,000 a year, including the salaries of the social workers. The magazine "Economist" suggested "Cutting the middle men" and concluded that "the most effective way to spend money on the homeless is possibly to give them the money" (2010).

In the next chapter we shall return to the issue of universal guaranteed income schemes. Not all agree with Bregman that they will pave the way for Utopia, and some warn that GMI outcomes will depend much on what political interests are behind (Battistoni, 2017). Parties on the political left are afraid GMI will be introduced as a strategy for the erosion of deserved and hard-won social welfare reforms and public goods provision. Parties to the right are concerned that "free money" will undermine job-seeking motivations and even the social ethics of society. Nonetheless, facing the enormous challenges of a green energy transition, land and wilderness protection, AI-competition for jobs, and the erosion of social sustainability, it is hard to find serious research references that do not contain mentioning or recommendation of some form of a guaranteed minimum income system.

In the present chapter GMI is mentioned as a prime tool for inequality reduction. In the next chapter it will be seen as an integral part of a reformed political-economic system.

Guaranteed minimum inheritance

After 40 years of increasing inequality in most societies, with environmental crises lined up, and prospects for AI and robotization-caused unemployment on the horizon, calls for GMI, or negative income tax systems, have multiplied. When referring to economic inequality, it may to some be a surprising fact that the differences in wealth are consistently greater than the differences in income. This holds for both rich and poor countries, and a main reason is inheritance. You may earn an income, but to a great extent you inherit wealth. And wealth generates additional income.

In "A brief history of equality" (2022), Thomas Piketty describes how those with the least resources are left without inheritance, property and wealth, moreover often with inherited financial burdens or family obligations. In addition to GMI, he therefore recommends the introduction of a universal, guaranteed minimum inheritance. A guaranteed minimum inheritance of US $100,000 would prepare all young adults for a fairer start in life. Similar to the negative income tax version of GMI, a guaranteed minimum inheritance would have to be truncated for whatever inheritance a person would have from family or beneficiaries, making the sum at least equal to the guarantee.

It is profoundly unfair that poverty is inherited and that some people are born heirs to absurd wealth while others emerge from start on bare feet. A guaranteed minimum inheritance will help young people acquire the basics for adult life, kickstart building families, getting access to higher education, or acquiring bank loans for housing or business start-ups. Unfortunately, the idea of a guaranteed minimum inheritance is very foreign in most political landscapes. There is not even an academic research tradition for the investigation of such schemes. Whereas the legal and economic technicalities of inheritance are discussed and re-defined at length, there is a conspicuous lack of concern for those who have no inheritance in sight, which ought to be the greater issue. And even Piketty's argument seems to have gone largely unnoticed. For the reduction of economic inequality, his proposal is very heartening.

Common arguments against GMI are discouragement of job seeking (reducing the supply of labor) and participation in other meaningful activities. Reduced self-esteem and problems of stigma are also mentioned, as is the fact that it will be expensive. GMI in the form of negative income taxation will not disturb the labor supply, however, and as for self-esteem and stigma, alternative systems of targeted social welfare, requiring detailed questioning of personal life situations, have been observed to add much more directly to vulnerability and shame than income statements for low-income groups. We may also foresee combined solutions. D'Orlando (2020) suggest a combination of guaranteed minimum income with a general reduction in per capita working hours, and the implementation of a "cap-and-trade" system for hiring – including a duty (quota) to employ people and allowing firms to trade quotas, paying other firms for employment of "their" unemployed.

Finally, added to the steps mentioned for inequality reduction from the economic top and the economic bottom, there is also a very important public sector role of service and infrastructure provision. Traditional public services, such as military service, police and law enforcement, monetary control, transportation infrastructure, etc., are of course imperative for the functioning of the whole society. But the addition of public education, health services and the like may be of greater impact for inequality alleviation – generally delivered for free or at low cost to rich and poor alike. Public services free of charge, or at a low charge are important for inequality reduction on par with a GMI (guaranteed minimum income).

There is an apparent, right-wing belief that real economic values are created by private business, whereas the public sector represents mainly needless appropriation and consumption. Forgotten is the value of public education and competence-building, infrastructure and defense, health and justice, social safety nets and road maintenance. One could just as well reason in the opposite direction, that real values are created when someone helps with childbirth, raises children, manages law enforcement or saves endangered species, and that the business world exploits such services for personal gratification. Of course, both positions represent nonsense. Both sectors, and in addition charitable and voluntary work at home and in neighborhoods, contribute to economic value in society. And both can contribute to both inequality and equity.

We have seen that the correction of internal and international economic inequality is a complex matter, in great need of attention, however. It is about the common good, taxes and social support systems, about executive renumeration and tech-billionaires political power, and also about average wages and availability of jobs. Attention to economic inequality is mandatory, both for social sustainability and for concerted environmental action and responsible development of AI. Under BAU conditions, the availability of jobs is ranked as the number one priority for the common good, political cohesion and social sustainability. That is an assumption that may be challenged in a future of AI, AGI, and robotization – to which we shall now turn.

Chapter 13

From AI Conundrum to Economic Security

There is a wonderful documentary film, "After Work", brilliantly directed by Erik Gandini and produced by Jesper Kurlandsky for Fasad of Sweden. The term "after-work" normally refers to life and activities at the end of a workday. In the film it mainly applies to a life situation where work and jobs belong to the past.

Most jobs of today will soon be history. Other jobs will be added – for a green transition, for new products and services, but more and more work is also being done cheaper and more effectively by robots and AI. It is time to rethink what work means in our lives, and what to do when we have excess time, or a workfree existence, or when possessing competence that is deemed "irrelevant" and "useless" (Harari, 2020), supported or not by guaranteed minimum income (GMI). The film describes a variety of life situations and cultures, work and leisure, on four continents. It is beyond my capacity to give justice to the powerful messages of the film. But at least, mentioning some content may stimulate thinking about the work and non-work universe. Like in the film, there is no logical order of themes.

A work rundown

- "After work" starts with the life of a Korean father. He wakes up every day at 06.00 am, meets at work at 07.00 – returns home at 23.00 pm, gets five hours of sleep before getting up again at 06.00 am. He claims that work is what adds meaning to his life. At work he can express his devotion to society, his contribution to the common good. South Korea is third on a list of global work intensity. His daughter suggests that he is wrong. He has no friends beyond work. He has no other purpose in life, and no time for family. He was never there when his daughter grew up, and now it is too late.
- A well-to-do Italian devotes his time to gardening, making work a form of art, precision, detail, and beauty – also offering time to think. His wealthy father cared only for possession – of riches and beautiful women. But the gardener looks to a statute of *Kronos*, God of time, reminding him of *carpe diem* and the fleeting nature of moments.
- An extremely busy American is observed in a gym – always reminding himself how busy he is – just unbelievably busy, always having many irons in the fire. For Americans it is all about jobs. He gets a huge kick from contemplating that Europeans take 4–6 weeks of vacation. For an American to take more than two weeks off is unheard of. 55 percent of American interviewees had access to more vacation but renounced. Among qualifications of applicants for whatever jobs, "performative workaholics" were chosen, able to demonstrate their devotion to the job. Positive work attitudes, ambition, and good appearance are rated highest in terms of work ethics.

DOI: 10.1201/9781003518815-15

- At this point in the movie, I was reminded of Alexei Stakhanov, a miner of Russian descent, who in 1935, in the Donetsk region of Ukraine, set a new work record, pulling 102 tons of coal in 5 hours and 45 minutes – 14 times his quota. The Stakhanovites movement for hard work in the Soviet Union, named after Stakhanov, was to initiate a new work culture, and the hero himself was to receive the Order of Lenin twice, the Order of the Red Banner once, making him Hero of Socialist Labor. I was also reminded of the Japanese Karoshi, for devotion to long work hours and even willingness to work to death. But this was America 2023, in a culture of "never satisfied" people, where finding the right job is an ultimate achievement.
- Since the long-hours work ethic in Korea is taking a hard toil on people, with depression, fatigue and illness, "PC-off policies" have been introduced in some organizations, shutting off data equipment at 06.00 pm, a time for sending people home.
- There is an interview with Pa Sinyan, Gallup's Regional Managing Partner responsible for Europe, the Middle East and Africa (EMEA). Data from 150 countries show that there are a bit more than one billion proper, formal jobs in the world. 15 percent – 150 million jobholders say that they are "engaged" at work. 85 percent – or 850 million either say they are "not engaged" or "actively disengaged" at work. They are frustrated, angry, and they spread misery.
- An Amazon delivery service driver is interviewed. She enjoys being alone in the car, engaged in her own thoughts ("time is everything"), and she often enjoys the scenery. She feels that the thanks and the joy in people's eyes when she delivers a package, make her feel good and add meaning to her life and work. But up to 300 deliveries in a day, no time to go to the toilet (she finds a bottle half-filled with pee in the company car), no rest periods, eating while driving, and five cameras in the car surveying her every moment and reminding her that a drink, a stretch, or whatever may disturb ones driving, all day long – it is all making the job a form of slave labor.
- Pa Sinyan of Gallup comes back, reminding us that making employees feel included and engaged is the ultimate of leadership – and he recognizes Hitler as a "great leader" – in spite of his terrible policies.
- Work is slavery when every part of the job is performed on the command of someone else, forced to be done out of fear or need. Only the market opportunity to find another job breaks with the norms of slavery, and only if there is a better alternative.
- Elon Musk reminds us that there are fewer and fewer jobs that a robot cannot do better. Irrelevance is worse than exploitation. Irrelevance is what we are waiting for. Political leaders will have no choice but to implement GMI – guaranteed minimum income schemes. It will be a platform for mass survival. It will also offer a chance to replace routine work with creative work and meaningful activity.
- We are reminded that a permanent, secure job and pay – for the rest of life – can be a nightmare.
- In oil-rich Kuwait, everybody is guaranteed to have a job. If not employed in private business, the government guarantees employment and income. 20 times as many people as needed are employed in the public sector. A newly hired person shows up for work first day. There is no office and no job content specification. He discovers that there is a basement floor, with lots of offices, some of them manned, with desks – but no computers, lots of people – but no regular office activity. They spend 6 hours of idling, reading, talking, watching movies, playing games. They are bored. They may show up three hours late, and nobody ask a question. Superiors are not around, ask no questions, give no orders and distribute no assignments. An interviewed person mentioned that she once had readdressed a letter to another public department – that was the work she had done the last three months. Over time, people get bored, depressed, and angered.
- The question is repeated. What would people do if they had enough money and time, free to choose?

- I think – a desolate basement office is not an ideal environment for creative action. But they could have started a choir, a band, a discussion group, a seminar, a jewelry workshop, or an in-door gardening experiment. Or they could even have moved outside to do gardening work, socialize, care for someone, walk a dog… Why don't they?
- Italy ranks first in a global count of NEET-youth. Greece follows next. NEET means No Employment, Education or Training. 30 percent of relevant groups in Italy are NEETs. Some just have a hard time finding work or educational alternatives, but the majority are simply free to spend time at free will. The average family wealth is 400,000 Euro. Fertility stands at 1.2–1.3 children per adult woman – meaning that the population size is shrinking, meaning again that young people on average will inherit more than the per capita wealth of the parent generation, allowing them to party and play, getting high or seeking meaningful dance and entertainment, go to the beach, engaging in sports or just passing the hours with friends. They engage in "voluntary unemployment", in "collective, decadent rituals" of discotheque and play. They distance themselves from a work-based society and create their own leisure-based societies. A leisured lady finds meaning in caring for a horse: a person needs at least one passion in life. Her husband devotes at least three fourths of his life to work.
- Back to Kuwait, immigrant workers do the real work. Families were used to have an immigrant maid – now they have two. The "workers" are treated as slaves or machines – and they are not seen as humans. People are lonesome, filling time eating and playing video games. They live in their own narrow comfort zone, and never go beyond. They become irresponsible, spoiled, and ignorant, unable to deal in a world of trouble.
- In Korea, a young man is exhausted from work. He gets more and more nervous and depressed. His heartbeat takes off. The situation is increasingly getting worse. He panics and starts to cry at his desk. He seeks help from his doctor and is told to take a rest. But he is unable to rest. He is offered pills to calm down. He quits work. His parents are both working, and they see him as lazy, and without a job, a loser. They have no concept of happiness in life. They work to live, not for happiness. If not working, they will die.
- The Italian gardener is back in his park labyrinth, contemplating the meaning of life, perfectioning the shape of every plant. There are seven traps in the labyrinth, and every time you go into one of them, you have to find your way back again, same way. There are also seven deadly sins, and for each one of them you have to find your way back. *Kronos* is watching you, and the filling of time offers insight and meaning. You add meditation, and you learn to find and appreciate your role in society, finding meaningful ways of spending your time. When death arrives, you can say to yourself: "I have lived a life worth living".

AI-technological unemployment

We noticed in Chapter 5 that the transition to green energy will create more new jobs than those that are lost in the fossil fuel sector, and other industries and service sectors will also need additional manpower in order to cope with new demands, including green energy and new products such as batteries. More work will also be needed for mitigation and adaptation, damage control and repairs. With too little money invested in mitigation, one may foresee a future where a significant portion of the workforce will be busy with adaptation and environmental damage repair. In addition, a considerable number of people will be recruited to care for the combination of aging populations and the lowering of birthrates.

On the other hand, while there are warnings about future labor shortages, there are also substantial warnings about climate-migration pressures. If advanced economies run short of laborers, there will be people at the borders begging for entry and work permits. Immigration control has become a key issue in both far right and centrist political debate. That may change if and when the health sector and other sectors run short of workers.

Beyond demographics and climate change, however, future unemployment caused by AI, automation, and robotization emerges as a main theme of social sustainability. Near full employment is a key factor in the stability and quality of society. If there is productivity growth (as promised with the implementation of AI and robotization), full employment and the absorption of the new "useless class" (Harari, 2020) can only be achieved if there is also correspondingly strong economic growth, enough to absorb unemployed labor. Strong economic growth without real decoupling will add to the environmental burden, however. Therefore, the greater the AI contribution to productivity, the greater the employment challenge, the growth challenge, and the decoupling challenge. In the present chapter the productivity-growth-employment-decoupling logic will be examined. In Chapters 15 and 16 outlooks for the "useless class" will be discussed.

Productivity and employment

An important question is simply what we can expect in terms of productivity gains from AI and robotization and the impact on employment. For specific industries and companies, we can often point to technologies, machinery, or managerial innovations that have led to specific productivity gains. For whole economies, however, we have to seek broader explanations in macroeconomics, in the organization and management fields, and in technology. And in the field of technology, there is again an important distinction between general purpose technologies (GPTs) and specific purpose technologies (SPTs).

Only GPTs have the capacity to elevate the overall productivity of the economy (Bekar et al., 2018). Steam engines were key to the industrial revolution, interrupting traditional production forms and impacting all spheres of society. The introduction of electricity was not just a means for lightbulb technology, but rapidly became the basis for all kinds of electrical tools and machinery. Later, electronics came to feed applications in a number of industries and served as a take-off of the information and communications technology. The personal computer was at first a product of special interest to data pioneers – today they are in every home, every office and every shop. Often it takes time to identify GPTs. Sometimes inventors or marketers claim to have come up with a GPT, and they are not always right.

As for AI, there seems to be an early consensus that the technology qualifies as a GPT, and already diffusion is proceeding at record speed. Still, estimates of future productivity impact are hard to make. A commentary article from the Brookings Institution (Baily and Kane, 2024) remind us that labor productivity growth in the US has been down from a previous level about 3 percent to 1,5 percent since 2008. Similar trends are seen in other high-income countries. As for specific examples of productivity gains, the authors mention several examples, from health care to computing to legal analysis and fraud detection. Interesting examples of pioneer AI implementers are offered by Davenport and Mittal (2023), still without any attempt to suggest overall productivity impact when generative AI is implemented on a fuller scale in the economy.

General purpose technologies (GPTs) have the capacity to impact a wide number of productive activities, but beyond that, they also have the capacity to change the ways we think about production and organizations. Business historians generally refer to the introduction of steam, electricity and information and communications technology (ICT) as three main GPTs that have fundamentally changed modern societies and the modern economy. The introduction of the steam engine was the basis of the First Industrial Revolution. Likewise, electricity spurred the Second Industrial Revolution, and ICT – combining digitalization and electronics, started the Third Industrial Revolution. And they all did more than just bring a set of new products to the market.

Alfred North Whitehead commented that "the greatest invention of the 19th century was the invention of the method of invention" (1925, p. 120). General purpose technologies enabled mass production and mass consumption, and railways and ocean liners set the stage for large scale transportation and communication, dramatically changing society while also encouraging a culture of inventiveness. In a like manner, the introduction of AI is by many seen as such a double function invention, enabling productivity and efficiency gains, but also stimulating invention and an overall transformation of society on a large scale and in most speres of life.

Estimating productivity gains

Clearly, with GPTs it is not easy to predict the exact productivity gains to be expected. While James Watt patented the first steam engine in 1769, and introduced it to the market in 1776, the full impact was only seen in the latter part of the 19th century. Thomas Edison started selling electricity in New York in 1882, but the full industrial impact came over a long period of time, with a great number of new electric devices, tools and machinery, and with the redesign of factories and assembly lines in the 1920s (Crafts, 2021). And while Intel introduced the first 4004 microprocessor in 1971, the full range of instruments for storing, retrieving, manipulating, processing, transmitting, and receiving electronic information is still emerging. The first smartphone, "Simon", was introduced by IBM in 1992, and the Apple iPhone was released in 2007. For automobiles, new electronics have been introduced over several decades and development is still on the rise. Similarly for AI, the overall GPT impact on society, and specifically on productivity and employment will only be seen in the long run. We can already conclude that the coming of AI, large-scale automation and robotization is bound to cause disruptions in job markets, causing work-life revolutions in both developed and developing countries. But no one can know how it will turn out, even for the next two or three decades.

In 1811, English workers formed the Luddist movement, named after its leader, Ned Ludd. In the years 1811–1814, workers attacked textile machines in a rebellion against the industrial revolution, arguing that machines were stealing people's jobs. It came to a "Luddite Rebellion" and to regular battles with the British army – and it ended with a mass trial in York in 1813. The trial led to the death penalty for many of the workers, and for many of them to be deported to penal colonies.

Short-term and long-term effects of technological development have been discussed at least since Aristotle, and an impressive number of solutions for job creation have been tested. Often "disruptive innovations" have created temporary job losses and short-term mass unemployment. In early industrial societies, economists and political leaders often feared serious and long-term job losses. But in most cases the pessimistic predictions did not come true. Luddism has been dismissed as a "Luddist fallacy" or a "myth" (Mincer and Danninger, 2000).

The third wave

Towards the end of the 19th century, classical economists concluded that technological progress would generally benefit society, including the working class, and that employment would rise as innovation and technological progress led to the development of new products, with economic growth, higher incomes, and higher consumption. With exponential growth in both innovation and consumption throughout the 20th century, technological progress has been identified with near-full employment and ever-higher standards of living – punctuated by temporary setbacks and crises. Today, however, there are several signs that the happy coexistence between technological progress, productivity growth, economic growth and full employment will be challenged.

Former technology waves have generally been followed by product development, changes in consumption, and economic growth, causing enormous job creation. Since the 1970s, four factors seem to have caused profound changes in the job markets (Andersen, 2020): (1) Fewer completely new mass consumer products were introduced - and those that came, required fewer industrial jobs; (2) economic growth in the industrialized world slackened; (3) under a political order honoring neo-liberalist globalization, large parts of industrial production were "offshore outsourced" to countries with significantly lower wages; and (4) many of the industrial jobs that remained were automated or "robotized" away. Andersen refers to the latter two, globalized outsourcing and automation, as a "job-killing tag team" – taking turns to eliminate jobs.

The number of employees in manufacturing industries as a proportion of the workforce reached its highest level in the US in the 1950s. In most of Europe, the peak came a little later. On both continents the number of jobs stayed at roughly the same level until the 1980s. About a third of families got their livelihood directly from manufacturing, and a great deal of other employment was a derivative of the manufacturing economy – before a gradual collapse began in the 1980s.

Still, in spite of the "job-killing tag team", mass unemployment did not occur. Instead of mass unemployment, the removal of manufacturing jobs was more likely to cause relocation of workers to the service sector, often for significantly lower wage pay. With the slide away from manufacturing came increased income differences, economic insecurity, and a stream of social problems. Similar patterns unfolded in other industrialized countries, but generally with less dramatic social and political consequences than in the US. The manufacturing sector now accounts for some 8.5 percent of the workforce in the US. The biggest reduction came in the period 2000–2010, when nearly six million industrial jobs were eliminated by the "tag team" – outsourced to "emerging markets" or automated away.

What we face today is a situation where we can no longer trust that "Luddist arguments" must be identical to "Luddist fallacies". It is often assumed that technological unemployment mainly hurts low-skilled workers. But historically this has not always been the case (Katz and Margo, 2014). Early industrialization tended to create jobs for low-skilled workers, able to mass-produce standardized, cheap goods. Skilled artisans, capable of producing high-quality handmade goods with a personal touch, often came out as the biggest losers. With AI, we are facing a similar phenomenon on a grand scale. Automation and globalization since the 1970s have hit blue collar workers particularly hard. In the near future, with automation and robotization based on AI, professional jobs and great parts of the service sector may be hit just as hard.

Gloomy predictions

Critical theory and critical economics, from Karl Marx to Herbert Marcuse, have for the most part looked optimistically at the consequences of automation. Robots have been seen as tools to ease work and set the worker free. The worker was to become the watchman and master of work, while machines were to perform heavy labor.

On the other hand, Kurt Andersen (2020) reminds us that the mathematician, multi-scientist and main initiator behind the cybernetics revolution and AI, Norbert Wiener, already at the end of the 1940s envisioned that with fully automated factories, it would be very difficult for humans to gain a livable income from routine jobs. A new generation of computerized machines would reduce the economic value of factory work to a level where the worker would no longer be worth employment – whatever the price, whatever the wage.

Another scholar who understood the development very well was the Nobel laureate in economics, Wassily Leontief, who noted in 1983 that while incipient automation and early use of computers had led to both increased productivity and higher wages, the trend would soon reverse. Higher productivity would be combined with stagnant wages (Leontief, 1983).

One who was astonishingly clear-sighted, but also quite a bit over-optimistic, was the economist John Maynard Keynes, who in an essay on "Economic Possibilities for our Grandchildren" (1930) envisioned that the standard of living in Europe and the US a hundred years hence would be so high that no one would have to worry about not making enough money. He foresaw that people would not have to work more than three-hour shifts, a total of 15 hours a week. But he also pointed to a "disease" that might accompany the shift, namely technological unemployment. With greater and greater efficiency, the problem of employment would increase. However, he optimistically envisioned that the citizens of the future would be able to distribute equally both work tasks and income. Increased free time and financial independence would give room for the development of other aspects of human life. Keynes believed that the so-called "economic problem" of sufficiency (about what should be produced, how it should be produced and who should produce it) would be solved. What remained was the challenge of how to distribute work tasks and wealth.

What Keynes could hardly imagine was that the new abundance would increasingly accrue to the richest in society, and that poor people would continue to work twice as much as necessary for the good life – to some extent by performing "bullshit jobs", also to pay for the health costs from working too much (Graeber, 2018).

Off-stage comments

Early signals have come from the top of the international business world – the so-called "Davos elite" of the World Economic Forum. In an article in the New York Times, on "The hidden automation agenda of the Davos elite", Kevin Rose (2019) wrote about what none of the leaders dared to say publicly, namely that they both wanted to, and were under great pressure to replace people with machines as quickly as possible. During the annual meeting that year, Roose had informal conversations with a few business leaders, and he noticed that the answers to the questions depended on who they were talking to and from what venue.

In public contexts, and on stage, the leader-billionaires complained about the possible negative consequences of AI for employees. They participated in panel debates on "human-centered AI" and on the need for safety nets for people who would lose their jobs. But in private contexts, and in conversations with consultants and technology companies, they were concerned with how they could automate their own businesses as quickly as possible to gain the greatest possible competitive advantage – without much concern for how the employees would fare. The aim was to create "lean, digitized and highly automated operations", while departments with thousands of employees were to be slimmed down to a few dozen. Former "rationalization" initiatives might involve slimming down departments by 5 to 10 per cent a year. Now they foresaw getting rid of most of the workforce in the shortest possible time. But they didn't say so publicly. Instead, they would mention how employees should be exempted from boring, repetitive tasks. Instead of saying that people were to be fired, they would admit that a "digital restructuring" was about to take place.

Already in 2019, the companies that supply systems and AI solutions to business were experiencing growth and good times. IBM's department for "cognitive solutions" in a short time became the second largest in the company. Investment banks anticipated that the entire international AI sector would grow by 20 percent a year. Kai-Fu Lee, author of the book "AI Superpowers: China, Silicon Valley, And The New World Order" (2018) estimated that AI would eliminate 40 percent of jobs during 15 years, and he was in particular aware that this time it would be to a great extent about white-collar jobs.

With the AI breakthroughs of 2022 and 2023, managers are under increasing pressure from shareholders and boards to go for short-term gains through the fastest possible automation. Kevin Roose found that Asian leaders were speaking more openly about the emerging employment conflict. In interviews, they talked about specific goals, such as replacing 80 percent of the workforce with machines in the next five years. Others could offer hope that the entire business operation would one day be 100 percent automated.

Practical issues

We have had decades to get used to computerized automation and robotization. The introduction of AI, however, represents a quantum leap in further development. The shock of the 2022 introduction of the ChatGPT series has taken AI from the dreamland of tec-executives to the business level of competition and personnel management.

In 1962, General Motors and AT&T were the two largest companies in the United States with a total of over one million employees, one in six of the total of workers in the American labor market. Today, Apple and Google have combined revenue and profits (adjusted for inflation) that are more than double what GM and AT&T had in 1962, but they employ less than a quarter as many people. Apple, Google, Facebook and Microsoft together account for nearly 10 percent of the value of all publicly traded companies in the United States, but they employ less than one percent of the workforce. They are in a sense the heroes of the economy and of the stock markets, but to the extent that other companies follow in their footsteps, there will be a long way between jobs.

Robots are world chess champions. They calculate within seconds whether Magnus Carlsen's last move was wise. In the USA, a duel was arranged in 2018 between 20 experienced business lawyers and the legal robot LawGeex, which specializes in contracts (McPeak, 2018). The task was to assess the quality of five unknown 153-paragraph confidentiality contracts, all couched in strict, legal language. This is a type of contract that is one of the most difficult to assess. The experienced lawyers spent on average 92 minutes for their assessment, with an average accuracy rate of 85 percent. The LawGeex machine took 26 seconds to do the job, with a precision rate of 94 percent.

Robots can produce a finished hamburger, with lettuce, tomato, pickle, cheese or whatever, in 10 seconds, 400 hamburgers with a significant variety of ingredients per hour. They can do most of what humans can do in the grocery store, including stocking items and helping people find what they need on the shelves. Staffing in the retail trade in the US falls by an average of 100,000 a year. The AI robots are already experts in online shopping. In an online bookstore, people can pack two or three orders a minute. It's lightning fast and very stressful. The robots that take over the jobs pack ten to twelve orders a minute, and they are neither stressed nor tired. They are also happy to work at night, without lighting, heating and night allowances. Similar stories are increasingly reported from a great number of economic sectors, professions and trades – including driverless cars, busses and lorries, 3D printer technology for housebuilding and many kinds of construction, etc. The AI robots are unusually willing to learn, but so far rather clumsy in their bodily movements.

Employment predictions

Previous predictions about technological unemployment have focused mainly on blue collar work. AI- and AGI-produced technological unemployment predictions will have to focus on the entire workforce, including white- and blue-collar work, professional jobs and handicrafts, creative industries and food production, even managerial work, leadership, and ownership functions.

Early contributions to the AI-unemployment literature recognized the far-reaching possibilities and consequences of automation (Autor and Dorn, 2013; Beaudry et al., 2016; Katz and Margo, 2014), but not the far-reaching consequences of advanced AI and potential AGI developments. Predictions for short-term and long-term job losses have varied a great deal, but a land-mark study by Frey and Osborne (2017) provoked great attention. The authors recognized the lack of quantified studies of unemployment in the wake of automation, robotization and AI. They introduced new research methodology and picked 702 specialized occupations for assessment of susceptibility in the US labor market. Their main finding was that some 47 percent of jobs registered in 2010 were at high risk. They referred to the Levy and Murnane (2004) study of machine difficulties in replicating human perception, and their conclusion was that "it is hard to imagine discovering the set of rules that can replicate a driver's behavior" – only six years before Google successfully presented the first self-driven car, a modified Toyota Prius.

Important clues for susceptibility assessment have traditionally been the distinction between cognitive and manual tasks, and between routine and non-routine tasks (Autor et al., 2003), generally discarding non-routine cognitive jobs from the high risk area. By 2017 it was evident that both cognitive and manual tasks were being automated, and that many non-routine jobs would have to be included in the high-risk group. Referring also to studies of susceptibility to offshoring of jobs, focused on needs for specific localization of job performance, and on the need for face-to-face personal communication (Blinder, 2009), Frey and Osborne added the face-to-face element to factors suggesting less susceptibility. The main finding by Frey and Osborne (2017) was that 33 percent of the US workforce faced a low risk of unemployment for the next two decades,

19 percent faced a medium risk exposure, and 47 percent faced a high risk. There was also a correlation between risk and timing, with the highest risk categories first to be automated.

The main insusceptibility factor was described as "engineering bottlenecks", more easily overcome in the high risk areas such as transportation and logistics, office and administrative support, and in manufacturing jobs. Also, in line for high risk were services, sales, and construction occupations, personal and household services.

In the medium risk category, tasks requiring "manual dexterity", "finger dexterity", "cramped work space", "complex perception", and "manipulation" (such as much repair work) were prominent. Here, human labor would be at a comparative advantage over robotic dexterity – at least for some time, due mainly to the combination of high intelligence, bodily movement and dexterity.

For the low-risk categories, problems of overcoming the engineering bottlenecks related to creative and social intelligence were the main factors. Such tasks included "fine arts", "originality", "negotiation", "persuasion", "social perceptiveness", and "assisting and caring for others". These all require knowledge of human heuristics and the ability to develop novel ideas and artifacts. In particular, the jobs of chief executives, requiring "conferring with board members, organization officials, or staff members", requiring negotiation skills and social intelligence, were seen as low risk positions, as were also other "generalist" jobs, such as in education, healthcare, arts and media, also requiring social intelligence.

Since 2017, AI has taken on roles for deep learning, self-coding, or self-programming, enabling modification of program codes, and the development of basically new programs, also for artistic, creative, and social purposes. While the number of jobs for a green transition may increase, humans may be at less advantage for securing the (low- and medium-risk) jobs.

AI-competition for jobs

A considerable literature on AI and unemployment has emerged over the last few years, and it seems to group into either a "replacement effect" or a "displacement effect" (Guliyev, 2023). Studies in the "replacement effect" group tend to focus on the negative impact of AI on the labor market, where AI takes away more jobs than it adds. It starts with Leontief's prediction (1983) of a near 100 percent loss of jobs, with Frey and Osborne (2017) prominent among scholars foreseeing deep cuts in the job markets, identifying jobs with high substitution risks (Felten et al., 2018; Fossen and Sorgner, 2019). The "replacement" literature refers mainly to prognoses and potentials. For obvious reasons, there is as yet little evidence of advanced AI impact on job markets. And since robot-automation of manufacturing jobs and computerization of clerical jobs have been at the forefront, the empirical consequences of AI on professional jobs and jobs requiring higher education have not been studied.

Like for latent climate change and damage, the early 2000s have been a time to talk about "latent" AI-job losses, while heading for "manifest" consequences in the 2030s and onwards. Some empirical work has been done under the AI-employment umbrella, seeking out trends based on early automation and robotization, however. The effects of robot density on local drop-outs from the labor market and on depression of local wage levels has been studied, in particular the impact on workers with low education (Acemoglu and Restrepo, 2020). A drop of 0.2 percent in the employment-to-population ratio, and a decrease of wages of 0.37 percent was found as a result of adding just one robot per thousand workers.

Another study confirmed that a significant number of jobs that had been picked as high-risk jobs in 2012, seven years later in fact tended to have been lost or subjected to less growth (Georgieff and Milanez, 2021). The precise numbers are not of particular interest. But the studies are reminders that unemployment is only one among the potential consequences of AI. People threatened with job-loss due to automation may have the option of seeking employment in another sector, or they may take a wage-cut in their present occupation (or not earning compensation for higher prices and higher productivity) – issues to which we shall return.

As for the "displacement effect" (when AI causes people to seek employment in newly added job offerings), there are studies documenting positive impacts of automation and robotization on employment, suggesting a positive outcome of a more general AI-transition. What seems clear, however, is the crucial role of economic growth for a long-term "job creation effect" of AI. With productivity improvement, meaning that fewer human jobs are needed for the same production quantity, or the same number of people may produce more, without economic growth there would be increasing unemployment. Automation may increase the migration between economic sectors, such as from manufacturing to business services (Dauth et al., 2018). Companies at the forefront of innovation and AI driven automation may also experience a net increase of employment, reflecting above average growth (Aghion et al., 2020; Koch et al., 2021). And nations at the forefront may have a higher positive employment than the follower-nations (Guliyev, 2023). But automation alone, without growth, will cause a decline in employment (Webb, 2019), and the greater the productivity gains, the greater the growth has to be in order to stabilize employment.

Decline in employment is a term hiding three alternative options – (1) direct job-loss, (2) wage decreases to make humans competitive with robots, and (3) transition to other sectors and occupations (frequently at lower wages). At each of these we may identify extreme points where (a) unemployment nears 100 percent, (b) wages fall below sustainability level, and (c) alternative employment does not pay a livable wage (forcing people to hold two jobs). These options for an AI-dominated future are seldom considered in the "displacement" literature, and they are essential to the prediction of political-economic outcomes of the imminent transitions.

In a sobering comment to the "hype" of both industry leaders and media possibly exaggerating the AI potential for extraordinary productivity benefits, economist Daron Acemoglu suggested a modest 0,66 percent productivity addition for the next ten years, or 0,06 percent per year (2024). Pointing to the great uncertainties surrounding AI economic impact, Acemoglu limited his prediction to the next ten years, when most economic activity will have to be based on present technologies. His estimate is based on a simple multiplication of the expected number of jobs to be affected times the degree to which they are taken over by AI-based technology (referred to as the "Hulten's theorem"). For labor cost savings he referred to various studies, suggesting a 27 percent average (with total cost savings of 14,4 percent). For estimates of shares of tasks to be impacted by AI he ended up with 4,6 percent of the total. This is a surprisingly low number, compared to previous estimates by Frey and Osborn (2017) and by business insiders like McKinsey and Goldman Sachs. Acemoglu is quick to admit that beyond the next few years, it is very hard to judge about productivity impact of AI and of combinations of AI, automation and robotization. Also, he confirms the main claim of the present chapter, that unless such productivity growth comes with strong economic growth, increasing economic inequality and unemployment will materialize (Acemoglu & Restrepo, 2018).

Below sustainability

Near full employment at a livable wage level is basic to social sustainability and a prime political-economic legitimatizing factor of capitalism and the neoliberal political order. Without the capacity to create jobs for people, and without livable wages in their present jobs or in relevant alternative jobs, political support for the present political order would erode. A question of greatest political-economic importance is therefore whether or not AI, AGI, and large scale robotization will support the employment/wage mix or undermine it.

There is little doubt that AI-guided robots will take over an increasing number of jobs, not so much during the present decade, but at an increasing speed towards mid-century. Besides direct job losses, there is also little doubt that jobholders, in competition with robots will come under pressure to work for lower wages, or seek employment elsewhere, often at a lower wage. At some point, in a major slice of the job market, robots will come to outcompete humans to such an extent that whatever the wage-cuts, humans will not be able to compete, even at wages below the existential minimum.

What then? Will AI-units and their owners become the new taxpayers (Abbott and Bogenschneider, 2018)? Will societies at least stop tax discrimination against human labor?

Growth potential and the environmental burden

AI stimulants to innovation and to new product and services development, will provide for added employment in some sectors and in sectors still to be discovered. This is the hope of AI proponents and technology optimists. But it also brings up two main new issues. Historically, technological unemployment has been more than compensated for by economic growth. New technologies may have caused short-term and local unemployment but added productivity and new product development have stimulated economic growth and thereby an expansion of the job market. Can the same thing happen now with AI stimulants to productivity, initially causing job losses and wage depression, but adding to new economic growth, causing new job-offerings and wage rises – more than balancing out the immediate job losses?

We have already, in Chapter 11, seen the main arguments for de-coupling, essentially that the global overshoot, and in particular the overshoot of the wealthiest societies and the wealthiest people have been carried too far, adding enormously to the environmental burden. Unless absolute decoupling is achieved, economic growth must be limited, even reversed, and mainly reserved for green elements and for societies that need growth to satisfy basic needs. Thus, if material growth is going to compensate for AI-caused automation, it would have to expand beyond levels compatible with sensible overshoot.

This brings up an issue of considerable political-economic consequence. If overshoot, de-growth and decoupling considerations combined with AI-productivity enhancement lead to mass-scale unemployment, what will be the economic outcome for people and for social sustainability? What kind of political-economic solution will be called for?

In Chapter 11, we referred to the relationship between economic growth and employment, and to Okun's law (that GDP must grow at about a 4% rate for one year to achieve a one percentage point reduction in the rate of unemployment). The exact numbers are open to debate and historical corrections. Several factors account for local and over-time variations, including public policies, governmental intervention, and productivity changes. In principle, any stable society, with a stable population size, a stable level of job seekers, and stable productivity can achieve stable employment without economic growth. With population growth, economic growth must at least keep up with the population numbers in order to sustain full employment. And with productivity growth (defined as output per employee), economic growth must at least keep up with productivity gains for full employment.

Disregarding all other impacts, future economic growth must at least keep pace with AI driven productivity growth in order to sustain near full employment. If the productivity gains will be closer to what AI-optimists hope for, extraordinary economic growth will be needed to secure full employment – or AI-driven productivity growth will sooner or later lead to mass-unemployment.

Today it takes some 30 man-hours to build a car. If robots are to take over 50 percent of the human work input, the manufacturer would need to produce the double number of cars to retain the same number of people in production. Extrapolating to the full economy, a doubling of everything would be needed to retain employment. Extrapolating to the green transition, a doubling of energy for production and use of materials for vehicles and batteries, would be required. For electric vehicles, instead of a 100-fold increase from present numbers, a 200-fold increase would be needed. Besides the obvious question of whether people would need all these cars (and whether there would be enough road-space and parking space), there is the more general question related to the green transition – of de-growth and decoupling, and the extreme added burden of making the electrification transition, the climate transition, and the nature protection transition, if feasible at all.

Unsustainable income

The economic growth level needed to absorb the full labor supply in a situation of greatly enhanced AI-productivity, will add massively to the environmental burden. And with failure to absorb the full labor supply, AI-driven productivity gains will also add significantly to social sustainability issues, threatening the collapse of the present political-economic version of capitalism and the neo-liberal political order.

Economic growth for the satisfaction of basic needs of poor societies can be defended – with or without AI-assistance. For wealthy countries, economic growth without de-coupling will add to the environmental burden and to all transition burdens, which can hardly be defended.

What is needed for social sustainability is, first of all, redistribution of economic means, not overall growth. Secondly, overall economic growth can be maintained at some level, provided an absolute decoupling of burdensome material growth from overall growth. Again, there is a "stickiness" problem, however. While manufacturing today represents a relatively small percentage (less than 10 percent) of overall employment, material products are key objects of trade and service, finance and even the health sector. Labor could potentially be transferred from manufacturing to other sectors, but these other sectors will also be subjected to AI productivity gains, and they are to a great extent coupled to material production. This all adds to the stickiness problem.

As mentioned in Chapter 11, Ted Trainer (2012) found that with a future global population of 10 billion people, if all in the world were to consume material resources at the level of the present generation of Australians, annual production would have to increase by a factor of eight over the present in a time span of 40 years. It would also take 80 billion ha of productive land – about 10 times what is available on the planet and put equivalent burdens on other transition factors such as GHG emission, sequestration, and green energy. To this may be added, that since wealthy people in wealthy nations over the last few decades have enjoyed considerably greater income booms than the average, if equal income at the level of the wealthiest ever were a goal of added AI-productivity, it is certainly beyond any reasonable expectation – requiring endless double-digit overall growth. The enjoyment of Quatar-level yacht sailing is only possible when others in the world do not achieve the same level of spoils and affluence.

The latter adds an important challenge to economic welfare discourse. Most economic models seem to assume full economic openness, such that the wealth of one person does not prevent others from seeking similar prosperity. Most models, including "trickling down" or "supply side" models, also gain legitimacy by postulating that tax relief for the wealthy will benefit the rest of society, without postulating that there will be some kind of ceiling on the benefits to the rest. Overshoot, circular, de-growth and decoupling theories, on the other hand, say that there are limits to how much can be consumed in total, and that when some human actors consume disproportionally much, this is directly or indirectly a limitation on others' opportunity to improve their condition.

As stated before, the limit may be set involuntarily by ultimate production peaks and the full emptying of oil-wells and by climatic and environmental collapses that may physically incapacitate further growth. Or limits can be set voluntarily by governments and by the international community, to mitigate future damage. Both trajectories represent definite limits, however. To this may be added the burden of AI-effected productivity gains, adding to economic growth, potentially bringing us closer to peak production and environmental tipping points, and putting increased pressure on mitigation-motivated governments. Also, potentially outcompeting human labor for most jobs, and bringing going wage rents below subsistence level.

An important political-economic conclusion is that if and when wages fall below existential minimum levels as a consequence of AI and robotization, the main instruments of neoclassical economics fail to provide answers, and we have to search elsewhere for analytical instruments, compensating forces, and intervention strategies (D'Orlando, 2020).

From economics to political economy

The discipline of "neoclassical economics" has contributed enormously to economic insight and policy formulation. There is a looming question, however, as to how much it can contribute to understanding and policy-making in a world of AI deployment, robotization, looming mass unemployment, and wage levels below existential minimum.

For comparison, D'Orlando (2020) refers to some main features of classical and political economics, allowing for subsistence wages and long-term unemployment. The classicists were also concerned with class distinctions and class struggles, economic surplus and undue profits, institutional impacts and political maneuvering. To this we may also add Harari's notion of competence irrelevance and the "useless class" (2020) – an important issue in the classical political-economic repertoire, but not in the neoclassical repertoire.

The role of class identification and contradictions in modern societies have been discussed mainly by intellectuals on the left. With the shrinking of manufacturing employment, and the growth of a wide variety of service and professional jobs, with a wide variety of ownership models and renumeration forms, including the emergence of a "precariat" class, there is no longer an easily identifiable contradiction between capital and labor, between a capitalist class and a labor class. There is still a lot of tension in modern societies, but not always an obvious "enemy" for targeting opposition or revolt.

Trapped

Looking ahead, we may anticipate a new massive line of demarcation, between capital owners and technological pioneers, benefiting from AI-generated wealth, on the one hand, and masses of unemployed, "irrelevant" and underpaid people on the other. I referred above to Lord Keynes (1930) who envisioned a future of abundance and little need for labor, but who also feared the "disease" of unemployment and failure to share -jobs and incomes. The alternative to having everybody working three-hour shifts, and equally sharing the bonuses of innovation and productivity, is that of a select few working long hours for very good pay, with capital owners and tech billionaires reaping enormous profits, while the rest are either underpaid or unemployed, living lives in misery.

Even a modern sceptic to the mass unemployment hypothesis, David H. Autor (2015) admits that if AI and robots were to replace human labor at a large scale, it would be a major challenge how to share the proceeds. If distribution is left to the market, equilibrium wages will fall below subsistence. And AI-pioneer, Erik Brynjolfsson (2022) warns about what he calls "the Turing Trap", an outcome where AI-unemployed are trapped in an alternative equilibrium without economic income and political bargaining power.

Brynjolfsson compares the trap of AI-generated unemployment, trapped without income and bargaining power, to that of workers in the wake of globalization. The economic arguments for free trade are textbook standards, but mainstream economists have always pointed out that while societies benefit from globalization and free trade, there will necessarily be some groups that end up with a loss, and these groups will need some form of financial compensation. The story over the last four decades is definitely one of national overall gain, but also one of failures to compensate those who stand to lose, stirring protests and despondency.

Brynjolfsson observes the same pattern in pre-AI digitalization and automation. Enormous gains from productivity were promised to everybody, and while owners, executives and other successful people prospered accordingly, workers and unemployed all too frequently were left in despair. Actually, recent research has demonstrated that technology discrimination has been a far more important reason than globalization for increasing inequality over the last twenty years (Tica et al., 2022). Similar mindsets and anxieties related to feeling trapped are evoked with the coming of AI and AGI. Great promises are in the air, with prospects for record wealth creation, but also for loss of jobs, income, and bargaining power for the many.

Imagine

In Part 3 we shall consider political-economic and social models for dealing both with environmental threats and energy issues, social sustainability and employment. But before diving into strategies and models, we shall bring the Turing trap issue one step forward. In a previous paper (Knudsen, 2021), I took the AI-unemployment issue to an extreme position. The philosopher, mathematician and universal genius, Gottfried Wilhelm von Leibniz (1646–1716) recalled a logical concern from antiquity, whether to allow for a form of argument based on extreme assumptions. *Reductio ad absurdum* or *argumentum ad absurdum* can sometimes be used to create truly absurd "proofs" or conclusions, but they can also be used to illuminate what may emerge under the most extreme conditions (Ferguson and Priest, 2016).

If we imagine an extreme future where almost all jobs are handled by robots equipped with artificial intelligence, what will people do and what will they do for a living? And what will happen to the market economy? A provocative answer is that if wage incomes disappear in total (*argumentum ad absurdum*), the demand curve in economic models will also disappear! Where there is no one to demand goods and services in the markets (except for robots and robot owners), there is also no money to put behind the demand side of economic models. In other words, no regular market economy can provide for human life support if all paid work is replaced by robots that do the job cheaper and better. In that case, all payment for work done will go to robots – for resource extraction and refinery, for production of robots and for robots to produce whatever is demanded by other robots.

A derivative sector

Increasingly greater returns to capital, to technology experts, and to non-replaceable human workers are predicted for the coming decades. At the same time, income erosion from both job losses and wage cuts are expected. At the extreme of uselessness or irrelevance we may talk about a growing "derivative sector" of society – of those unable to make a living in a regular labor-market economy. However, if more and more people are unemployed, without income, or working at below subsistence level wages, less money will be available for consumption, until at the stage of *absurdum*, there is no demand, and for the companies no sales and no income.

The only way a market economy can work, is when consumers have at least enough money to keep themselves alive and for capital to keep production going. In the extreme case of *reductio*, where owners and a minority of experts and technicians acquire the full value of what is produced, and the unemployed majority receives nothing, there will be no use for most manufactured or agricultural products, or for the long chains of supplies, services, transportation, warehousing, financing, and insurance that go into providing for such consumption. Only products of interest to the minority of super rich people would be worth producing, such as private jet planes, super-yachts, luxury food, and castle-size mansions. As for the rest of the population, the vast majority, they would either starve or freeze to death, or they would need some form of economic compensation at above subsistence level. In a sense, the emergence of a "derivative" sector caused by AI developments, takes us back to Brittain of 1795 and the Speenhamland legislation introduced for "the right to live" – to which we shall return in Chapter 16.

The term "market failure" is normally used for special cases of market inefficiencies, as when rational behavior by individuals give less than rational outcomes for a group or society, as when there are lasting imbalances between supply and demand in a market, distorting externalities, or monopolization tendencies. In the imagined case of *Reductio ad absurdum,* a market failure is not about special cases, but rather about the market system as such. We are back to the Keynesian challenge of sharing the benefits of productivity, here at an extreme imagined case of mass unemployment and masses of unemployable, "irrelevant" people.

There is a long tradition in classical economics for thinking in terms of "surplus value", later excluded from the neoclassical economics canon. In the 18th century, French physiocrats were concerned with the "net product" accruing to business owners, paying workers little while extracting solid profits for themselves. For the British classical economists, from Adam Smith (referring to "net product") to David Ricardo, and for critical political economists of the 19th century, such as Karl Marx and some socialist followers of David Ricardo, the concept of "surplus value" was much debated, generally with reference to William Thompson and his book with the appealing title: "An inquiry into the principles of the distribution of wealth most conducive to human happiness: applied to the newly proposed system of voluntary equality of wealth" (1824).

Whatever the many definitions and qualifications, the term "surplus value" generally refers to income from sales of a product minus the cost of production including materials, plant, production equipment, depreciation, and labor. Having been disregarded as a useful analytical term in neoclassical economics, replaced with legitimate profit, assumed to represent the marginal cost of capital, for our purposes the term may again be useful. In the *extreme* situation where labor costs are to be eliminated, but where profits still will be made, such profits will represent surplus value, either to be taxed and redistributed for a guaranteed minimum income (GMI) in an economy of voluntary occupation, or retained by capital investors for investment and luxurious spending, in a shrinking economy or eventually a dystopian society of mass elimination.

Reductio - not quite ad absurdum

Arguments *ad absurdum*, like postulating that nobody pay taxes, nobody holds a normal job, nobody can afford anything unless supported by the government, are meant to be absurd, in the sense that even the person who presents the argument believes that it will not be like that. The point is to help us visualize what will happen if there is a trend in the direction of the absurd state, help us see what we may be up against.

What if we imagine a long period of turbulence, with wage-cuts and rising unemployment, eventually reaching stabilization and social sustainability, but where "eventually" can take generations (Berg et al., 2018)? As referred above, Tica et al. (2022) found that technology was more important than globalization in accounting for economic inequality. They also found that employment and low inflation were the most important factors in defense of equality. It seems fair to say that we are already into an era of growing *absurdum* – where record wealth and technology bonuses are collected by owners and upper echelons, while workers at best have obtained inflation-adjusted wages equal to 40 years ago, and more realistically below that level. While most AI commentators seem to agree that at least some productivity bonuses will be achieved at societal levels, few seem to think that it will come entirely without major job losses, or at least without substantial turbulence in job markets, or that technology will bring increasing economic equality.

Of course, if we allow for unlimited material growth, and we foresee a period of active governmental job creation, re-education, employment bonuses and quotas, and tradeable (maximum) robot quotas, the period of near full employment may be prolonged, and the turbulence of job markets minimized. But there is also the question of increasing "*ressentiment*" – of alienation and disintegration, calling desperately for economic security as the most essential factor for integration and concerted action, also in dealing with the many environmental and societal threats of our time.

As mentioned, for people to be kept alive, the future productivity bonus from AI and robotization will have to be shared through some form of GMI system. As for the near future, the technological unemployment threat in parallel with increasing despair and desperate hopes for economic security are sufficient enough reasons for starting an era of sharing and building safety nets. Beyond that, there is also the question of life occupation, not equal to wage-paying jobs. What will "irrelevant" and "useless" people of the derivative sector do, if they receive a guaranteed income without having to work for pay?

Guaranteed minimum income

Detailed historical examination of experiments from different corners of the world, for social welfare and economic development, have documented that cash in hand is an effective and efficient instrument for poverty alleviation (Bregman, 2018). A detailed historical listing by Wikipedia (https://en.wikipedia.org/wiki/Guaranteed_minimum_income) includes social GMI-type innovations from around the world, including the minimum wage/child support scheme by Persian monarch Cyrus the Great, 590-529 B.C.; the *Cura Annonae* of free grain to the poor, introduced by Gaius Gracchus of Rome in 123 B.C.; a guaranteed minimum monetary income system introduced by the first Sunni Muslim, Caliph Aby Bakr, in 632 C.E.; and the *citizen's dividend* to all citizens of the US who had lost natural inheritance, introduced by Thomas Paine in 1795.

An equal "share in the Earth's produce" was advocated by Emperor Napoleon Bonaparte of France. American economist Henry George advocated a "dividend paid to all citizens" from the revenue of a land value tax. Conservative economist Milton Friedman, author of "Capitalism and Freedom" (1962) and "Free to Choose" co-authored with his wife (Friedman et al., 1990 [1980]) advocated a negative income tax – probably the most rational and effective model of a GMI design. Another conservative voice, the Austrian economist Friedrich Hayek, in his "Autobiographical Dialog" from 1994 (2012), expressed strongly that he always had been in favor of a minimum income for all. At the centerstage of neoclassical economics, James Tobin, Paul Samuelson, and John Kenneth Galbraith were joined in 1968 by 1,200 economists signing an appeal to the US Congress, calling for a system of income guarantees. Most famous among GMI initiatives is perhaps President Richard Nixon's "Family Assisstance Plan", from 1969, with a minimum income guarantee for poor families. The proposal was passed in the House and was only one week away from being adopted by the US Senate.

Robert Theobald's "Free Men and Free Markets" (1965) is perhaps the most important historical book marker for GMI. Martin Luther King advocated GMI in his latest book, "Where Do We Go from Here: Chaos or Community?" (1968). In 1987, New Zealand's Finance Minister, Roger Douglas from the Labor Party, announced the introduction of a Guaranteed Minimum Family Income Scheme and above that a flat income tax – again an economically rational approach. It was stopped by Prime Minister David Lange, however, and Douglas was sacked from the Government. On January 1, 2023, Germany introduced "Bürgergeld" as a substitute for social security and unemployment benefits. The time has come for learning from experiments and experience, initiating further testing and implementation of GMI and also guaranteed minimum inheritance in practice. We don't have to wait longer.

More time for reflection

Traditional societies learnt to cope with local, short term and directly observable threats. Our list of modern issues are of a different kind, provoking different challenges: They are both local and global, they have long-term and to some extent unforeseeable consequences, and they are a result of mainly complex and interdependent cause-effect relationships (Beck, 2005). Such problems and challenges have been referred to as "grand challenges" (Reid et al., 2010). The challenges are not only formidable in terms of size and consequences, but they are also complex, also referred to as "wicked problems".

A wicked problem is one that is difficult both to formulate and solve, and sometimes impossible to find a clear solution to. Climate change has even been characterized as a "super wicked problem", enormous in both time and space, and growing in complexity and severity the longer mitigative action is delayed (Lazarus, 2008). Implausible contradictions and often unknown elements affect the end result and sow doubts about claims to have solved the problem. Solving one side of the problem can also quickly create entirely new problems elsewhere (Churchman, 1967; Rittel and Webber, 1973).

The wider concern with climate change and nature depletion as related to patterns of overall global development has become more accentuated in recent UN publications related to climate and nature, in particular the Climate Change 2023 Synthesis Report (IPCC, 2023), repeatedly referring to CRD or "climate resilient development" (see also Eriksen et al., 2024). It is increasingly recognized that just knowing about climate change and nature depletion, while necessary is not sufficient to bring resolve and concerted action to deal with the problems. Concerted action takes human agency, cooperation within and among nations, recognizing disparity of interests but also commonality of vulnerability and fate.

It seems that the combined problems linked to climate change, nature depletion and the spread of environmental toxins are of the wicked kind. The coming of AI adds hope for easy solutions, but it also comes with added grand challenges. Basically, we know what needs to be done for climate and nature. The wickedness accrues to problems of concerted action, which again accrues to a set of disintegrative forces and the lack of "glue" in societies. At the extreme, the wickedness exceeds the capacity of society to deal with the challenges, and the result may be some form of societal collapse or dystopia. In order to handle the green-plus-AI transition, we need to know more about the forces of delay and lack of concerted agency.

In Part One of this book, we dealt with the environmental burden – with environmental threats, green energy alternatives, energy stickiness, and the impact of AI and AGI. In Part Two we have dealt with factors that interfere with the capacity for resolve and concerted agency.

From theory about change, we may expect the coming of manifest damage caused by climate changes and nature depletion to stir an increasing sense of crisis and urgency in populations, hopefully outclassing the roles of denials and doubt, psychological and practical barriers to the green transition.

Manifest environmental damage sends strong messages that it is time for action and change. But the political-economic landscape may not be ready for concerted action. In Chapters 9 and 10 causes of *ressentiment* and disintegration, inertia and paralysis were examined. Immiseration and elite rivalry were identified as significant hurdles. In order to deal effectively with environmental threats and also with threats linked to AI and artificial general intelligence, AGI, we need both to improve and modify the present political-economic playing field, and to seek "possibilist" solutions beyond present boundaries, and we need to open up for a "breakaway" political order more conducive to dealing with the threats.

Chapters 11, 12, and 13 has focused on the combined challenges of dealing with growth and decoupling, economic inequality and security, and the potentially disruptive labor market effects of automation and AI-supported robotization. Reading across the three chapters, and accounting for *argumentum ad absurdum*, the reasoning is summarized in Figure 13-1.

As shown, there are two main entries to the figure – the environmental and the digital. As for the environmental entry, economic growth has historically been strongly linked to resource depletion and environmental problems. Also, the present global economy is an enormous burden on nature. The Paris Agreement, the Montreal Protocol, the UN Sustainable Development Goals and other agreements and target formulations aim for a "net zero" of environmental destruction around 2050. That means first of all to bring present levels of pollution and depletion to near zero. Secondly, it means that economic growth will have to be fully de-coupled from the environmental burden. The main green challenge is to achieve an absolute decoupling between growth and nature destruction.

The second entry point in Figure 13-1 is about AI and expectations for AI to add significantly to productivity growth, allowing again for significant overall economic growth – potentially adding to relative decoupling, but also to the overall environmental burden. Without economic growth, increased productivity will lead to unemployment. With strong productivity growth over a long period of time, enormous economic growth will be required to maintain full employment. Economic growth without absolute decoupling will significantly increase the environmental footprint (related to climate, nature, pollutants).

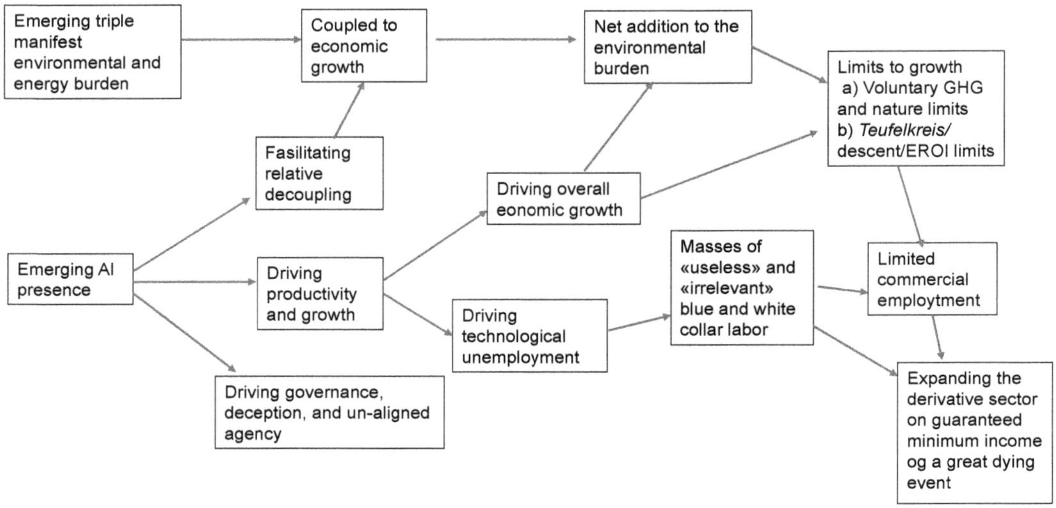

Figure 13-1: The derivative sector.

In summary, we must achieve several things at once. First of all, for a net zero, there must be an absolute decoupling between environmental footprint and economic activity – both concerning present activity and new, growth-related activity. The imperative of decoupling will reduce the growth potential of the economy and the productivity effects of AI, even allowing for "de-growth" or "negative growth". AI development may be used for green efficiency and overall greening of the economy, but AI may also produce challenges related to the environmental burden and to "unalignment" and dystopian outlooks, for which strong regulations are needed. AI will facilitate robot-driven production of most things, disrupting labor markets, causing joblessness, wage levels to drop and returns on capital to increase, causing the emergence of a "derivative sector" of "irrelevant" and "useless" people.

Livable conditions and meaningful lives will have to be secured by means of guaranteed minimum income and inheritance, allowing for the "derivative sector" to thrive, with meaningful activity absolutely disconnected from environmental destruction, and freed from the repressive norms of the commercial sector.

At the extreme, the environmental burden, combined with the productivity benefits of AI, with the need for decoupling of economic growth, and for redistribution of wealth and income, may in sum bring the political-economic system to a precipice of potential system breakdown. Respecting the ageless "right to live" imperative, a guaranteed minimum income (GMI) for all will become necessary, even for saving market-economic capitalism. At a time when AI and robots can take over a significant share of human work, at the same time reducing the purchasing power of "derivative section" people, wealth and income will have to be transferred to the needy, in order to support their "right to live", but also to uphold a modified consumer role in the system.

Adam Smith's reasoned that market egotism serves the society by aid of "the invisible hand". But the logic of "the invisible hand" does not work under premises of a limited Earth system, where for the first time in history, we have reached the outer boundaries of the resource system, and where machines take over the roles as both producers and consumers. Kenneth Boulding replaced "the invisible hand" metaphor with "the Spaceship Earth" metaphor. As a follow-up, we have seen that Trainer's observation (2012) is an extremely provocative one, telling simply that if passengers in the business class continue their partying, there will be too little of everything left for passengers in the economy seats, and no chance that economy riders will ever attain the level of consumption longed for, not to mention the level now enjoyed in business class.

Since the days of Rousseau, it has taken a less severe message to stir aggressive *ressentiment* around the world. As people take in the reality of a limited Spaceship Earth and what it means for economy class passengers and for people still in the waiting lines for getting onboard, the outlook for peace and concerted action are not ideal. From the Spaceship Earth metaphor, we shall in Part Three examine three metaphors for warnings and for change and hope.

PART THREE— IS THERE HOPE?

Part Three is about our final question of what needs to be done. Scientists are warning that the environmental threats of our time are of a planetary and potentially devastating reach. Much uncertainty is also linked to the coming of AI and potentially an AGI singularity. Topping the concerns and uncertainties are geopolitical rivalry, wars, uprisings, anger and ressentiment, potentially opening for dystopian developments even in traditionally democratic societies.

In Part Three, three metaphors are introduced to support the notions of political-economic "possibilism" and a political order "breakaway" – as signaled in Chapters 9 and 10. They are all aimed to inspire thinking related to issues of economic de-coupling, inequality and employment – as signaled in Chapters 11–13.

The war economy metaphor is presented to inspire thinking about political-economic alternatives, preserving democracy while allowing for strong regulations and public interest priorities of production and mega-projects. It is basically up to people in the roles of voters to go for harsh priorities, regulations and subsidies to promote the necessary mitigative and adaptive investments.

The world music metaphor is used to inspire creative engagement by people in all roles and at all levels of society for green and safe solutions – to GHG emissions, extinctions, and pollution. The world music metaphor inspires learning across cultural divides, innovation built on variety of experiences from around the world. It is up to people in the roles as workers and care-givers, professionals and artists, to seek green solutions and implement them, for the betterment of their own lives and the lives of their neighbors and descendants.

Finally, the *dugnad* metaphor, borrowing from old Norse tradition, is about work forms built on mutuality (of give and take) and reciprocity (of being in it together). The *dugnad* metaphor inspires the search for cooperative and commoning initiatives and is linked to the notion of hope and betterment of life from whatever position, under norms of volunteering. It is up to both politicians and people in their everyday roles to create hope through action and action through hope – soundly anchored in the love of people and all creatures of nature.

At the end of Part Three a "final reflection" is added, underscoring the dimension of hope.

CHAPTER 14

The War-economy Metaphor and Commands From Above

Scientific investigations and technological development are going on at an accelerating speed, changing contingencies of life at an exponential rate, making it harder and harder to know exactly what kinds of problems people will face the next hundred years, and what solutions will be available to them. This also makes it harder to suggest societal and political-economic solutions to challenges of the future. Even for short-term challenges, the complexity of modern societies makes it difficult to design policies and structures for reliable positive change. As for deep-seated transformation of societies, interventions are still more risky – and results still more uncertain.

As an alternative to specific advice for long-term intervention and transformation, I am inclined to go for metaphors. Three metaphors will be presented in the next three chapters. In the present chapter, the war-economy metaphor will be used to convey the need for top-down solutions – major governmental engagement in regulations and mega-projects to structure the society and the economy for major steps to deal with the environmental burden and hedge against dystopian AI development. The main challenge to ordinary people is about choice of political support and rejection. In Chapter 15, the world music metaphor is used for engagement of common people of good intent in innovation, global learning and creative search for environmental improvement and social sustainability. Here, the main challenge is about engagement, agency and popular action. Finally, in chapter 16, the "*dugnad*" metaphor is used to stimulate imagination of alternative structures for societal transformation, also allowing for hope. Here the main challenge is about the formation of both traditional and experimental structures of interaction and cooperation, for meaningful activity and for mutuality and reciprocity.

The term "metaphor" is from the Greek *meta* (over) and *pherein* (to carry). Translated literally it suggests a manner of language likened to that of a moving van. The word is "carried over" to represent something else than the original meaning. In everyday talk we use metaphors all the time (the *foot* of the mountain, *drowning* in work, a heart of *gold*). In the present context, the use of metaphor implies a step back from literal precision and narrow commitment and a step forward for unlatched thinking and creative searches, more akin to what Donald A. Schön (1979) called a "generative metaphor".

War-time economics

When thinking about political economic systems, there is a tendency for both politicians and lay people to think in terms of the grand distinctions between command economies and market economies, between capitalism and communism. In reality, most political-economic systems tend to be of the mixed category, with considerable leeway for more or less command and more or less market economics. The US today, as of 2024, is definitely a mixed system, leaning strongly towards

DOI: 10.1201/9781003518815-17

neo-liberalism and market solutions and by 2025 open for radical change of course. China also has a mixed system, leaning more towards command economics.

While much has been published about the command-market mixes, remarkably little has been reported in the academic press about the quite deep-ploughing political-economic reforms in the US and the UK during the Second World War. The experience of war-time economics in basically market-based, capitalist systems, may help us envision some called-for steps and system consequences of a green transition. The choice is not between a Soviet style command economy and a neoliberalist market economy, but between continued mitigation delays and concerted economic action at a scale approaching war-economic levels.

The idea of a war-economy metaphor for environmental mitigation and adaptation in this book is not original. William James wrote the classic essay "The Moral Equivalent of War", first delivered at Standford University in 1906 (James, 1911). William James was concerned about political unity and civic virtue in the absence of war. Inspired by the essay, President Jimmy Carter built his "Moral Equivalent of War Speech", delivered on April 17, 1977, for handling the 1970s energy and oil crisis. Our concern here is about handling of the accelerating environmental crisis.

All countries that are actively involved in war must expect far-reaching changes in the economy. They are forced to make grand-scale priorities and commands trumping individual consumer preferences and the invisible hand of markets. It is seen in the Ukraine today: National defense requires a balance between the full involvement in war, and the economic priorities related to the war effort, on the one hand, and the needs of social sustainability on the other. Nations outside of the war zone may deliver weapons and economic support, but with less need for the most demanding priorities. To the extent that the war economy becomes a dominant part of the overall economy, it will require new organizational and institutional solutions, and changes to legislation and logistical systems. Planning and control of the economy will take a greater role, including rationing of food and basic necessities, price regulation and inflation control, tax increases and the sale of wartime bonds, confiscation of property and logistics control, regulation of labor markets, including wage control and redeployment of manpower, enlistment in the military and mobilization of labor for arms production, also sometimes direct government involvement in production and distribution. It all belongs in the toolbox of a war economist, and some of it may come in handy for the green and the digital transition.

As the possibilities for wilderness crash-landings and climate tipping points come closer and increase in probability, the world community must prepare itself for thinking in war economic terms. If instead of 1940 we think 2040, and instead of war economy we think "earth overshoot" economy, we may learn from historical experience. The example from the Second World War shows that even capitalist USA went far in the direction of a planned economy. We shall start with a summary of "The American Economy during World War II" as presented by Christopher Tassava (2008).

Shock opening – Pearl Harbor

In the USA, careful planning in the form of increased "preparedness", was already underway towards the end of the 1930s, based on experiences from the First World War and motivated by the outlooks for wars in Europe and Asia. The country was still struggling with the consequences of the crash of 1929. Under the auspice of the "New Deal" policy, new administrative institutions had been created to stimulate economic activity and reduce unemployment. But even after ten years, the economic crisis was not over. The level of production in 1939 was only marginally higher than in 1929, but picking up at a high rate, while the unemployment rate was still above 17 percent. The "New Deal" policy had, however, served to educate both politicians, the federal bureaucracy, and the general public about economic command elements for intervention in favor of employment, economic sharing, and social sustainability.

Crucial for the transition from "preparedness" to full "conversion" of the economy was the attack on the American fleet at Pearl Harbor in Hawaii, December 7, 1941. What happened in the weeks after the attack was a rapid transition from a market-based civilian economy to a near command-based war economy. The federal military budget and the production of war materials were considerably increased. So was also economic and material support for allies. And a significant part of American consumer goods production was converted into production of military material.

Many of the manufacturers opposed the transition from production for the consumer goods markets to production for government purchases of war material, mainly for fear of losing market share and customer loyalty. But thanks, among other things, to the "New Deal" experience, large-scale "conversion" had considerable support both in the federal bureaucracy and in major trade unions. Even before the United States joined the war, in 1940, Walter Reuther, one of the top leaders of the United Auto Workers labor union, had advocated that the big car manufacturers should be able to produce war planes. The proposal caused people in the USA to become more aware of the seriousness of the situation, and for "the Reuther Plan" to be launched in 1942.

Other important industries that came under state control were shipbuilding (which had been near dormant since 1921) and which, under the leadership of the U.S. Maritime Commission (USMC) was to gain increasing importance during the war years. The "Lend-Lease" program for aid to the Allies was decisive for both Great Britain's and the Soviet Union's efforts in the war in Europe, even before the United States became directly engaged. The most important content of "Lend-Lease" was the delivery of warplanes, ships and ship repairs, military vehicles, trucks and weapons, but also food to the Allies.

A broad mobilization

The first step after Pearl Harbor was to establish an effective war bureaucracy, with a whole series of directorates, councils and administrations, several of them with the authority to direct production and distribution in private enterprises and with budget funds for direct intervention in the economy.

An important characteristic of the American model was that the most important war-economic bodies retained their civilian character. Similar bodies in both the UK and Germany were more or less directly controlled by the military. In the US the civilian administration went to great lengths to limit the production of consumer goods such as textiles and clothing, non-essential foodstuffs and cars. Also, civilian administrative units went a long way in organizing the economy for the production of aircraft, tanks and ships and in securing access to steel, aluminum and personnel.

For the administration of "preparedness" programs, from 1939, and for "conversion" during the war years, President Franklin D. Roosevelt initiated a series of new administrative units, some of them very effective – others less so. A main issue was to balance civilian needs against military needs, which was not only about production and logistics, but also about the mobilization of female and minority labor for war production and military service. The challenges were not totally unlike those to face us in relation to the green transition.

Balancing military demands against social sustainability was an important concern. Too strong military prioritizing might undermine long-term social sustainability and defense capability. In order to meet the combined civilian and military needs, a War Production Board (WPB) was established in January 1942, followed by a Controlled Materials Plan (CMP), with particular responsibility for the distribution of steel, aluminum and copper between military and civilian production. Subsequently the Office of War Mobilization (OWM) was to become the supreme authority in the war economy, controlling the distribution between civilian and military purposes.

New administrative bodies were erected for recruitment to the military and for the distribution of the workforce between civilian and military production. For the rapid development of the shipbuilding industry, the U. S. Maritime Commission (USMC) was created; for provision of food and administration of foreign aid programs, there was the War Food Administration (WFA); for control of prices and inflation, the Office of Price Administration (OPA) was erected. In order to secure financing of the war effort, The Department of the Treasury introduced the USA's first general, federal income tax system, with tax payment deducted directly from the payroll. Also the mass sale of government war bonds was in some cases deducted directly from wages as a means of saving for workers.

The unionization of workers received a major boost during the war. In practice, all businesses that received assignments for the defense were unionized, and since these had strong growth throughout the war years, the scheme led to a significant growth in trade unions, from 10 million members in 1941 to close to 15 million in 1945.

The sum of measures led to significant productivity improvements and economic growth. In particular, great advances were made in production planning and process technology for mass production of military vehicles, aircraft, ships, and all types of equipment and munitions. Particularly important was the total transformation within shipbuilding, a sector which had been mostly fallow for almost 20 years. Of particular importance for efficiency was the development of modern tools for project management.

While the American practice of "war economy" administration is worth learning from, we also need to recognize the level of conflicts and chaos involved in the transition. Martin Hart-Landsberg (2023) notes the quite significant contradictions generated by the enormity of military spending, with federal expenditures for military procurement doubling from 1939 to 1944, accounting for approximately one-half of all goods produced in America in 1944. While improved productivity played a part in that build-up, curtailment of civilian production, suppression of product offerings, and direct rationing of supplies were also parts (such as prohibiting the production of cars and tires for civilian purposes). From 1941 to 1944, in absolute terms, non-war industrial production declined by more than 10 percent.

The government often faced strong resistance from the business community. Overall, private investments fell during the 1941–1943 period, with business investments down 37 percent from the 1940 level. As a consequence, there was a great deal of opposition from business owners and executives, even from companies enjoying quite lucrative government contracts. Also protesting were people with a great deal of political influence, with strong representation in advisory and executive government agencies. Of government contracts between June 1940 and September 1944, more than half were secured by the top thirty-three corporations.

Also, as the war economy heated up, inflation became an increasing problem, and while wage control was quite effective, getting control of food prices and overall inflation turned out to be much more troublesome. In the end, however, the government prevailed, and from early 1943 to April 1945, the consumer price index rose less than two percent, and food prices fell by more than four percent. Hart-Landsberg attributes success to ongoing learning processes, involving grass-roots participation and the involvement of unions, business corporations and administrators. For a full war-economy mobilization in a capitalist society, business interests (and outlook for rewards) have to be fully involved, as do also worker organizations and regional and local communities.

The full "conversion" of the American economy came after Pearl Harbor, December 1941, following an immediate and massive "sense of urgency" (cf. Chapter 8). The historical lesson is not to copy the American war-time experience for solving the environmental threats of today. What we can learn is, first of all, to think in terms of political-economic flexibility and possibilism, using command economic instruments where needed, leaving much else to markets and even to tradition. Secondly, we learn that command elements, used sparingly and for necessary prioritizing and regulations, can be used as instruments for the preservation and defense of democracy and freedom. Command elements do not threaten democracy, when graded according to intervention needs and when democratically installed and practiced in ways that require open processes, due diligence, and broad participation – even when significant interests rise to protest. Related to environmental threats, we observe that some involvement areas require direct and obtrusive command-economic intervention, such as governmental involvement in investment and production, while others are more fit as frameworks for general mitigation. Also, as we shall see, there is room for both general regulations, with a new set of rules and managerial implications, but also for governmental involvement in megaprojects.

Broad approaches

The war-economy metaphor is a reminder that the full mobilization is not just about the military and about weapons procurement, but also about overall manpower needs and employment, finances and investments, inflation, price, and wage control, and about administrative challenges and institutional solutions. Similarly, concerted environmental action will have to include a number of social, administrative, technological, and economic causes and considerations in addition to dedicated targets for climate, wilderness, and the broader environment.

The "war-economy" metaphor is also a reminder that not all societies respond with equal fervor in times of mobilization. In the following we shall identify three forms of intervention, commitment, and resource mobilization, ranked as (1) a BAU level of minor market-economic regulations, then (2) a Green New Deal type of public, multi-sector involvement, and finally (3) Manhattan style megaproject approaches.

BAU and market regulations

Most responses to climate warnings have so far been cost-based, with an emphasis on carbon taxes and cap-and-trade policies, including also some green technology investments and subsidies, and measures to arrest "free riders". The general economic approach is to address problems of negative externalities (such as pollution) and "free rider" issues (companies that avoid paying compensation for pollution damage), by forcing enterprises to "internalize" the costs (such as paying carbon taxes).

Meckling and Allan (2020) studied how economic ideas shaped climate policy and advice of international organizations, such as the World Bank and OECD, from 1990 to 2017. They found that while a rather narrow neoclassical approach was followed in the 1990s, in the mid-2000s a more diversified policy discourse emerged, with a broader set of policy advices for a green industrial policy with more active uses of subsidies and regulations. Still, climate policies were mainly seen as a quite narrow GHG-emissions problem and a matter of technology innovation, isolated from broad societal concerns. The whole set of "stickiness" issues was largely overlooked, and while the need for mitigation investments for climate and issues of social discount rates were hotly debated, matters of broad societal resilience through systemic transformation were generally not part of the discourse. Also, the triple content of the environmental burden was largely overlooked.

In terms of "war-economic" content, the BAU position barely qualifies as an option. But dependent on the rigor and radicalness of measures such as carbon taxes and market interference, with environmental product classifications and controls, with governmental GHG emissions measurements and international reporting, and with subsidies and government participation in green technology projects, the role of government has become more pronounced – even within the BAU framework. Also, as the numbers of people suffering from extreme weather has gone up, there has been an increasing concern about human vulnerability, displacement, and resource availability, and a second level of engagement, with the emergence of broad, multi-sector initiatives, such as in America the Green New Deal and in Europe the European Green Deal – aiming to make the EU the first climate-neutral area in the world by 2050 and also to cut pollution and restore ecosystems.

Still, in relation to environmental threats, by the 2020s subsidies of fossil energy consumption vastly exceed those of green energy. Carbon tax incomes are significantly short of carbon externalities. And in relation to potential technological unemployment, taxation of labor is in most countries significantly heavier than the taxation of physical investment – by wide margins.

National, state, and city level initiatives

Within the BAU framework, it is not always easy to compare and rank societies. For the US, there have been a steady change of national policies, as evidenced by the support and rejection of – most visibly – the Paris targets for temperature rise and climate policies. Sometimes policies at state, city and local levels have come to directly countering national policies. Thus, even as Donald

Trump in 2017 announced that the US would cease participation in the Paris Agreement on climate change mitigation, several states and cities around the country were actively continuing work to meet the targets. At other times local and business initiatives policies have served to complement national policies. It remains to be seen what the final outcome of the Trump II backlash on the green transition will be.

In a somewhat similar way, the European Union has taken a leading, more persistent role in promoting the green transition, while gross differences of engagement exist on national, regional, cities, and communities' levels. European Green Capital & European Green Leaf Awards have been granted since 2010 and include cities from most of the continent as pioneers in the transition. Similarly in the business community, BAU positions camouflage an enormous variety of engagement and sincerity.

The Green New Deal and the Multi-sectoral approach

Among movements and initiatives, the Green New Deal Resolution (GND), presented by Senator Edward J. Markey and Congresswoman Alexandria Ocasio-Cortes in 2019, still remains what is for a major nation the most comprehensive "war-economic" type of approach to date. It has also served as a model to other nations and movements. It refers specifically to the New Deal policies of the 1930s, for assumptions about governmental and public responsibilities and radicalness of intervention, and it offers a solid listing of "war-economy" initiatives for a peaceful and sustainable green transition. For a broad approach, it is a good place to start (Boyle et al., 2021).

The Markey/Ocasio-Cortes resolution starts with the recognition of climate change and temperature rise as a cause of future mass migration, loss of annual economic output, and specific consequences of extreme weather, such as wildfires, loss of coral reefs, heat stress, food scarcity, damage to public infrastructure and coastal real estate. To keep temperature-rise below 1.5°C (Paris target), immediate GHG emissions reduction and a US net-zero target for 2050 are taken as starting points. The authors also demand that the US take a leading role in global GHG reduction and technology development.

Societal contexts

Beyond the environmental challenges of the Green New Deal (NGD), the Resolution includes a detailed listing of societal crises, in the US and other parts of the world. Declining life expectancy and lack of basic needs satisfaction for the poor are recognized. Clean air, clean water, healthy food, and adequate health care, housing, transportation, and education are listed as inaccessible to poor people in America and in much of the world.

After years of wage stagnation, deindustrialization, antilabor policies, and loss of bargaining power for workers, inequality has increased between rich and poor, and also along racial, gender, and social divides, in America and much of the world. According to the Markey/Ocasio-Cortez resolutions, the level of "systemic injustices" has increased affecting disproportionately frontline and vulnerable communities of "indigenous peoples, communities of color, migrant communities, deindustrialized communities, depopulated rural communities, the poor, low-income workers, women, the elderly, the unhoused, people with disabilities, and youth".

Added to social arguments for an NGD is also a recognition that years of underinvestment in relevant infrastructure, and in climate damage repairs, mitigation, and adaptation, at national, state, and local levels, has weakened the nation's preparedness for handling climate challenges. Climate change is also recognized as a "direct threat to the national security of the United States", while also impacting the "social stability of countries and communities around the world", acting as a "threat multiplier".

Summing up the needs for intervention, Markey and Ocasio-Cortes proposes a "a new national, social, industrial, and economic mobilization on a scale not seen since World War II and the New Deal era". The Green New Deal is to include "millions of good, high-wage jobs in the United

States… to provide unprecedented levels of prosperity and economic security for all people….. and to … counteract systemic injustices". A 10-year "national mobilization" is envisioned for implementation, and a quite detailed listing of intervention areas is offered, including climate and pollution measures, electrification, transportation, green industrialization, health and education, public works and vulnerable community development, participative democratic processes and international leadership.

An important learning in the present text has been that climate, energy, and environmental challenges represent "sticky" areas of the economy, presenting us with "wicked problems", and that concerted action will be needed to effectively deal with the threats. Also, that a lack of social sustainability tends to cause disintegration and unrest in societies, and thereby undermine the potential for concerted action: In order to deal with climate first, we have to secure social cohesion, "glue", and commitment first – first!

Not all societies have the same social issues as the US, but everywhere there is tension linked to mitigation and environmental investments, including energy pricing and inflation problems, costs of living and level of income – all influenced by environmental policies. And while enthusiasts have reminded us that the green transition and circular economics will bring new jobs and added incomes, less fortunate people and pessimists have reminded us that those jobs will be for the select few, and that AI and robotization, decoupling policies, and austerity measures will tend to eliminate jobs.

Paying for environmental defense

There are no definite numbers for the costs of a GND scale intervention. It would be about trillions of US dollars – some of it recovered by ending fossil fuels subsidies and tax benefits, some of it paid for by better health, less conflict and violence, better jobs, and a safer and sounder society. As said, some are scared of making the payment commitments – others are more scared of not doing it. Much depends on the scope, timing, and prioritizing of intervention areas.

For a better grasp of the economics involved, we may again learn from the war-economy metaphor, more precisely from a concept of material resource accounting instead of fiscal and cost accounting, about project costs versus taxation, public revenues, and public debt. Nersisyan and Wray (2021) refer to John Maynard Keynes book "How to Pay for the War: A Radical Plan for the Chancellor of the Exchequer" (Keynes, 1939, 1940). Keynes suggested that instead of looking at financial figures, we should look behind for the "real resources accounting" and for taking stock of the real resources of manpower, plant and equipment, raw materials and productive capacity versus needed resources to fight the war. If the needs of the war exceed what would be available, inflation would be the result. Increased taxes and higher interest rates might then be used to counter inflation.

As related to the green transition, if the resources demanded are in balance with existing capacities, there is less need to worry about expenses. Instead, we should be concerned with the effectiveness of specific strategies and projects. Instead of looking for better financial estimates, Nersisyan and Wray seek out estimates of the real manpower and material needs for implementation of GND-type policies, including the transfer of manpower and resources to new uses, the utilization of presently fallow resources, new employment through job creation programs and the like. They also examine financial tools for mitigating inflationary impacts – just in case. To this must be added, however, that all kinds of mitigative and adaptive instruments, carbon taxes and green energy subsidies, austerity programs and recycling initiatives will have distribution effects that will need to be compensated. Even for proper "real resources accounting", manpower and capital used for mitigation will not be available for other consumption.

What implementation then hinges on, is the availability of human and physical resources – not finances, and what the success depends on, is the effectiveness of projects in reaching GND targets and similar priorities in countries around the world. Calculations of real resource needs for various elements of such steps for the green transition have to be produced, and final answers are hard to arrive at. But even so, "real resources accounting" should be recognized for project planning and also as an educational instrument.

From multi-sector to megaprojects

Green New Deal (GND) type reforms for the environment and for social reform build on multi-sector, broad approaches. Also included are prospects for relatively small actions from large numbers of people, such as reducing meat intake, use of heat pumps in homes, reducing textiles shopping, switching to electric vehicles, using public transportation, to be enacted by millions of people. At the opposite end of the scale, there are searches for special, huge projects, with a potential for major leaps in dealing with environmental problems and threats – and sometimes for causing further destruction.

For the multi-sector approaches we may ask whether they are truly transformative, eliciting engagement from multitudes of people. For megaprojects we may ask whether they are truly pivotal for the "environmental war economy effort". Tonn and Stiefel (2014) listed six levels of intervention to avoid climate catastrophe, ranging from doing nothing to the complete re-organization of the economy. In the middle they mentioned Manhattan scale megaprojects.

Historically, there have been a number of Manhattan-type projects, and presently a number of ideas and proposals for megaprojects are on the table. The historical Manhattan project (to develop the nuclear bomb) was never part of the "ordinary" war-economy repertoire. Still, it came to be a major element of the overall war effort, producing the bomb before Hitler, and forming a historical tipping point for a cold-war future and the deterrence form of terrorism – which is still with us. Another truly mega-size megaproject, of course, was initiated on May 25, 1961, with President Kennedy's speech to Congress: "I believe that this nation should commit itself to achieving the goal, before this decade is out, of landing a man on the moon and returning him safely to the Earth." Around the world, there are few projects that in terms of investments, size and impact, will match the main American ones, but adjusting for size and timing, many nations have prestige projects beyond annual standard budgets, sometimes in support of the green transition.

Mega-style for the environment

A variety of ideas for real breakthrough environmental technology projects have emerged since the first climate-reports were presented. Hopes and ideas for new forms of clean energy have regularly appeared in the news and in technical publications, including ultra-thin film photovoltaic cells plastering buildings, a long list of new principles for wind turbines, wave and tide electricity generators, clean, safe and profitable nuclear energy and recycling systems, also better automobile batteries, and more efficient electricity grids. New principles for Earth conservation, carbon storage, and plastic handling have appeared. To the list of potential megaprojects, geoengineering of gigantic sunscreens has been included, seen by some as a last resort against climate collapse. Some of the ideas have been implemented, and some have turned out to be unrealistic or uneconomic.

The main characteristics of megaprojects are complexity and size – with megaprojects generally classified as costing more than one billion US dollars (Ahmed et al., 2022). Added to complexity and size are significant risks and uncertainties related to technology development, environmental concerns, and economic unpredictability, with the potential for cost and time overruns. Accordingly, megaprojects are mainly reserved for rich developed nations.

Examples

Recently, a number of megaprojects related to climate, energy, and preservation have been suggested. Some are still at the idea stage, and others are ahead in the pipeline of exploration, analysis, and planning.

Among ideas for energy, "gold hydrogen" is enticing, with prospects for a decarbonized energy future based on naturally occurring hydrogen. Hydrogen burns without producing carbon dioxide. But industrial production of hydrogen is both energy demanding and polluting, with a need for expensive and not highly effective carbon capture and storage (CCS) technology.

Referred to as "gold" or "white" hydrogen, geological hydrogen (H2) exists with other gases (methane, helium, hydrogen sulfide and/or nitrogen) and is available for exploitation directly from the ground. The challenge today is to assess availability and exploitability, demanding megaprojects for further exploration and verification. At this stage there is limited evidence of significant reserves, and further exploration comes with significant commercial risks and uncertainties (Dinneen, 2024; Stalker et al., 2022).

For an even more "airy" green energy idea, proposals for launching space solar power stations (SSPS) have been around since the late 1960s. So far, no such power system has been launched, and researchers are still analyzing control systems and technological requirements for elliptical orbital navigation and electricity production, and for navigating the energy transfer back to Earth (Hu & Deng, 2023; Yermoldina et al., 2019). On Earth, solar power is burdened with the darkness of night and the greyness of clouds, combined with wind energy potentially adding to risks of "*Dunkelflaute*". With SSPS in place, neither nightfall nor cloudy weather will interfere with power production. A voltaic panel, 10 km wide, would supposedly produce 570 terawatt-years of energy, enough to supply 10 billion people at a level six times present US per capita consumption (Cartwright et al., 2024). Beyond the costs of launching the solar panels (considerably lowered with reusable rockets), concerns with the technicalities of bringing the power back to Earth have been a major hurdle. Using microwave technology for transfer to "rectennaes" on Earth is now deemed possible after successful experiments by the Space Solar Power project in Pasadena, California. Another concern has been that the construction and launching of panels would be so energy consuming as to cause energy and materials "cannibalism" – consuming more energy in the process than what can be gained from later production. Again, research indicates that over 60 years of operation, there would be a net energy payback with huge positive margins. Also, how to receive the enormous amounts of power on the ground has been a concern, with a need for large "collecting areas", each serving an SSPS in space. The rectennae areas will require considerable space, perhaps best done offshore – as an alternative to offshore wind power turbines. To go on with experimentation, analyses and pilot testing will require multi-megaproject commitment. But besides the Space Solar Power project in Pasadena, Japanese and Chines projects are being launched, and the European Space Agency is actively testing out the potential through the Solaris programme (Cartwright et al., 2024).

Other climate-related megaproject ideas range from the establishment of energy islands with green hydrogen production; to the speeding up of direct air capture (DAC) installations for sequestration of carbon already in the air, and carbon capture and storage (CCS) facilities at production sites; to measures to halt the melting of the West-Antarctic glacier to halt a sea level rise of some five meters.

Fewer ideas and plans for megaprojects seem to be on the agenda for nature conservation and control with environmental toxins and plastics. Conservation purposes are normally better served by smaller and local initiatives, and by regulations limiting the negative side effects of megaprojects such as large highways and infrastructure installments. Worth a grand international public megaproject for nature would be the implementation of standards for nature conservation, preservation of pristine forests and ecosystems, landscape and ocean controls, and fresh water dispute settlements. For environmental toxins and pollution, another international public megaproject is needed for waste handling, for chemicals toxicity assessment and health impact control, and for plastic and micro- and nano-particles handling.

Megaprojects have the potential for mobilizing resources for great achievement. Sankaran et al. (2020), realizing that time was running out for nations to achieve the 17 UN sustainable development goals (SDGs) by 2030, also noting severe shortfalls of initiative from governments, NGOs and business have suggested the use of megaprojects to motivate increased and improved cross-sectional performance, mobilizing politicians, the technology and science community, the business community, and the general public for greater efforts.

On the other hand, there are also megaprojects causing great damage - from hydro-electricity dams, roads, ports, urban developments, pipelines, petrochemical plants, mines, and vast industrial agriculture and plantations. There is also critique of megaprojects representing neo-colonialist development and globalization that threaten indigenous communities and landscapes. Gellert and Lynch (2003) refer to Schumpeter's concept of "creative destruction" for a conception of megaprojects that have caused transformation and destruction of landscapes and communities. Of course, there can be no general estimation of goodness or badness of megaprojects. Each projects has to be evaluated on its own, often including climate, nature, and wildlife elements, and great uncertainties tied to both.

Like war-economic strategies, megaprojects can be used for defense and solidary action. But they can also be put to subversive uses. In the case of the green transition and the digital metamorphosis, they may serve to undermine the public interest in security and well-being. The following is a presentation of mega-projects with a non-green signature, including mega-thinking of the super-rich, potentially using war-economic command-economic instruments for dystopian purposes.

Megaprojects for dystopia

In Chapter 8, the concepts of "sense of urgency" and "sense of crises" were introduced. A sense of urgency or a sense of crisis were seen as necessary for deep change and concerted action. The transition from a stage of latency and prognoses to one of manifest change and damage was seen as a main trigger of urgency.

On the other hand, forces of denial and "merchants of doubt" were recognized as significant obstacles to action, as well as psychological and practical barriers to action. In order for war-economic policies and projects to be implemented to save the planet, the magnitude of environmental challenges has to be recognized, and the authorities behind mega-budgets and mega-projects need to act in the interest of the environment and the global human community. After the German Blitzkrieg on London and the UK, and after Pearl Harbor in the case of the US, both nations were unified for a war-economic response.

Unfortunately, there is today no corresponding unified political climate for the environment and for global social sustainability. First of all, wide and increasing economic inequality is stirring tensions and adding to a great number of dividing lines, causing anger, *ressentiment*, and disintegration, and making concerted action difficult for whatever cause. There are also strong indications that the wealthiest people will seek to create safe havens for themselves, leaving the rest of us to pay for their sins. Secondly, we shall see that present-day megaprojects are being launched to undermine a sense of urgency and to let us dream our lives away from involvement and action. Instead of war-economy level mitigative action, powerful agents seek escape.

Physical escape

In his book, "Survival of the Richest: Escape fantasies of the tech billionaires", digitalization expert Douglas Rushkoff (2022) tells about how he was called up and invited to a private meeting with five tech billionaires. At first, he thought of turning down the invitation – asking himself why he should help greedy people turn the internet one more time to the advantage of the one percent wealthiest. But curiosity and the promise of a lavish fee made him change his mind.

The meeting took place at a private resort in a desert somewhere in the USA. The five participants all represented what Rushkoff refers to as a special "mindset" - originating in a Silicon Valley tech-culture, where some people believe they are entitled to break the laws of physics and economics, as well as moral codes, in order to avoid the disasters created in part by themselves. At the forefront of the mindset are dreams about endless economic gains and bottomless crypto-coffins, of a virtual, lucrative "metaverse" orchestrated by themselves, or of colonization of Mars in search of endless richness from the universe.

Also included among war-economic size megaprojects is the physical escape from angry crowds by creation of private, artificial colony islands somewhere in the Pacific. To have a vision of escape at a time of earthly apocalypse is a privilege of the superrich, not on the agenda for common people thinking about the environmental burden and earthly collapse.

The participants at the meeting in the desert resort were concerned about "the event", seen as an inevitable societal collapse in a not-too-distant future. And they wanted inputs from Rushkoff on how to best protect themselves and their loved ones from the consequences - money was not an issue. For the meeting, Rushkoff had prepared a presentation about the consequences of the tech elite's decisions for the rest of society. But instead of delivering the speech (no interest was expressed), he was asked direct questions, first about some technicalities, and next about their real worries: What would be the safest area on Earth to build a survival bunker - New Zealand, New Mexico or Alaska? What would be the biggest threat of "the event" - global warming, nuclear war or biological warfare? How long could you expect to survive in a luxury bunker without outside help? Should the bunker have its own air supply? What was the probability of groundwater contamination?

One of the five stated that he already had an underground bunker, but he wondered how to ensure that the security forces on site would be loyal after "the event". The event could be one of environmental collapse, social upheaval, nuclear war, solar storm, an engineered pandemic, or a data infection that would bring all computing to a standstill. Threats to the bunker could be angry crowds or burglars, and it could be about countless different challenges to both the owner and security guards. But how could he be sure of the guards' loyalty?

Another admitted that he had already contracted with a dozen "Navy Seals" - an elite group within the US Navy's "Special Warfare Development Group", a group which, among others, was responsible for taking the life of Osama bin Laden. The agreement was that the commandos were to appear at the bunker site on a signal from the owner, which could mean overriding a presidential order! But how was he to pay the guards at a time when cryptocurrencies could be worthless? And what would prevent the guards from following their own military leader rather than the billionaire? Maybe he would have to install locks that only the owner knew the code to? Maybe the guards would have to wear collar locks? Perhaps he would rather invest in artificial intelligence and super-smart bodyguard robots? But could these be produced in time to replace the human guards?

Rushkoff likens their concern to that of building a car that would go fast enough to escape its own exhaust. He suggested means for securing the loyalty of the guards, treating them like friends rather than investing in ammo, collar locks, and electric fences. They were not convinced - they saw his advice as a philosophy for a hippie society.

Rushkoff ended up feeling sorry for his clients. He came to see them as the great losers of their time, as victims of a system of perverse rules and norms, where the winners are those who have enough money to insulate themselves from the countless victims of "the event". They seemed to be locked into a mindset that drives them to prepare and seek some kind of end game, where some are left as winners while the rest are losers, victims of climate disasters, refugee flows, wars or other destruction. The disaster event is something they expect in their own lifetime. They hoped to be saved by technology and innovation on a level that presupposes new laws of nature and undiscovered economic and political playing grounds. Billion-dollar megaprojects for bunkers, and evacuation from the rest of human society will come in handy. They reckon that large masses of people cannot be saved, and they bet their fortunes on how to isolate themselves and their families "from the coming wrath" (Matthew, 3.7).

Multiplying escape plans

As mentioned in the Introduction chapter, the building of bunkers and escape routes has seen explosive growth since Rushkoff wrote about "survival of the richest" (2022), and in particular since news were out about Mark Zuckerberg's 5,000 square foot bunker on the Island of Kauai. The Covid pandemic experience was also a significant trigger for wealthy people to think about isolating themselves from the dangers of the world. Since first observing "How tech's richest plan to save themselves after the apocalypse" (Rushkoff, 2018), news about protection against a possible "catastrophic event" by building "doomsday bunkers" have become increasingly frequent. In interviews with Ingrid Schmidt (2024), specialized bunkers entrepreneurs as well as general contractors refer to armed conflicts and geopolitical rivalry, mass shootings and political disintegration, civil unrest and cyberattacks, nuclear bombing threats and power grid failures, as well as climate change as reasons to build world war style survivalist installations. Facilities include underground survival shelters, but also (vintage) cars and airplanes that can operate independently of power grids and computer interface. Power grids are seen as crucial to vulnerability, but for long-time survival in a bunker, advanced air filtration systems are also in high demand. Besides underground installations, there are also "island fortresses" – on shore and on land properties. For defense of the bunkers, great creativity is being displayed to withstand attacks from the air and from the ground, including natural fortifications, concrete and steel shelters, and in one case a deep lake around the compounds, with a surface of flammable liquid (for creation of a ring of fire), with access to the "island" limited to a (medieval style) swing bridge, and with water cannons for shooting down helicopters or land-based intruders. Also included are tunnel systems and advanced door and lock installations for secret access and escape. On super-yachts there are helicopter decks and submarine terminals. In up-graded bunkers and facilities there are state-of-the art decontamination booths, operating tables and medical supplies for extended survival. Many of the bunkers are equipped for playful entertainment of residents and select visitors, so not to get depressed with boredom while waiting for the "event".

One may ask whether the building of bunkers is just a fad or a wild idea of people with enough money to spend, or whether they are expressions of sincere, rational concern for the life of families. At a micro level, there is little doubt that bunkers may offer at least short-term protection against most kinds of "events" – whether natural disasters or armed conflicts. On the other hand, local people (and soon everyone on the Internet) will know about the bunkers, about locations and much of what is considered secret. In case of grand societal collapse, there will be no general "law and order" premises, and even the best defended bunkers will in the long run be vulnerable to violence and war or whatever form of attack, with little bargaining power left in the hands of the owners. Owners may not at such a time be among the most popular of neighbors, and even when surviving in a bunker for a long time, the occupants may have a hard time adjusting to life outside of the bunker upon release.

The virtual escape

In 1996 there was not a single tech company among the top five companies traded in the US stock exchanges. In 2021 there were not a single non-tech company among the top five on the same lists (Chafkin, 2021, p. xiii). The most profitable companies and the wealthiest owners are now all hi-tech, engaged in some form of computing or media business, and they more or less control the US government while exerting increasing pressure on governments around the world.

Venture capitalists, owners and developers understand that the worse the situation gets, and the more ordinary people see that the richest are getting away with it, while ordinary people live with floods and heat waves, with hurricanes and rows of refugees, and with food shortages and war, the greater the chance of rebellion. Therefore, it is important to lull and deceive people into a virtual world where they forget the problems of the real world and their own capacity for agency. In "The End of Reality", Jonathan Taplin (2023) describes how four tech-billionaires are engaged in trillion dollar mega-projects for escape and escapism rather than for assuming responsibility to solve the real problems threatening our planet.

First on the list, for a "mental escape", Mark Zuckerberg is the chief architect and visionary for a future Metaverse. "Facebook" has been renamed Meta, and the Metaverse (Web3) is to be the preferred virtual world of the future, where people for several hours a day seek virtual escape from a dull, depressing, and demanding real world, and from the calls to concerted environmental action. Putting on a virtual reality (VR) headset, or an augmented reality (AR) headset, people enter a world of games and dreams, with personalized avatars performing whatever acts that add arousal – be it truth or lies, violence or sex, sensible instruction or deceit, paid for by advertising and crypto tokens. VR headsets are for sale, costing a few hundred dollars for last year's models to $ 3,500 for the Apple Vision Pro mixed reality version.

According to game developer and tech billionaire Gabe Newell, we are closer to "the Matrix" (a film about a dystopian future) than most people realize (in an interview with Gault, 2021). The virtual world of the Metaverse might be used to make viewers more conscious of the environmental burden, more prepared and motivated for effort. But such content may not be the best for generating incomes to tech billionaires.

The economic escape

The second escape covered by Taplin (2023) is the escape from poverty by means of crypto currency. Like all pyramid investment schemes, crypto currencies like Bitcoin may create huge premiums for early investors and insiders, as rumors spread and more people want to get in and capture their share of the bonus. Losers are generally the late entrants – young people and poor people who hear about enormous profits and hope to leave their world of poverty and destitute, who end up being the ones paying for the winners.

The promises behind Blockchain technology and crypto currencies went far beyond the promises of former schemes, like the Ponzi schemes, pyramid schemes and economic bubbles. Crypto currencies were supposed to put an end to corruption, democratize national governance, empower people, help migrant workers sent remittances to families, avoid the fees charged by banks for wiring cash, register property and land ownership, offering financial services to lay people – combining profit-making and public service. Started mysteriously by a person or team, using the pseudonym Satoshi Nakamoto, and seconded by tech billionaires and libertarian ideologists Peter Thiel, Elon Musk, and Marc Andreessen, promises of instant prosperity were backed by the spread of rumors that central banks were about to go bankrupt and that now was the end of time of fiat money.

The "whales" in the system are the largest crypto holders (or "bitcoin cartels"), with enough holdings to move the market on their own, buying tokens when the price is low and selling when it is high, using media control to lure the market (Chernoff and Jagtiani, 2024). Five or six "whales" may be able to stir the waves of the crypto markets, with insiders profiting whether prices go up or down. FOMOs ("fear of missing out") investors and HFSPs ("have fun staying poor") are nick names for the latecomers. Taplin notes that Peter Thiel in October 2021 praised Bitcoin as a hedge against inflation and against the collapse of the dollar. Shortly after, Bitcoin fell 30 percent against the dollar, and overall crypto currencies fell 70 percent. Overall, crypto currency has made some people incredibly rich, while others have lost all their savings. No real wealth has been produced, but lots of wealth has been shifted into the pockets of mainly the insiders. The "wealth pump" has worked remarkably well.

The Martian escape

A third escape covered by Taplin (2023) is the escape to Marchi to be orchestrated by Tesla founder, X.com. owner, and Head of the US Department of Government Efficiency (DOGE), Elon Musk. Musk is known for his venturesome visions and business plans and for his ruthless leadership style. In 2001 he joined the Mars Society, a nonprofit organization concerned with human exploration of the planet Mars. His initial ambition was to put a greenhouse growth-chamber for growing plants on Mars, but after great success with commercial space travel, a vision for the human colonization of Mars has taken hold.

As for the colonization of Mars as an escape strategy, Musk has warned about a "Terminator-like" AI earthly apocalypse as the number one existential threat to humanity, about climate change as a threat next in line, and about declining birth rates on Earth as another long-term threat. His main argument is that to stave off planetary threats, humanity should become a "multiplanetary species", installing institutions for fertility and multiplication of colonists on Mars.

For scientific historical-geological exploration of Mars, modern rover technology (using small, mobile robots to take pictures and temperature readings, and collect rock and soil samples) would generate much more information at a fraction of the price of human space travel. Taplin (2023) reminds us that to land a rover on Mars would add one metric ton to the cargo, while landing a minimal crew of human researchers would add more than one hundred metric tons. And unlike the AI-supported rovers, the crew would definitely ask for a round-trip ticket. The idea of Mars colonization is simply a stupid idea. At Mars, there is no water and no soil (only rock dust). Average night temperatures fall to –140°C and average daytime temperature is –20°C. Space travels to Mars are only possible once every 26 months. Of (unmanned) missions so far, more than fifty percent have failed to reach their destination, and 65 percent of those reaching the destination have crash-landed.

For overall survival and livability of a Martian settlement, the Biosphere 2 experiment in the Arizona desert the early 1990s showed that the most likely result will be total disaster (Goldsmith and Rees, 2022). As Taplin sums up (2023, p. 222): "The only lesson to be learned from Biosphere 2 is that creating an artificial atmosphere inside a sealed chamber is really hard on Earth. On Mars it is probably impossible". As for long time livability and reproduction at a level qualifying for (Toby Ord) "human long-term potential" or (Elon Musk) a "multiplanetary species", Jonathan Taplin refers to science fiction writer Charlie Stross, suggesting that a population of a hundred million people would be needed. With no local food production or resource availability, that would be a hairy target, even for Musk.

Only one argument in favor of a Martian colonizing project remains, that of earthly profits to earthly greedy dreamers. For the environment, putting trillions of tax dollars (or euros) into an insane space project would be one of the worst possible additions to the environmental burden. The US-flag on Mars goal has very little to do with science, and much to do with the (Turchin) "wealth pump" – pumping tax payer money from common people into the pockets of the tech-billionaire contractors.

The transhumanist escape

The transhumanist escape is about the escape from limits related to the facts of being human. Taplin (2023) suggests that in a sense all the four mega-projects mentioned (physical barriers, metaverse, cryptocurrencies, and Mars colonization) in a misguided way strive for a fifth mega-project of transhumanism. It is about dreams for a world of longevity, happiness, god-like intelligence, and the bending of the laws of nature.

In the article, "A History of Transhumanist Thought", Oxford philosophy professor and transhumanism-researcher Nick Bostrom (2005) points to cultural and philosophical, religious and technological antecedents to modern transhumanism. Ancient religious writings and ceremonials point to the desire for longevity, afterlife and to an expanded understanding of the world, for the purpose of enhanced well-being. Intensified searches have been on the agenda since the dawn of Enlightenment, while the term "transhumanism" was coined by Aldous Julian Huxley in 1927, a prominent biologist, founder of the World Wildlife Fund, and the first director general of UNESCO. He was also the brother of Aldous Huxley, famous author of "Brave New World" – perhaps the most important work related to subsequent transhumanist discourse. Aldous Huxley described a dystopia where psychological conditioning, drugs, biotechnology, sexual promiscuity, and manipulation is used to breed a conformist, caste-type population.

Transhumanist ideas includes the position of Australian philosopher Peter Singer, author of "Animal Liberation: A New Ethic for Our Treatment of Animals" (1975). Singer applies utilitarian ethics to the animal kingdom, relativizing the lives of people to the value of animals by ranking according to intelligence, and in due course launching the animal rights movement. Among tech-billionaire transhumanists, Peter Thiel, is a main supporter of longevity and superintelligence research, also ready for a considerable investment in a cryotic afterlife.

For the enhancement of the human race and the extension of life and capacity, several branches of research are being pursued, including parabiosis (extending life by getting regular blood transfusion from young people), gene-editing (as with the CRISPR-Cas9 tool), cryonics (freezing a deceased person in liquid nitrogen for a wakeup to continued life when technology is ripe), superintelligence computation (see Chapter 7), computer "uploading" (cyborg-style transfer of a human mind to a computer), and molecular manufacturing. To these we may add a long list of innovations and developments that have been seen by transhumanists as signs of progress (Bostrom, 2005), including Metaverse-style virtual reality computers; genetic engineering and preimplantation of genetic diagnosis; memory-enhancing drugs and pharmaceuticals for concentration, wakefulness, mood, performance-enhancement, and anti-aging; cosmetic surgery; sex change operations; and prosthetics. Much of this is by now on the market or under development. As for molecular manufacturing, Eric Drexler in "Engines of Creation" (1986) argued that non-biological molecular nanotechnology assemblers would become a future reality opening for any kind of separation and combination of molecules at the nano-level, enabling the transformation of coal into diamonds, removing pollution from the air and tumors from healthy tissue, in the long run abolishing disease and aging, reanimating cryonics "patients" and enabling revolutionary space travel – while also producing entirely new weapon systems.

All of this has been heavily criticized, among others by Richard Smalley (2001) and a host of bioethically concerned scientists and practitioners. But the dream of transhumanism still keeps a strong hold on believers. There is also a growing attachment between transhumanism and libertarianism, conservatism and MAGA ideas in America, generally facing considerable internal contradictions and elite rivalry, stirring great confusion while opening for totalitarianism – the ultimate Huxley, warning.

The political escape

Many of the ideas here seen as potentially dystopian have sprung from sources otherwise classified as libertarian, a term of much confusion. The modern libertarian movement has roots back in history. The term "Libertarianism" (from the French *libertair* and the Latin *libertas*, meaning freedom), was first used by William Belsham, an English historian and supporter of the Whig party in 1789. Libertarianism was at birth a movement on the political left – defending liberty, personal freedom and autonomy, civil rights and freedoms of association, speech, and thought. Libertarians also often supported socialist, anarchist and communist movements around the world, while opposing feudal and capitalist state power and suppression, war and violence, nationalism and private ownership of natural resources and means of production.

Philosophical roots were traced to John Locke, the father of classical liberalism, and to Enlightenment movements such as the English "Levellers" and the French "Encyclipedists". For libertarianism in America, Thomas Jefferson and Thomas Paine were important forebears, and for terminology, there was in both Europe and America an overlapping use of the words libertarian, liberalist, and liberal.

There is today great confusion about the meaning of libertarianism. Common to all libertarians is the support of individual rights and freedoms. But for the protection of such rights, there is great controversy. There is also a fundamental distinction between left-libertarianism and right-libertarianism. Traditional left-libertarianism sought reduction of state power in societies of conservative, repressive governments. The front was against royal governments instituted by acts of

God and more directly by wealthy landowners. Modern right-libertarianism also seeks reduction of state power, but for entirely different reasons and mainly against governments generally perceived as democratic and benevolent. Right-wing libertarian purpose of state power reduction has been mainly for deregulation of business, with tax reductions for the wealthy, and for downscaling of social welfare for the many.

The difference between left-leaning and right-leaning libertarianism is an important reason for the present confusion. In the Chilean presidential election of 2022, Gabriel Boric, a self-described "libertarian socialist" won, while the Argentinian election of 2023 was won by openly "right-wing libertarian" Javier Milei. Both libertarians?

Of greatest interest today, in relation to politics in general, to transhumanism, business regulations and democracy, and also to the green transition, AI-regulations, and the broad issue of social sustainability, is the rise of right-wing libertarianism in the US and in parts of Europe since the 1950s. Early libertarianism in the US, like in Europe, was left-leaning and on par with being liberal. Following the New Deal reforms during the 1930s, however, many liberal politicians in the US came to jump ship, rejecting the interventionist policies for economic recovery and the enlarged role of the state in social welfare. As for the present transition to authoritarian tech-billionaire and DOGE (Department of Government Efficiency) demolition of benevolent US governmental and educational institutions, sometimes legitimized by direct reference to libertarian ideology, it would be better classified as an extreme contrast to libertarian roots.

The political escape takes many forms and routes in present societies. Leading the way in high-risk and high-uncertainty directions is Trumpian America. At the individual level, as we have seen there is a growing business of bunkers-building, adding a sense of security to wealthy people scared of the consequences of present policies, and diverting attention away from environmental mitigation and adaptation. At the collective and national level, there is the MAGA-ideological support of American geopolitical hegemony and supremacy, rejecting the environmental burden and the threats of uncontrolled AI developments. In the political-economic sphere in between there is the crypto-currency escape route, technologies for dreamworlds, and transhumanism. At the political level there is escape from the rule of law and from democratic institutions and traditions.

For all areas of development, including environmental concerns and AI and automation initiatives, there are path dependencies and lock-ins in operation. In both business and politics, the emergence of path dependencies is entirely rational. When a decision is made to drive on the right side of the road, endless decisions are rationalized. Supply lines are streamlined, behaviors are made more predictable, infrastructure is adapted, and investments are made more profitable. The same holds for most areas of business and politics.

The down-side of path dependencies and lock-ins is that fundamental changes become more demanding, and for practical purposes sometimes impossible. Among proponents of uncontrolled digital development and environmental neglect, the present regime of due diligence and business regulations becomes a burden and a barrier to creative expansion and experimentation. Unable to break significant societal lock-ins and bring the whole of society to a new stage of development, political escape from laws and institutions is now being explored in "freedom cities" (also linked to "network states", "start-up nations", and to "free trade zones" or "special economic zones" in China).

While the topic of freedom of cities and metropolitan areas is an old one (Brooks, 1915), and while there are recent cases of planned cities in the US such as Irvine in California, Reston in Virginia, and Walt Disney World in Florida, the idea of building ten new high-tech, innovation-focused "freedom cities" in the US was introduced by Donald Trump in March 2023 (Lutter and Allen, 2024). The idea was to stimulate American manufacturing and product development, and encourage more of a frontier spirit and entrepreneurship in a nation of (supposedly) stagnation.

Historically, cities in America have been growing and expanding mainly organically. Around the world there are famous examples of planned cities, most famously St. Petersburg in Russia from 1703, built to create a modern capital, opening Russia to the Western world; Brazília – capital of

Brazil since 1960, promoting modern architecture and administrative efficiency; and Shenzhen in the south of China from 1979, sometimes called China's Silicon Valley as a leading global technology and computer hub. Shenzhen is actually much more than a technology hub, as China's number one special economic zone, a center of large scale manufacturing and a main recipient of foreign direct investment, also a leading international trade and finance center housing China's leading stock exchange and figuring among the top ten city-economies in the world, with the second largest number of skyscrapers, the fifth highest number of billionaires, and the seventh largest density of Fortune 500 companies in the world.

Common to previous planned city establishments has been government involvement and combinations of business and government financing and implementation. In contrast, the present American "freedom city" initiative is predominantly a business venture. First out has been the Starbase City establishment in the desert of Texas (formerly known as Boca Chica), housing Elon Musk's Space X company, where in May of 2025, 97 percent of a population of 283, most of them X employees, voted to build a city. The town's new mayor, 36-year-old Bobby Peden, has worked at SpaceX since 2013 and is the vice-president of test and launch operations of X in Texas. While Starbase City (and also freedom cities to come) are promoted as pioneers of new forms of democracy and freedom, initial signals points toward authoritarian (plutocratic) government by billionaire founders, with minimal restrictions on "anarcho-capitalist" power and greed – reminding more of 19th century "company towns" or "private kingdoms" than of liberal democracies. The powerful messages from Merle Travis 1946 company town song "Sixteen tons" comes to mind: "You load sixteen tons, what do you get?", and even stronger: "I sold my soul to the company store".

While Starbase City is to have a minimal municipal administration, including its own fire department and police force, the main focus will be on spaceship technology and in the future defense contracts and perhaps nuclear power generation - without governmental restrictions. Great potential for growth is envisioned by Musk, mainly based on billion-dollar governmental X-contracted space travel developments. Rumors are out that sites for American "freedom cities" abroad are sought in Greenland, in Honduras, Guantánamo, and on a rebuilt Gaza (without a Palestinian presence and including casinos) – all in the future subject to digital colonialism, and all evading democratic controls and the traditional rule of law. Outside neighbors are in the meanwhile concerned that access to public beaches and environments may be denied.

War-economy mobilization

Pervasive mobilization of people and taxpayers, sacrificing convenience and well-being for a common cause, is not a BAU preference or a normal occurrence. In Chapter 8 the "sense of urgency" or "sense of crisis" hypothesis was presented. Manifest damage caused by man-made climate change, ecosystem and food production breakdown, or dramatic consequences of environmental toxins and pollution may have the momentum, similar to that of Pearl Harbor or the London Blitz, to move the political landscape for acceptance of war-economic measures.

On the other hand, psychological and practical barriers, denials and delays, forces of *ressentiment* and disintegration, reinforced by inequality and indifference, and by escapist maneuvering, may serve to undermine the effort. Still, the challenges remain. There is a lineup of war-economy style candidates for investments and budget adjustments, from small-scale tax and subsidy adjustments to large scale public mega-projects, related to the environmental burden and energy provision, to direct cooling initiatives and carbon capture technology, to AI developmental regulations and political-economic adjustments.

Overall, the war economy metaphor tells us that the command elements needed to guide the economy and society at large for the green transition can be introduced without jeopardizing democracy and civil rights, and without abolishing markets and freedom of consumer choice. But the metaphor also tells us that the content of the commands, and the level of threats to our freedoms will depend on who will be behind the commands, and what their ideologies will be.

Looking at the global map, and being reminded of *ressentiment*, there is little doubt that democracy is under threat from the far right and from tech billionaires with unlimited funds and undemocratic ideas. Even so, corruption researcher Sarah Chayes has expressed hopeful views for democracy, suggesting that while "they have the money – we have the numbers" (Chayes, 2020). We do not believe that a majority of people want to destroy our planet and the democratic societies, but at some point even democracy itself may not work for the majority of people.

The war economy metaphor is about government and political level action – with agency from the top of society. The role of ordinary people relative to war-economy level engagement for a green transition, is mainly about the political choice of support or resistance. Before turning to the issue of hope in a world of increasing uselessness, we shall in the next chapter consider the role of initiative and innovation from below, with agency located in the hands of ordinary people, small businesses and local communities.

Chapter 15

The World Music Metaphor and Engagement from Below

In Chapter 14, we considered what can be done to speed up and make the green transition more efficient by learning from war economic history and from megaproject initiatives. For macro-economic maneuvering and megaprojects, what ordinary people can contribute is mainly political influence for support or criticism.

In the following we shall tap into the world music metaphor to learn about grass-root engagement, from historical experience and from exotic cultures, from the presently irrelevant and from future membership in what I have called a "derivative sector". Where the war-economy metaphor was about high-level expert, governmental and business engagement, the music metaphor is about personal involvement, mitigative innovation, and adaptation. Where the war economy metaphor was about the engagement of science, technology and industrial excellence (and also about the meg counter-forces), the world music metaphor is about learning from basic practices, from local and indigenous communities, through cross-cultural interaction and the engagement of common people, bottom-up.

World music has existed for the full length of human history. According to Tendera and Rubiś (2016), a first step to make it an academic discipline is credited to Robert E. Brown, professor of ethnomusicology at Wesleyan University in Connecticut. He developed undergraduate and graduate coursework and research of musical styles from all around, and he invited performers from Africa and Asia to teach and perform at the university. He also encouraged similar program development at other institutions in the US and abroad, with still wider recruitment of participants.

According to the Wikipedia (https://en.wikipedia.org/wiki/World_music_(term)) based on "Minutes of Meeting Between the Various "World Music" Record Companies and Interested Parties", a less academic presentation of the history of "world music" was presented on 29th of June, 1987, in London. Paul Simon had released "Graceland" in 1986, stirring curiosity and interest in non-Western music. Others, like the British singer and composer Peter Gabriel and the South-African musician Johnny Clegg, had helped bring non-Western records to a wider Western public. At the meeting of the record companies and "interested parties", there was a recognition that while specialist music stores had traded music trans-continentally for many years, there was now an emerging market of interest to the record industry and outlets, encouraging a broad presentation of "world music" for a wider audience.

Classifying and naming

According to the minutes from the meeting of the Record Companies and Interested Parties, there was an even more pragmatic concern behind the naming of "world music". With increasing volumes of music from distant places, London store owners found it difficult to categorize and rack the

DOI: 10.1201/9781003518815-18

new titles. All other music forms had their categories (classic, jazz, rock…), properly placed in some order, and they had their alphabetical rankings of famous composers, bands, soloists, and conductors – easy to file and organize on shelfs. But then there was the rest, a stack of exotic music from the world over, with not so famous names and a mix of categories. Supposedly, someone just decided to call it "world music" without further classification and organizing.

The problems of record stores to come up with classification of "world music" is parallelled by the industry's problems with definition of what exactly is meant by the term "world music". Clearly, to include "all the music of the world", is to make the term meaningless. "Local music from out there", or "folk music from different corners of the world", are part of it, but misses out on the innovative new trends of bringing musical roots and traditions to an increasing list of hybrid subgenres, combining various local music forms, or local and traditional forms with popular, modern music. A lack of precise definition has also caused some experts and commentators to dislike the term and even suggest abandonment.

The term is still with us, however, and there is a growing professional interest in experiments and new forms of production. From a small beginning of recording traditional music, world music has emerged to generate entirely new creations, such as ethnic fusion (Enya) and worldbeat – blending traditional and folk music with pop and rock, and also blending folk music genres and whatever traditions may be thought of. There are also now great "world music" festivals around the world.

In this sense, world music may serve as a generative metaphor and guide for an emerging new political order, for political-economic frameworks for saving the world, and for meaningful living in a future AI-framed, job-stealing world. As we shall see, the metaphor is even useful for tracing the roots of the greatest philosophical, political and scientific turn in history – that of Enlightenment in 18th century Europe.

For the timeframe of a few decades ahead, we may perceive of a futuristic society where most of what we think of as typical activity of markets and production systems is undertaken by AI-guided robots and data-units, where people for most commercial tasks are uncompetitive – "irrelevant" or "useless", and where life support come not from wages but from some form of guaranteed minimum income, allowing for generous, but not luxurious living according to an Aristotelean "natural wealth" standard. Some people will still find occupation in the commercial sector, in jobs where humans still are competitive. But a sizeable "derivative sector" of people freed from commercial activity will have to generate other activity.

For inspiration to the absolutely necessary green transition and for filling the "derivative" (natural wealth) sector with content and meaning, I suggest use of the world music metaphor – representing historical culture and modern innovation, distinct traditions and global exchanges, preservation and innovation, voluntarism and mutualiaty. Thus, the present chapter and the choice of metaphor, also serves as an introduction to the final chapter of the book and to the theme of "commoning".

A bottom-up metaphor

There is as yet no definite definition of world music. But there are a number of characteristics – and more are still emerging. Sometimes it is easier to say what something is not. World music is not for the production of musical pieces as definite, monumental, and inalterable works of art, with inalterable orchestration, scores, and arrangements. It may include musical score, but not rigidly, and not a priority. It is mainly improvised, and open for complex modal harmonics, dissonance and polyrhythms. Its essence is to take seriously the great varieties of musical roots and the blending of folk musical traditions from distant places.

From an initial interest in combining parochial musical traditions with western music, the genre has matured to equally experiment with the blending of whatever local traditions with any other (non-Western) tradition. Thus, at the heart of the genre is an interest in the variety of traditions, and in the outcomes of combining traditions. The latter also includes an interest in the history of music, where even genres generally assumed to be solidly of national or local origin, are found to represent hybrid innovations and trans-culturalism, sometimes combining forms over wide distances. In the following we shall first consider some examples of world music, travelling, combining and innovating, and thereafter adapt the metaphorical use of world music to our wider interest in the role of combining and influencing for innovation and learning for a future "derivative sector" natural wealth economy and for relief of the environmental burden.

Examples

The world music literature is ripe with examples, including the origins and later developments of jazz, Latin, world fusion, worldbeat, and the variety of stimulants, such as Indian and Asian, Arabic, Jewish, Gypsy, African folk and European music. Also included are histories of instrument inventions, refinements, and adaptations to new musical expressions.

In world music history, the merger in jazz of traditional European melody and harmonic patterns with African rhythm and polyrhythm, allowing also for spontaneity and dissonance, is perhaps the most spectacular example of world musical creative achievements. A result of the Atlantic slavery trade, emerging musical forms came to include church spirituals, ragtime piano, guitar and vocal blues, further merging into traditional jazz, merging again with Latin rhythmic and instrumental influence from the Caribbean and South America, and emerging further into gospel, swing, be-bop, and mainstream jazz – further inspiring the coming varieties of rock, modern pop and hip-hop, while also bringing inspiration and influence back to the roots of European symphonic music and to African pop music, as well as energizing American country style music. Quite an accomplishment! Quite a reminder of great results from a humble beginning in slave quarters!

Radio music reporter, Sigbjørn Nedland (2023) has traced a great number of such trans-oceanic and transcontinental "music travel" stories, with innovations and combinations of all aspects of musical instrumentation, singing and playing modes, melody characteristics, rhythms and harmonics. Fascinating examples are from the beginning of human visual artistic expressions, formerly believed to have emerged in Europe in the shape of cave paintings, lately having been traced to Africa 20,000–30,000 years earlier. From Africa visual art migrated north to Europe, with musical instruments shown on some of the paintings, and with 22 bone-flutes found in the Isturitz-cave in the Basque area of France – showing that music and musical instruments were already there.

It has even been found that tone production at certain positions within the caves, marked in red on the walls of the cave, add to the acoustics of the cave as natural amplifiers. Some such caves (including the Nerja-grotto in Malaga, Spain) are today used as concert venues because of superior acoustics. And some of the limestone walls can be tapped by a metal object or a hard rock to produce harp-like tone quality. From India and elsewhere, there are reports about "ringing rocks" producing special timbres and pitches (Spitzer, 2021).

Flutes made from bone are thought to be the first musical instruments. While the beating of rocks and tree trunks have been used for both rhythm and sound, the first man-made drums were supposedly produced in China some 7,500 years ago, and the first harps may have been invented alongside the agricultural revolution, some 10,000 years ago, definitely in common use more than 5,000 years ago. In America there were supposedly no string instruments before the arrival of Europeans, while both the Maya, Aztec and Inka people had quite advanced music and song traditions, with wind instruments and percussion.

Instrumentation

Wind instruments have come in a great variety of horns – first from animal horns, then from silver and bronze, from wood and ceramic, and lately plastic. They also come in endless shapes and sizes, appropriate for a wide variety of scales, pitches, and timbres, and for a variety of occasions, for wartime signaling and religious and political ceremonies, for postal service and herd protection. Only a few hundred years ago was the trumpet first used for purely musical performance (for the greater history, see Wallace and McGrattan, 2011). And only in 1818 was the first valve trumpet produced – by German horn player Heinrich David Stolzel in cooperation with Friedrich Bluhmel, capable of handling the chromatic scale with near perfect intonation.

Of particular interest is the development of string instruments. Nedland (2023) has traced the history from ancient harps and lyres to modern guitars. Harps have a hollowed soundbox at the base, whereas lyres have a flat soundboard with two projecting arms. The harp was supposedly created first, in ancient Mesopotamia (now Iraq) more than 5,000 years ago, and the lyre later, in ancient Greece, some 3,600 years ago. For unknown reasons, in much of the world, and even in modern orchestras, harps have been played by women, and lyres by men. The term "lyrical" is derived from the lyre, as a name for poetry to be read to the accompaniment of that instrument. 3,000 years ago in Israel, the young David played a "kinnor" lyre to please and calm King Saul who most likely suffered from some form of bipolar disorder.

From the base of harps and lyres, both with fixed strings limiting the tonal expression to the number of strings, a next step of instrumental development happened in Mesopotamia, with the addition of a neck, making for the Mesopotamian lute some 5,000 years ago, useful for near unlimited variety of tones. From the present Iraq, the lute spread to Egypt in a matter of 1,500 years, and from Egypt to the Kush kingdom of the Sudan, where it stayed until the dissolution of the kingdom around 400 AD, spreading then from Sudan westwards to West-Africa, where it was observed around 1300 AD as "ngoni" or "akonting" , and then north through the Sahara to Morocco, Algeria, Tunisa, and Libya, transformed and named the "Moroccan ginbri".

From Mesopotamia, the wonders of the lute also spread east, north and west. To the east and north, the flute fostered the Japanese "biwa", the Chinese "pipa", the South-Indian "veena", and the Central Asian "tanbur" and "dombra". To the north it fostered the Turkish "saz". To the west it fostered first the Middle Eastern "oud" and the Greek "pandoura", before rounding off with the European "guitar". And from the short-necked guitar stems innumerable varieties of acoustic, steel, and electric guitars, and all sorts of instrument varieties influenced by lute-family standards. To this we may add innumerable influences, exports and imports of songs, instruments, playing styles, band and orchestra compositions, and genres.

So, what can we learn from world music innovations to use for a world in despair and for a future of natural wealth?

The innovation paradigm

Innovation is at the heart of intellectual capital, and intellectual capital is at the heart of commercial value creation, but also of societal and soft value creation. Innovation is also a historical phenomenon, in musical and cultural development, in farming technologies, building skills, and ultimately wisdom. It is basic to the learning of new things, new ways, and to bringing new life content. A classical definition of the content of innovation was provided by Zaltman, Duncan, and Holbek (1973, p. 10), stating that an innovation is "any idea, practice, or material artifact perceived to be new by the relevant unit of adoption". The core building material of an innovation is anything already known. The innovative leap is to find new combinations and applications of things or knowledge or practices that already exist. That is also the secret of the world music metaphor. The lute was produced by combining the idea of a neck with the existing artefacts of harps and lyres. And the guitar and all other varieties of instruments in the lute family were innovations combining the lute artifact with local materials, traditional handicrafts, and musical traditions, further allowing for innovative musical expressions by combining the adapted artifact lute to local song traditions, and to rhythmic instruments and harmonics, at our time allowing for infinite varieties for guitar-sustained music.

In this sense, and multiplied by the speed of communication and learning, with digital information available in real time, and physical combinations of artefact and practices within reach at the speed of jet travelling, the world music metaphor can stimulate innovation in all aspects of living. The innovative process even for a time of AGI means bringing forth and adding new value to traditional knowledge and practices, while stimulating human life expression and adding to life meaning in the process of generating new ideas, practices, and sustainable artefacts. It is about what people can still do for satisfying living, even when AGI supposedly takes over most of our present jobs and the world community puts brakes on material consumption. And it is not just about the future. Cross-cultural learning and local innovation is going on all the time. And enormous gains are being made for environmental protection and improved living.

The world-music metaphor points our focus in a pragmatic direction, however. A basic premise for the following pages is the pragmatism-philosophical recognition (inspired by William James) that finding variety in the past is a sound ground for anticipating variety in the future, in other words opening for possibilism. The neglect of science, the loss of freedom and equality and the erosion of democracy in modern societies, need not be a straight-jacket. Demands for truth, freedom and democracy are emerging bottom-up all the time.

Exegeses

The world music metaphor is a reminder that for handling the green transition and for dealing with AI challenges, we need knowledge about the environmental burden, about green energy provision, and about regulations for AI development. And we also need knowledge about the contingencies for concerted action and political support. The metaphor is basically about learning and bottom-up (and some periphery-centrum) knowledge creation – about learning to learn.

In the following exegeses, the world music metaphor will be revisited, first to explore the roots of our concept of theory and scientific investigation, secondly to trace the roots of modern democracy and political self-correction, and thirdly to recall inspiration from Non-European sources behind the Enlightenment mantras of Liberty, Equality and Brotherhood. Inspiration for the three areas of high-level insight did not originate among the philosophes in the salons of Paris. As seen in previous chapters, those same mantras are presently at stake for disintegration versus concerted action for a planet under threat.

The birth of science

Linguists and philosophers generally trace the concept of "theory" to the Greek verb *theorein*, meaning to "behold" and "contemplate". A more enticing story links the word theory to religious and sport festivals at the time before Greek antiquity – from which also the origins of the Olympic Games may be traced. At the start, such festivals were small, local insignificant events, inviting neighboring villagers to participate in sport games and religious ceremonials. Over time they grew into national events of substantial standing, with the Olympian and Pythian games at the apex, attracting participation from every Hellenic community. Attendants included a group of "envoys" (delegates, deputies) called *theoroi* (from the singular *theoros* – meaning sent to see God – *Deos*). Their duties at the festivals included both the administration of sport events, sometimes in roles as competitors, and the overseeing of religious rituals and sacrifices. The term *Theoria* came to represent the group of *theoroi* as a sacred legation.

The innovative aspect of the games was brought out by the British historian George Grote, famous for his voluminous "History of Greece". In the third volume (1924. vol., p. 28-29), Grote traced the history of the festivals, and noted how *theoria*, at the start was not about abstract thinking or theorizing, but about the assembly of delegates. The missioni of the delegates, besides various roles at the festival, was to also bring news and ideas back to the village – including innovative ideas stimulated by the *Theoria* experience. The early conceptualization of *Theoria* was built upon by philosophers and proto-scientist of antiquity to generate the vocabulary underlying much of modern science.

Besides sport activities and religious ceremony, bard storytelling (often retelling the stories from the Iliad and the Odyssey) was a main attraction at the games, and Grote notes the challenge to the bard, added to by the diversity of people in the crowd. There was among the bards competition for attention, by use of articulation and artistic exposition, and by demonstrating intimate knowledge of story details to be presented in improvisational verse form. With both farmers, mountaineers, warriors and sailors in the crowd, a superior performance required intimate knowledge of trades and customs.

Adding to the critical evaluation of the performance, and to the learning effect, was the fact of shared language. Grote makes a comparison to the Phoenicians, who were more advanced at sailing, traversing longer distances, being exposed to more varieties of people, but lacking the opportunity for intimate communication across the lines of diversity. Their communications across experiential borders were mainly about purchase and sale and did not foster development in the kind of refinery that helped bards at Greek festivals develop storytelling to the level of The Iliad and The Odessey, works that after near 3,000 years are still ranked at the top of world literature (Grote, 1924, vol. 3, pp. 15-16). Two main qualities stand out from Grote's observation – the role of variety of experience, at individual levels and at the level of crowds, and intimacy of communication, to pick up the finer points and the essential quality markers. Variety of experience arrests "in-house blindness", and intimacy of communication allows for exchange of knowledge.

Grote's understanding of "*theoria*-qualities" have later been confirmed and expanded by other historians (Rausch, 1982). It has also been linked to modern business innovation (Knudsen and Holbek, 2007), and to thinking about ancient Greece as a miniature Europe, in the global-historical treatise of Eric Jones (2003). Jones traces European rise to global economic leadership to what we may refer to as "*theoria* qualities", including variety of experience related to topography, living conditions and tools innovation, and to learning from border-crossing, and from combining local knowledge and practices with traditions and practices of neighboring peoples.

For an even more exotic application, we may also refer to the "sagacity philosophy" of Henry Odera Oruka, a Kenyan philosopher devoted to the preservation of knowledge of indigenous African thinkers (1990). Oruka recognized the important roles of sages in traditional African societies, known to be intimately familiar with tribal and village history, rituals, and expressions of meaning.

Oruka made a distinction between the "cultural sage" and the "philosophical sage", however. The "cultural sage" was loyal to tradition and to inherited belief systems, often achieving great personal influence based on traditional political and spiritual guidance and power structures. "Philosophical sages", on the other hand, were people of curiosity, who might travel outside of tribal borders, to learn about foreign customs, beliefs, and values, later to bring home new perspectives and solutions to practical problems and for the renewal of society – often facing resistance and rejection in the process. The "philosophical sage", in other words, is an African distant relative of the Greek *theoros*.

Rebirth of democracy

Where the *Theoria* ancestry is about knowledge generation, innovation and new standard-setting, democracy is about societal self-correction on top of knowledge creation. Authoritarian governance systems may allow for self-correction, but there is no built-in mechanism for ensuring that authoritarian leaders (historically kings by the grace of God) will not let personal interests interfere with the interest of society. In cybernetic terms, negative feedback is needed for correction. Democracy is the only governance system devised to ensure that feedback from the society works for effective correction. This is not to say that all democracies work equally well, or that only societies with democratic formal constitutions can be true democracies. Informal and small-scale democracies, marked by freedom and equality may serve the democratic interest as well or better than modern regimes (as will be seen below).

Greek democracy in antiquity is often referred to as a root of modern democracy. As far as we know, however, there was little or no such impact on the democratic undercurrents slowly emerging in Enlightenment Europe. The great charter of political liberties, Magna Carta, forced upon King

John by rebellious barons in 1215, is to the English a watershed event in democratic evolution, but it was basically about power-sharing at the top of society.

An interesting account of democracy in Europe and around the world was presented by Robert D. Woodberry. His basic thesis is that while economic and political elites tend to perpetuate class distinction and autocratic rule (conversionary), Protestantism served to promote democracy mainly by way of alphabetizing. With the protestant call for individual salvation by faith followed a logical demand for everybody to read the Bible and the scriptures, then to decide what to believe, and to erect religious organizations outside of state control. With Protestantism followed mass education, mass printing and the fostering of civil society. From alphabetization and the spread of information followed power dispersion and an increased chance for democratic initiatives to survive (Woodberry, 2010, 2012). While Jewish communities in Europe traditionally have enjoyed high levels of education, the social and political consequences have been near exclusively within the community. And while Catholic tradition has called for scholarly excellence, the impact has been mainly within the clergy and church hierarchy. With protestant education, common people were involved in making the Gospel widely available, and the political and social consequences were spread in wide communities, gradually to transform whole societies. The idea of a "universal priesthood", or the priesthood of all believers, independent of Holy Orders, was in Protestantism a key platform, a true mechanism for self-correction and a true inspiration for democracy.

First out among the protestant reformers was John Wycliffe (1328–1384), an English scholastic philosopher, priest, Bible translator, and theology professor at the Oxford University. He insisted on frugality and radical poverty of the clergy and on repentance and ruefulness for the many, and he came to sharply criticize the clergy and priests for corrupt living, for the collection of annates paid to the pope, the practice of indulgences, the preaching of worldly punishment for sin, and the trading of church offices. For years he was involved in ongoing disputes between catholic and worldly powers, and between king and pope. He was loyal to the church, but critical of the clergy, and he increasingly came to voice his opposition in books and tracts. Wycliffe died peacefully, but at the Council of Constance in 1415, 31 years after his death, he was declared a heretic, and his writings were banned. As a heretic, his works were to be burnt and his body to be removed from consecrated ground. The verdict was confirmed by Pope Martin V. The exhumation was carried out in 1428 and his remains burnt.

Beyond the impact in England and Brittain, Wycliffe's writings were made available to the Czech reformer Jan Hus (1369–1415). Hus was a theologian and philosopher and later rector at the Charles University in Prague. He was perhaps more of a church reformer than Wycliffe, preaching the protestant revival message, also confronting some central church doctrines and practices, and starting what was to become the Bohemian denomination. He was subsequently excommunicated from the Catholic church and exiled from Czechia, but invited to attend the Council of Constance of 1415, with a promise of safe-conduct. Upon arrival, he was put in prison, and failing to recant his views, he was burned at the stake for "heresy".

Both in his homeland and abroad, his being burnt at the stake added to the impact of his teachings. The Bohemian denomination became a forceful organization. Perhaps most important was his paving the way for Martin Luther in Germany.

Martin Luther (1483–1546) was a German Catholic theologian and priest, also an extremely productive author and hymnwriter, Bible translator and preacher. Luther is generally recognized among the most prominent of European religious and political reformers. For concrete criticism of the church, he is mainly remembered for his condemnation of indulgences and for his Ninety-five Theses nailed on the door of the Castle Church in Wittenberg – starting the Protestant Reformation on October 31, 1517. He was excommunicated. His teaching focused most of all on salvation by faith and the grace of God, recognizing only the authority of the Bible and honoring the priesthood of all baptized. He translated The New Testament and made it available to the laity, which was to dramatically change German literary and political culture. His hymns (and the subsequent musical accompaniment of Johan Sebastian Bach's compositions) brought the message to masses of people.

As for overall political impact, the personalities and role model effect of the reformers, including the tragic endings in the cases of Wycliffe and Hus, and the imposing figure of Martin Luther, have contributed to the spread of democratic ideas. Also, on par with the protestant manifestation of the right to disagree, the anti-corruption content of the message, and the reading, printing and communicative practices among followers were of significant impact, as underscored by Robert Woodberry. While some of Woodberry's statistics on the international spread of Protestantism and democracy has been criticized (Nikolova and Polansky, 2021), there is no denial that the early protestant democratic awakenings in north-west Europe, also spreading throughout the English-speaking world, but running against more of Catholic cultural and political barriers when spreading east and south, and against the much more impenetrable barriers when entering into Russia and Orthodox regions.

Calls of the wild

Our concern is how the world music metaphor can enlighten modes of awakening, both for practical bottom-up green transition initiatives and for the principles and political orders necessary for agency. The learning-to-learn model of pre-antiquity Greece inspires thinking about variety of experience. The protestant-democracy connection inspires thinking about the roots of democratic self-correction. Reports from the periphery are instructive for learning about freedom and equity from people of humble circumstance.

The notion of learning about such advanced topics as liberty, equity, and non-punitive jurisprudence from indigenous people is by Graeber and Wengrow (2021) seen to be "almost a heresy" among mainstream intellectual historians today. There has been no denial that European explorers, settlers and missionaries spent years learning native languages and developed competence to engage in conversations with locals (as much as – taken for granted – that locals picked up French and English, Spanish or Dutch). But there has been a general tendency to deny that there was anything beyond trivial practical things to learn from the natives, and also a general tendency to "forget" that the letters and literature produced in the periphery and about the periphery was widely read by both common people and well-educated Europeans.

The colonists brought along missionaries and scholars who learnt native languages and engaged in exchanges with the natives. Their reports were proof that "sawages" in French Canada, Inuits in Greenland, and indigenous people across Africa and the Americas were able to eloquently present their critique of European values, institutions and behaviors. Hard to take in for later arrogant European scholars was that the indigenous critique came to form the basis of the Enlightenment critique, with a main focus on the issue of freedom. Graeber and Wengrow do a marvelous job of bringing forth the critical voices from the periphery.

Graeber and Wengrow dive especially into early accounts from Canadian New France (before turning to examples from around the world and from distant epochs in subsequent chapters). Already in 1608, a report from Father Pierre Biard, a professor of theology assigned to evangelize among Mi'kmaq people in Nova Scotia, caused surprise back home. Biard made clear that while feelings of disdain were mutual, the Mi'kmaq tended to complain that the French were "always fighting and quarreling" whereas they themselves lived peaceably. The French were envious and slandering; they were thieves and deceivers; they were covetous and neither generous nor kind – whereas the Mi'kmaq, if they had a morsel of bread would share it with their neighbor. Biard admitted that this could not be denied. The natives would claim to be richer than the French, even as the French had more material possessions, but the Mi'maq had more of ease, comfort and time – the greater good.

In 1628 a "Recollect Friar", Brother Gabriel Sagard, wrote about similar observations among the Wendat nation of Québec. The good friar was perhaps even more critical of the indigenous culture than Father Biard, recognizing in particular the unheard freedom of Wendat women (including pre-marital sex and women's right to divorce). But he also had to admit that the Wendat culture was in many ways superior to that of the French. The Wendats had no lawsuits and they expended little pains to acquire the goods needed for a good life – where the French, with their "excessive and insatiable greed" had to torment themselves for what they needed. Friar Sagard also admitted to being "justly and with reason" reproved by the Wendat's quiet life and tranquil disposition, also by their great generosity to one another. All necessities were provided for on a reciprocal basis, and there were no beggars in the villages. The Wendats were strongly offended when hearing about beggars and lack of charity in France.

Bringing the message closer to the heart of Enlightenment, Sagard was especially impressed with the Wendats' eloquence, and their ability to form coherent, logical arguments, cultivated through near-daily collective settlement of everything communal. The Wendats, on the other hand were equally unimpressed with the French style of argument, with their constant scrambling, interruptions, and weak arguments – indications that they were not especially bright. Much as they were grabbing material goods when they had a chance, the French would grab the stage and deny others a chance to present their case.

Graeber and Wengrow (2021) report that Sagard's "Le Grand voyage du pays des Hurons" became a bestseller in France and across Europe, studied carefully by leading philosophers like Locke and Voltaire. It was followed by the publication of the much more extensive, 71 volumes "Jesuit Relations of New France", covering the 1632–1673 period, produced by a great number of Jesuit missionary authors. Perhaps the greatest impact was achieved bythe 1703 publication "Curious Dialogues with a Savage of Good Sense Who Has Travelled", by Baron de la Hontan. The young La Hontan had served as a deputy to the Governor-General of French Canada, but he came to disagreement with the Governor and was ousted from office and upon return forced into exile from France. After the publication of the "Curious Dialogues" he become a person of great celebrity – in Amsterdam. The book is about a series of dialogues with the chief of the Wendat Confederacy, Kandioronk – an unequaled master of eloquence, logical reasoning, critical observation, and a great supporter of freedom and equity.

Underlying much of the indigenous critique, was generally a native negative reaction to French competitiveness and selfishness. Also, the French hostility to freedom and equality was basic to the critique, including the French fear of their captains. The only real power of native chiefs was "in their tongue's end", in their ability of eloquence, and their ability to generate followership. The Montagnis-Naskapi considered the common people among the French more like slaves, living in constant terror of superiors. The Jesuits of 17th century French-canadian Québec observed that native chiefs gave no orders. Holding a political office did not give anybody a right to command others. Participation was voluntary.

The natives of French Canada practices much of what is 400 years later called "transformative leadership". Peter Senge, author of the systems theory classic "The Fifth Discipline" (1990) presented a vision of traditional Amerindian decision-making. The chief would call upon heads of families to come to his tent. There would be a fire in the middle. A pipe would be lit. The chief would comment on the present situation. Another tribe might be approaching. There might be problems of hunting and food provision. He would state the problem and send tobacco smoke out into the room, then pass the pipe on to the person next. His neighbor might add some more facts and observations, smoking the pipe, and handing it to the next in line. The process would continue until they all had their say. The words swirled in the air much like the smoke from the pipe and the firewood, both words and smoke being inhaled by all. No command was given. No explanation was needed. They all knew the outcome. The decision was already made. It was in the air.

As for the indigenous penal philosophy, the Jesuits were shocked to observe that there was no direct individual punishment for crime. Even murderers got off scot-free. Instead of punishing culprits, Wendat custom called upon families and clans to pay compensation. The whole community would assemble and agree to the number of gifts that was to be bestowed on grieving relatives of the victim. The settlement and the process behind ensured that all parties felt that justice had been done, and due respect was restored.

Interestingly, as observed by Graeber and Wengrow (2021), there was no disagreement between natives and missionaries about the fact of greater freedom in indigenous communities. There was disagreement, however, about the desirability of freedom – natives advocating the blessings, and the Jesuits pointing to sinfulness (not entirely different from present-day criticism of European, American and generally urban liberalism in areas of homosexuality and LGBTQ rights – generally expressed by Russian Orthodox and American MAGA and evangelical denunciation).

A great line of publications premised on outsider views of Europeans was to follow. Montesquieu in 1721 wrote "Persian Letters" about the Paris experience of two Persian noblemen, Usbek and Rica; Marquis d'Argens wrote "Chinese Letters" about a Chinese traveller in Paris; In "Supplément au Voyage de Bougainville" Diderot used a dialogue between a Tahitian man and a chaplain about views on sexuality; Chateaubriand wrote in "Les Natchez" romantically about the Natchez people in a "state of nature"; finally Voltaire wrote about a Huron "child of nature", "L'Ingénu", torn between the demands of a French lifestyle and native impulses, also disclosing corrupt practices in his new homeland.

Learning for societal and green transition

Extending the "world music" perspective to also include social organizations and circular economies allows us to see solutions and sustain hope. Roman Krznaric is a political scientist and historian who in "History for tomorrow: inspiration from the past for the future of humanity" has dug into history and exotic contexts to find solutions matching our present threat areas (2024). For social sustainability and harmonious co-living, he refers to the city of Cordoba and the Moslem kingdom of al-Andalus in Spain. The kingdom lasted from 711 to 1492 and provided a peaceful and prosperous arena for life and business, for both Moslems, Jews, and Christians – quite instructive for our time.

Inspiration for a circular and re-generative economy was found in the Tokugawa Shogunate of Japan, from 1603 to 1868. Exhaustion of natural resources caused the Shogunate to install harsh rules of re-use and repair. A Kimono was to be used until it was worn out. But by that time, it had also become very soft, ideal for re-use as a pajama. After it was worn out as a pajama, it was to be used for baby diapers. Too worn for use as diapers, it could still be used as a rag for washing and cleaning. And when too worn for cleaning, it could finally be used to start a fire for heating.

For the taming of social media, Krznaric refers to the role of British coffee houses following the invention of modern printing. It is generally accepted that the printing press brought a revolution to publishing, discourse, and news diffusion. It is less recognized how important the printing press was to the protestant reformation, mainly to the spread of Luther's thoughts, but also to the protestant movement as such, and to the resulting spread of all kinds of ideological and political messages, leading sometimes to polarization, unrest and wars. Modern social media are of course more potent than 16th century spread of printed material, but history is a reminder of the power of such messages, and also an awakening signal for what to do about it to promote truth and maintain peace. By the late 1600s, printed material became more responsible, less extreme and less prone to sensational reporting of witchcraft and the like, and more involved in the spread of education, science, and democracy-building. And an important instrument in that development was the growth of coffee-houses. While formerly a Turkish institution, the idea rapidly spread to England, where the first of its kind was built in Oxford in 1650. Soon there were more than 2000 coffee-houses in London alone. Mainly seated at long tables, there were rows of people reading newspapers, drinking coffee, smoking pipes and engaging in endless debates and exchange of information and news. A whole new concept of a public sphere and civil society emerged – serving also the roles of factfinding, inventive thinking, and social bonding.

From different corners of the world, Krznaric has collected social and political solutions to the sharing of freshwater and common land and resources (to which we shall return in the next chapter). From the ancient city of Djenné-Djeno in Mali (250 BCE – 1400 CE), he brings in experience with direct democracy and alternative governance forms. When archeologists started digging in the late

1970s, they found remnants of an advanced urban civilization, with ceramics, metal works, and sculptures of high quality, but no traces of hierarchical governance or royalty. Instead of hierarchy they found traces of "heterachy" – of horizontal reporting and self-governance, with no grand palaces or government buildings.

Krznaric reminds us, like Graeber and Wengrow (2021) that history is not linear, moving from less to more progress, also that the geography of modern progress is not one way, with the arrows pointing from Europe to the rest of the world. We may find treasures of hope from ancient times to the present and from the world over to anywhere in the world.

The message of all this is that world music learning across cultural divides also served an extremely important role in the conception of the Enlightenment revolution, an important reminder that for a green transition and the possible transformation of modern societies into something more in harmony with circular economic and democratic ideals. Inspired by William James' "Pragmatism – A New Name for Some Old Ways of Thinking" from 1907, my short version of the "learning from the periphery" experience is simply that the recognition of variety in the past causes us to see more variety and more possibilities in the future. That is essentially what the world music metaphor is about - possibilism.

A turn to goodness

My exegesis of the world music metaphor so far has focused on bottom-up and periphery-centrum learning. Behind such learning, and generally behind mobilization and bottom-up learning and invention, lies a fundamental striving for goodness. Striving for some kind of goodness may be expressed in musical beauty, in the search for knowledge, in promotion of the common priesthood or in freedom and equity for all. In the promotion of a green transition and disciplined development of AI, the idea of goodness needs to be an anchoring point. Love of nature and the planet is at the heart of environmental engagement. Not all innovation is for the good, and not all political systems are targeted for cultivating the good. But at least we shall see that most people have positive intentions behind their behaviors.

We noted in Chapter 9 a pre-historical backdrop anchored in the Book of Genesis, about an initial state of human innocence and a subsequent fall from grace, pointing to a final stage of redemption and restored harmony. I also referred to Machiavelli's "The Prince" for leadership authority by use of cunning and deceit, and to Tommas Hobbes for dictatorial authority to prevent the "war of all against all". It is all about the control of supposedly unruly people. Unfortunately, the pessimist heritage, while unscientifically based, has tended to gain powerful influence.

The pessimist heritage

Few books of philosophy and political science have been studied more intensely than Hobbes "Leviathan". The threat of all against all in the absence of illiberal leadership has been a cause of authoritarianism. Like "Leviathan", Machiavelli's "The Prince" was read by kings and great men of the Renesance, and in more recent times by Napoleon, von Bismarck, Stalin, Churchill, Mussolini, and Hitler. As Bregman says (2020), the book has been perceived by many as "realistic" – assumed to relate "how people really are", and "how leaders must behave in order to rule". A number of modern books on management have revolved around the same theme, such as "Machiavelli for managers" and "Machiavelli for moms". The film "The Godfather" and TV series such as "Game of Thrones" and "House of Cards" have followed suit.

The French psychologist Gustave Le Bon, wrote "Psychologie des foules", or "The Crowd" (1896) to become one of the greatest celebrity psychologists of his time. He postulated that when state control weakens, people will react with panic and violence and fall down many rungs on the ladder of civilization. Man's true nature will be expressed when the Leviathan power apparatus is not in place. His "Psychologie des foules" was read by the most powerful leaders of the last century, including Hitler, Mussolini, and Stalin, but also Churchill and Roosevelt. The book is also perceived even today as a main reference in the psychology of masses.

An important topic of the book was how people react to crises in society. Hitler believed in Le Bon's theory and reasoned that England would collapse as a civilized society, when only the bombing was shocking enough. Churchill thought the same, and feared what would happen when the bombs started to fall on London. He continued to believe in it towards the end of the war, when he aimlessly ordered mass bombing of German cities and densely populated places, culminating in the bombardment of Dresden with incendiary bombs. Putin may not have read the book, but he seems to believe the same when bombing apartment buildings, hospitals, kindergartens, and vulnerable infrastructure in Ukraine.

They were all wrong, however. Bregman (2020) reminds us how the London population kept their composure even throughout the "blitz". In the midst of bombing, postmen continued to make their usual rounds, shops remained open, and neighbors helped each other the best they could, all in a peaceful and orderly manner. It was admiringly argued at the time that the English were a special breed of people, with a stiff upper lip and commendable determination, sustained British humor, unrivaled self-control, and social discipline. Thus, it was even more surprising that the Germans displayed exactly the same capacity for helpfulness and discipline even after several years of war, under a dissocial leader and under a rain of bombs that created ten times as many casualties as the London Blitz. More than half of Germany's cities were razed to the ground by the end of the war. Churchill was still convinced, however, that the bombing would break the German nation's unity and morale. As it turned out, the Germans were, if possible, even more disciplined, solidary, and caring than the British. In one single night in Dresden, the firebombing led to more loss of life - children, women and men - than in the much larger city of London throughout the entire war. There was still no disciplinary breakdown. Neighbors helped each other and made unimaginable sacrifices.

Unfortunately, post-war military leaders were among the last to realize this. The Americans dropped three times as much bombing power over Vietnam as the sum of all the bombs dropped during the Second World War - and had to pull out with great loss of influence and prestige, three decades after the end of the World War. People are not built for war of all against all, not even after a blitz.

As for Dawkins' book on "The selfish gene" (2016 [1976]), outr interpretations of survival selection seems to have turned 180 degrees. It is an irony, noted by Bregman, that two years after the publication of "The selfish gene", it was shown that the selfish gene was not necessarily dominant in the animal world, and that in human development the fittest" might as well be "the friendliest".

At the level of genes, bacteria, fungi, plants, and animals, Silvertown (2024) has shown that cooperation tend to be ubiquitous in nature. There is competition and there is selfishness, but for survival, species have learnt to cooperate, and in order to propagate, they have learnt to be friendly where friendliness makes a difference.

Anthropologist Douglas Fry (2013), historians Graeber and Wengrow (2021), and a long list of archaeologists and anthropologists have found is that the hunters and gatherers who lived more than ten thousand years ago, lived relatively peacefully both locally and in relation to strangers. They were rarely at war with each other, they avoided conflict and violence, and they were good at resolving conflicts by simply talking together. They had extended networks of friends and acquaintances, and great social activity and festivity. They entered into marriages across tribal boundaries, and they had frequent changes of affiliation on the basis of kindness and goodness. Our genes are not developed for evil. It is easy to explain. People who are able to form friendships, establish relationships of trust and treat each other well, will have greater reproductive capacity on the basis of kindness genes, and they will also have better conditions for developing language and complex communication, also for establishing societies with room for innovation and trade, ultimately for creation a better life. Procreative ability is not just about muscle mass on the part of the male part. There is little reason to believe that our grandmothers wanted children with the most vicious and violent of men.

We know that not all people are equally kind and good. We also know that good and kind people can be provoked to do stupid and ugly things, and that in certain situations and in close contact with violent movements and charismatic leaders, people can come to carry out horrible acts. But it does not mean that humans have something fundamentally evil within them. All it says, is that most people who are basically kind and good, in vulnerable situations, and assisted by vile leaders and malicious groups, can be seduced to do nasty things.

A friendly experiment

The Dutch historian and social scientist, Rutger Bregman, in a stimulating book about human nature, "Human Kind – A Hopeful History" (2020), postulates that people are predisposed for kindness, love and social community, evolutionary selected on the basis of "survival of the friendliest" more than "the fittest".

Bregman finds a model for the evolutionary cultivation of virtues in an experiment on dog breeding. Dmitri Belyaev at Moscow State University had an idea for better understanding of evolutionary development in the animal world. Taming by selection is a process that in nature takes thousands of years. Belyaev wanted to see how fast it would work in a controlled experiment. Together with his assistant, Lyudmila Trut, he in 1958 established a large-scale laboratory, with several thousand cages, in the far east of Siberia. For an experimental animal, he chose the silver fox, one of the most asocial, angriest, and wildest animals of the *canis* genus. The aim was to see for how long, and for what genetic traits one had to breed, in order to produce an animal with a dog's characteristics.

Only one genetic trait for selection was chosen, that of kindness or friendliness. The researchers kept minimal contact with the animals and used no behavior modification techniques or rewards. Only when they spotted one that showed the slightest sign of kindness (such as not biting when it had a chance), it was pulled aside for further breeding. The result turned out to be astonishing. Already in 1964, after only four generations, Lyudmila saw for the first time an adult fox wagging its tail. And with each new generation of puppies, it became more difficult to maintain minimal contact. More and more, the foxes began to ask for contact and attention, and they became more and more playful.

In the wild, the foxes are eager to play at the puppy stage, but already after around 8 weeks they become markedly more aggressive. After a few generations of breeding for kindness, the animals in the Belyaev experiment continued their childish play throughout adulthood. Ideally, they would play and waddle all day. They also underwent a change of physical appearance. The ears began to hang down, the tail got curls, the fur got patches of a different color, the nose became shorter, the legs became thinner and there was less difference between males and females. Instead of howling, they began to bark like dogs. They also began to recognize and respond to their own names. Still, only friendliness was a criterion for breeding.

Dmitri Belyaev initially had to keep his basic hypothesis of breeding for kindness a secret. The Soviet authorities were dogmatic opponents of research in the capitalist, Darwinian tradition - only the "new society" could produce the "new man" - and the same had to apply in the animal world. Belyaev's own brother had been executed a few years earlier for working on forbidden theory. Officially, the breeding program was about producing silver foxes with better fur!

The result of the experiment was presented to a large international gathering of geneticists in Moscow in 1978. Meanwhile, the scientific climate in the Soviet Union had become milder, with more openness for genetic research. Belyaev's presentation at the conference became a sensation. There was a video of an adult silver fox wagging its tail, never before seen anywhere in the world! Data about hormone changes were also presented - with less of stress hormones and more of serotonin (joy) and oxytocin (love). The conclusion was that this type of change hardly applied only to silver foxes!

The philosophical footing

Aristotle was challenged to define an ideal regime, but he abstained from doing so. Instead, he argued in *Nicomachea* that ideal forms would vary among societies. Common to good regimes was that they enticed good behavior among people. The specifics of the law might vary. What was common to good regimes, was not a universal set of laws, but the character of rulers and lawmakers. Where the population perceived the lawmakers to be righteous, intending for justice and equity to prevail, people would tend abide by the laws (also fearing punishment in *Hades* if they did not comply). Aristotle was more optimistic and realistic than Hobbes, and better at assigning responsibility for the morals of society than Machiavelli!

Since Aristotle, there has been a recurring discourse about ideal regimes and ideal laws – mainly focused on our understanding of natural law. Aristotle made a distinction between natural law, anchored in the physical contingencies of life – *to physicón*, and conventional law, anchored in practical considerations – *to nómicon*. Conventional law applied to areas where one might choose one or the other, such as for a religious ceremony to sacrifice two goats or one sheep (or in modern society to have right hand side driving or driving on the left). Natural law, on the other hand, applied to areas where basic human interests were at stake, including issues of life and death, fairness and justice. Still, even the formulation of natural law was for Aristotle in the hands of humans, to be settled by political processes (Yack, 1990). For conventional law, simple agreement would be sufficient for the law to work well. For natural law, however, agreement was not enough. The quality of the outcome depended on the character and morals of the lawmaker and leaders as perceived by the citizens, providing for trust in societies.

Natural law arguments were present in early political-economics discourse, most visibly in the writings of Adam Smith, and most strikingly in the finding that thanks to markets, people might serve purposes of the common good, even when acting for their own well-being. Adam Smith did not thereby conclude that people are generally selfish – only that acts performed for their own good, and given favorable institutional arrangements, also may serve the common good. Ahead of Adam Smith and the classicists, there were also a generation of influential moral philosophers sorting out natural law issues – including Hugo Grotius, Richard Cumberland, Alberico Gentili, and Samuel Pufendorf. Among them, Samuel Pufendorf may have been the most important. In a doctoral dissertation about Pufendorf, Arild Sæther (2017, pp. 59–60) summarizes the Pufendorf position. First of all: "Every man cherishes his life, his person, and his possessions", but in addition "there can be observed, deep-seated in his soul, a most sensitive self-esteem. If someone undertakes to impair this self-esteem, he is rarely less and often more disturbed than if someone tries to injure his person and property". The source of self-esteem is located by Pufendorf in human nature, belonging equally to all men, and "no one can live a social life with a person by whom he is not rated as at least a fellow man". It therefore follows as a precept of natural law, that "Every man should esteem and treat another man as his equal by nature, or as much a man as he is himself" (Sæther, 2017, p. 60).

Practical impact of beliefs

In the aftermath of hurricane Katrina, in New Orleans, August 2005, one of the worst natural disasters in US history, with almost 2,000 dead, massive news came out about collapses, about dangerous gangs, shootings in the streets, attacks on rescue helicopters, burglaries, rapes and murders – also of small children. Bregman (2020, p. 4) sums up the initial observations: The police chief declared that the city was headed for anarchy. The governor of Louisiana was afraid of the same thing – she declared that such a tragedy brings out the worst in people. A commentator in the English "The Guardian" wrote that as soon as the usual prerequisites for civilized life break down, we are back to Hobbes' "struggle of all against all".

Bregman, however, also notes what really happened according to research reports a few months later: What had sounded like gunshots was in reality a gas tank that had exploded. In the great Superdome, where 25,000 people had sought refuge, there were six dead - four had died of natural causes, one due to an overdose and one due to suicide. It was confirmed that there had been break-ins in several places in the city, but almost always to find food for themselves or to give to others in need. The police chief admitted that there had not been a single case of reported murder or rape. The mentioned "gangs" mostly consisted of people who had banded together to survive, or to assist the police and rescue workers, or to form voluntary rescue groups. Private boats flocked from as far away as Texas to rescue people from the flood and to help recover bodies from the water.

The conclusion was that when the usual prerequisites for civilized life break down, spontaneous reactions occur which are mostly of a "prosocial nature". One of the things that unnecessarily cost lives was the authorities' handling of the situation. Thousands of soldiers from the National Guard were in place. Completely unmotivated, they shot six unarmed black residents, also killing a 17-year-old boy and a 40-year-old disabled man on the Danziger Bridge in East New Orleans. As for consequences after the disaster (Bailey, 2009), the murder rate, the robbery rate, and the larceny rate were lowered. The burglary rate, the rate of aggravated assaults and the motor vehicle theft rate were first lowered and later back to normal, and (for burglary and motor theft) again increasing. Overall, violent crime rates were reduced.

To the stories of disaster and violence , Bregman (2020) add war stories from far and wide of young soldiers refusing to kill their enemy peers (by aiming over or by busying themselves with carrying ammunition or helping with the wounded), about officers who have had to threaten the soldiers to get them to shoot, about pilots who deliberately miss the targets. In the US Airforce, a report from the Second World War showed that less than one percent of pilots was responsible for 40 percent of the shootings. Findings from previous wars, from the American Civil War (The Battle of Gettysburg), the Spanish Civil War and other wars show similar results. Soldiers pretend to be involved, busying themselves with fetching ammunition, loading weapons, helping the wounded, aiming and moving about - to avoid having to kill.

After studying all the documentation and hundreds of photographs, the sociologist Randall Collins (2008) found that only 13–18 percent of the soldiers in a war actually shoot to kill, and that humans by nature are predisposed to solidarity – which also makes the use of violence difficult. Violence is usually not contagious, and the use of violence usually does not last very long.

Psychologists leading astray

Bregman (2020) highlight three psychologists who have disproportionally reinforced our skepticism of human nature. Muzafer Sherif managed the "Summer-Camp Experiment" in Oklahoma, June 1954, and led us to believe that innocent boys in competition during a boy-scout type summer camp can turn into violent antagonists, if not kept in check and restrained by the camp administration. Philip Zimbardo was responsible for the infamous "Stanford Prison Experiment" and made us believe that in a prison-like environment, innocent university students on summer jobs, acting in the roles of "prison guards", will in a matter of two or three days turn into vicious bullies, beating and abusing fellow students in their "prisoner" roles. Finally, Stanley Milgram, responsible for "The Milgram Experiment", made us believe that volunteers from the street outside Yale University in New Haven in a matter of minutes would turn into willing killers if only guided by people of academic research authority in a laboratory-like venue, implying also that participants would come to believe that killings at some scale took place inside the Yale university. In all three cases, however, the experiments were biased and misrepresented in research reports.

As for the "Summer-Camp" experiment, the positive intention behind the experiment was for Sherif, who had observed much undue violence in his home country, Turkey, to understand the sources of violence. The experiment was reported by Sherif in his book from 1967, "Conflict and Co-operation. Their Social Psychology", reprinted several times and cited in 3034 psychology sources (2015). Critical examination has later uncovered an entirely different story from the one told by Sherif. Australian psychologist, Gina Perry, in her book "The Lost Boys: Inside Muzafer Sherif's Robbers Cave Experiment" (2018), referenced by Bregman (2020) uncovered that violent tendencies were not initiated by the boys, but orchestrated by research assistants, acting as camp leaders, manipulating the games for the purpose of eliciting conflict and violence, under strict, direct command of Sherif. Perry also reminds us that Sherif, a year before the Summer Camp Experiment" did an experiment, meant to demonstrate the same claim about violent tendencies, but this first experiment was called off, because the assistants, in Sherif's view failed to achieve the expected results.

In the case of the "Stanford Prison Experiment", first reported by Haney et al. (1973, cited by 2353), the original explanations again have been up for correction. Bregman (2020) says that he first became skeptical when he read in Zimbardo's later book, "The Lucifer Effect" (2011), mentioning that Zimbardo, on the Saturday before the experiment, had a meeting with the prospective "prison guards", instructing them about what he expected in terms of guard behaviors. More details were revealed when the French sociologist, Thibault Le Texier, after hearing Zimbardo on a talk show got the idea for a film, and began to dig into the original material, uncovering the full details behind the experiment. Le Texier found that as many as eleven out of seventeen "rules" for the guards had been thought up by Zimbardo's assistant, Jaffe, in advance and used to instruct them about the entire scheme for bringing out brutality, drilling them for six hours on methods to break down the "inmates", with direct instructions from Zimbardo about sleep deprivation, physical punishment, and isolation. Le Texier also uncovered that a majority of the "guards" resisted the methods. Two-thirds of them refused to participate in the brutality. A third participated, but many tried to treat the "prisoners" in a friendly manner – to Zimbardo's frustration and with subsequent reprimand. One of them backed out the day before start, protesting the brutal instructions. Others followed suit. But many followed orders. They were ordered to be brutal – it was not spontaneous and not an expression of human nature. Even before the result of the experiment had been analyzed, Zimbardo was active feeding news media about the result, with shocking film cuts and comments, adding 40 years of fame to his career. There was no latent cruelty at play, only direct instructions, subverting scientific principles, manipulating young students to act against instincts.

Bregman also informs that in 2001, two psychologists, Alexander Haslam and Sephen Reicher, were commissioned by the BBC to set up a reality show modeled after the original Stanford prison experiment. The series had to end prematurely, however. Nothing happened - nothing at all. There was no enmity, no violence, and no powerlessness. At the end, scenes of "prisoners" and "guards" in good company were shown to be hugging and exchanging uniforms. TV viewers who had hoped for greater drama felt cheated, but the world received a new message about what happens when good people end up in bad circumstances - very little. They sort things out among themselves. There is no battle of all against all. Unfortunately, almost no one has heard of "The BBC Prison Study", while the spirit of "The Stanford Prison Experiment" still keeps haunting us.

The "Milgram Experiment" is perhaps the most famous of the three experiments. In June 1961, Stanley Milgram was a young researcher, Jewish by background, with a great interest in finding out about the world's evils in the aftermath of Holocaust. Unlike Sherif and Zimbardo, who claimed to observe people "as they really are", Milgram's study was set up to study the effect of manipulative authority figures. The standard narrative tells us that Milgram advertised in the local newspaper for 500 people to participate in an experiment on memory training. It would take an hour, and the participants were paid four dollars for the job. On arrival, they were told that they would contribute in pairs, one of them in the role of "teacher" and the other as "student". If the "student" answered incorrectly, the "teacher" would give him an electric shock. The shocks started at 15 volts, adding more volts if the "student" continued to answer incorrectly, upward to a red light "danger zone" of allegedly 450 volts. What the participants did not know was that the "student" was always one of Milgram's regular assistants, and that he was never exposed to any electric shock. Instead, he howled and screamed as the "teacher" approached the danger zone. After 350 volts, the "student" became quiet – presumably dead.

The message that came out from the experiment was that the vast majority of us follow authorities' orders blindly. Such is human nature! The message was reinforced when at the end of the experiment, Milgram decided to film an entire session. The short film turned out to be incredibly convincing - and it is still available on YouTube and used as instructional material both in academia and at professional leadership seminars.

Bregman has noted a number of weaknesses, however. First, around half of the participants did not believe in the set-up and chose to leave. The rest stayed on, and many expressed directly that they hoped to contribute to something good for people. Contrary to the main hypothesis, they did not seem to respond to direct commands to enter the high-volt zone. They did not allow themselves to be manipulated to do evil. As reported by Gina Perry in "Behind the Shock Machine: The Untold Story of the Notorious Milgram Psychology Experiments" (2013) they responded more positively to "please continue", or "it is absolutely essential" – all framed as appeals for contributing to knowledge for the good of people. When, on the contrary, appeals came in the form of direct order, there was a tendency for the session to stop, for participants to disobey, protest, and withdraw (Gibson, 2013; Haslam et al., 2014). Contrary to the message that 70 percent of ordinary people simply follow the orders of authority figures, even with deadly outcomes, the opposite actually happened!

People are not by nature inclined to do evil things, nor do they just follow the commands of superiors or scientists. Most of us do so only when we are convinced about good intentions. Dangerous leaders are the ones who convince people to do nasty things while making them believe they are doing it for the good.

The question in this chapter is whether the engagement of people around the world, through interaction and innovation, have the capacity for adding positively and decisively to a green transition and to a secure technological and AI development. People are not by nature evil or egotistical in a way and at a scale that threatens human society and nature. People may not always be friendly and good. Some are less good-willed and empathic by nature. Some may turn cruel by upbringing and experience. Some may end up in bad company. But the goodness of the average men and women is enough to compensate and on balance contribute to the betterment of society. That is also the message of the world music metaphor.

Not all leaders, however, have the Aristotelian character for bringing out the best in people.

The paradox of power

There is a main factor threatening the marvelous project of betterment for all of humanity. It is a tempered threat that may eventually turn ultimate – the threat of malicious leadership and human powerlessness. World music is only possible when engaged musicians are free to develop their art and to interact in innovative ways across cultural borders. World music-level engagement for a green future, and for meaningful "derivative sector" activity, is only possible when concerned and engaged people far and near can work and interact freely, building on cultural roots and cross-cultural exchange to invent ways and means for safety and blooming.

The psychologist, Dacher Keltner, author of "The Power Paradox - How we gain and lose influence" (2016) grew up in a poor, small town outside of Sacramento, California. In the introduction to his chapter 5, "The Price of Powerlessness", he explains that his mother got a job as a lecturer at Sacramento State University, a step up the academic ladder. The family had to move from Los Angeles, and they chose to settle in the small town of Penryn at the foot of the Sierra Nevada mountains, a rung down on the social ladder.

The nine-year-old Dacher quickly made friends in the neighborhood, and he learnt early to appreciate the solidarity and generosity of poor neighbors. But he also learnt about the widespread powerlessness that characterized the small community, and he got an idea of what drives some people to gain and hold on to power, sometimes abusing it, while others come to live a life of powerlessness. He ended up as one of the world's foremost power researchers.

Preludes to power

Many powerful people have come to believe that the Machiavellian need for power exertion and for manipulative ways to accrue power is basically correct. Relatively few have tried to test the truth of his basic thesis. That is what Dacher Keltner has sought to do. He chose to observe behavior in situations where competition for leadership and dominance can typically arise, such as in summer camps, student housing - places where people come together for the first time and where it is common for someone to position themselves for leading roles. The result of his investigations shows that typically those among the participants who behaved most "Machiavellian" quickly were identified and excluded from important positions (2016).

The great paradox, says Keltner, is that we tend to achieve leadership positions and gain the power to influence the world by virtue of what is good in human nature, such as empathy and care, but we lose position and spoil influence for the good by virtue of what is the worst of human nature, which largely follows as a side effect of gaining power and money. Keltner defines power as "capacity to make a difference in the world, in particular by stirring others in our social networks" (2016, p. 3). Machiavelli's understanding was that power is something you grab onto, fight for, and defend by all means. In contrast, Keltner sees power as something that is given to us by others when we are acting in ways that others perceive as useful. The person who is bestowed power must preferably be seen as reliable, honest, and caring. From historical societies, good leaders are generally selected for ability to empathize, be generous, express gratitude, and tell good stories. If a person is known for selfishness and deception, it weakens the possibility of approval and power.

People who have had lasting impact on significant changes in society, such as the abolition of slavery, women's rights, civil rights, freedom of expression and democracy, the rights of gay and LGBT people, the abolition of apartheid, and much more, have been people with an undeniable emphasis on care and generosity, with an ability to express gratitude and convey effective messages. We generally want to choose managers who are good at the qualities we value most - what Keltner refers to as "the greater good".

Keltner (2016) did one of his first experiments with first-year students at the University of Wisconsin, Madison. At the start of the semester, he asked students in a dormitory to indicate how much influence each of the other students in the same location had, and how they scored on five different personality indicators, those of enthusiasm (reaching out to others), friendliness (generosity and cooperation), focus (promoting shared goals, rules and values), calm (bringing calm and perspective to stressful situations) and openness (being a listener, open to other people's ideas, reactions and feelings). After 4 months and again after 9 months, the participants were asked to assess the influence of their fellow students. The main conclusion was that the enthusiasm factor was the most decisive. Students who were active and outgoing in contact with others were seen as those with the most influence. But the other "big five" factors also came into play - friendliness, focus, calm and openness.

The survey was repeated among other student groups and among groups in finance, factories, hospitals and the military. The consistent observation was that colleagues who were promoted to leadership positions were people who scored high on the five factors and similar measures, such as being generous or brave, being wise decision-makers, good at conflict resolution, good at speaking and telling stories, upright, fair and impartial, reliable, calm and tactful, and also ethically reflective (Judge et al., 2002). Leaders can be strong and persistent, but they must also show humility. This is by now also widely recognized in the mainstream business literature (Collins, 2005; Collins and Porras, 1994). Some of the characteristics may be more dominant than others in different types of institutions (such as in the military or in business). But the basic conclusion is the same.

Postludes to Power

The great paradox of power is that, given such positive preludes, a great many leaders, supposedly under influence of the power experience itself, and of the money that tends to come with it, tend to develop characteristics that are often the opposite of those bringing them to power. At some stage there seems to be a shift in the balance between the power holder's devotion to friendliness, openness and similar qualities, and – on the other hand – concern about personal privileges, status, power, wealth and the like.

The paradox of power is about the tendency for leaders, once they have gained power, to easily explain away their own faults and shortcomings, exaggerate their own skills and qualities, and take it for granted that they are entitled to prerogatives, including the right to more power and increase of wealth. While the ability to empathize and show care is very important for gaining the approval of others, and for getting into a leadership position, the paradox is that leaders may soon develop antisocial tendencies, more and more blind to the poverty and powerlessness of others, and to the "greater good" of societies.

Perhaps those who bestowed powers upon the chosen person made the wrong decision in the first place. It is not always easy to form a true imipression of people's inner qualities. Also, a general problem with decision-making is that even a decision that from all considerations, at the time it was made appeared to be the best and wholly rational, can turn out to be wrong. It is a problem that no decision-maker, leader, or politicians can properly guard against. If the decision was rational and sound at the outset, we should not hold it against a decision-maker that it ended up being wrong.

The special thing about the paradox of power is that it is not about contextual change and change of job requirements, but about changes in the person who was selected leader. Of course, there may have been deception. Some people are extremely good at presenting appearances – appearing to be enthusiastic and friendly, using such appearances to gain support, even while at the outset planning for power and greed. Generally, people are quite good at disclosing deception and manipulation, and they use such understanding to give approval, and thus power, to people who inspire trust. People also benefit from gossip, inside information, and personal history to build an impression, and in public life we tend to rely on mass media for making up our minds. But that is still not at the heart of the power paradox. The real tragedy of the paradox is that even leader candidates with a sincere devotion to the greater good may quite dramatically undergo a change of values and character once they are in power, and once they gain access to (Aristotelean) "money getting"!

At an individual level, the experience of power leads the person towards a turning point - either in the direction of abuse of power, with impulsive and unethical behavior, or in the direction of awareness of responsibility and humility, and of empowering others to work for the greater good.

Cases of brain damage

A fundamental human capacity for promoting a range of positive leadership qualities is the capacity for empathy. At times, people may willingly choose leaders perceived to be tough, even brutal, with a capacity to "speak out", "setting things straight", "turning things around". But under normal circumstances, we go for trust, loyalty, and honesty, and we pick out the ones who demonstrate enthusiasm, friendliness and so forth.

Power, if not grabbed by more or less suspect means, is given by others. But to be able to offer others what they want, and make a real difference that others appreciate, prospective power-holders must have the ability to "read" other people. It requires quite specific qualities, at the heart of which is the ability to empathize – qualities that can be sharpened or dulled when a person comes into a position of power.

The experience of having power – the capacity to be able to make a difference – is for most people extremely inspiring. It can be experienced as a special energy and "flow", a source of joy and optimism, perhaps euphoria, which spurs effort and the will to "move on". The absurd observation of the power paradox (and what Chayes (2020) referred to as the Midas disease), however, is that it also can come with a direct loss of empathy, and with more of recklessness, power-hunger, and greed. At the heart of empathy is the ability to notice, interpret and understand the feelings of others.

This may involve noticing small changes in facial expression, voice use, body posture, gesticulation, gaze and touch, often only within seconds.

A series of experiments were done with pictures and sometimes cartoon-type illustrations of 8–10 different facial expressions. It could be faces that showed anger, embarrassment, fear, pride, sadness, shame, or admiration. Or it could be about awe, enjoyment, disagreement, or motherly love. The participants in the experiments were asked to first read a list of such keywords and then link each word to a picture or cartoon symbol. The conclusion from a lot of experiments was clear: People with high empathy are better at interpreting facial expressions, which is important for being able to act empathically (Keltner, 2016, pp. 80–82). The critical conclusion was that the quality of empathy in all too many cases was lost or reduced as a function of power. In many cases, gaining power was like injecting a numbing drug directly into the brain. Keltner says that many powerful people behave almost as if they are brain damaged.

It is a shocking observation. Psychiatrists and neuroscientists have come a long way in identifying abnormalities in the brains of psychopaths (Phillips, 2013) and for the identification of traits for violence and moral disorders (de Oliveira-Souza et al., 2008; Gregory et al., 2012). While the idea of leadership "brain damage" may be used metaphorically to explain the power paradox, there is also a direct physical and literal link. For there to be empathy, a number of brain parts have to be active, and the links between them can be tracked by brain scanning (Eres, 2016).

With the help of MRI (magnetic resonance imaging), it is possible to trace how different parts of the brain are activated through a kind of "relay" in empathetic individuals, in turn causing changes of body language, mannerisms, tone of voice, and facial expressions (Carr et al., 2003). It has long been known that we all have a tendency to imitate those we are with, as an expression of compassion and understanding. Yawning is contagious, and so is the use of voice, touch, facial expressions and body language - as expressions of both joy and pain. Generally, the stronger the empathy, the stronger is also the relay and the imitation.

The interaction involves more or less automatic muscle tightening and movements, but the footprints of mimicry and empathy are visible in the brain. Riess (2017) mentions a study in which sixteen female volunteers were subjected to quite painful (but not dangerous) electric shocks to their hands, while their brains were scanned. The pain experience produced a typical activation pattern in the brain. Afterwards, they underwent a new test, receiving a signal that their partners were exposed to a corresponding electric shock. Again, their brains were scanned, and it turned out that almost exactly the same areas of the brain were activated - a directly measurable empathic reaction.

Absurd results

In fiction and modern film, Charles Dickens' character Ebenezer Scrooge (in "A Christmas Carol") appears as the prototype of avarice, ingratitude and lack of concern for others. Scrooge stands out as an inborn egocentric, miraculously converted to philanthropy after a bad dream (being visited by the Ghost of Christmas Yet to Come), and in meeting with Tiny Tim, the ailing son of Bob Cratchit, Scrooge's underpaid clerk.

In the power paradox framework, the story is the opposite of that of Dickens. A number of observations and experiments have shown that the shift from an empathic attitude to a much more demanding and actually arrogant attitude, that all too often comes with leadership positions and power, can take place in a short time and manifest itself in many different ways. Experiments are well known that show that as soon as the "leader" gets behind the wheel of a "premium" car, the driving becomes rougher and the consideration for fellow road users deteriorates. The tendency is also reflected in real traffic behavior.

Power corrupts, and Keltner reminds us that it produces the most absurd results. It has been shown that people with power are worse than average when it comes to impulsive eating, bad eating habits, sexual adultery, violation of traffic rules and safety (when driving prestigious car brands), lying and cheating (about travel bills, tax data, bus tickets, or bribes), shoplifting or stealing candy from children, use of coarse and profane language, name-calling, and generally lack of respect for others (Pitesa and Thau, 2013). It comes with a particular form of brain damage, caused by power-seeking and money-seeking – the Midas disease.

Poor and powerless people can profit from shoplifting, but for a person with power and means, the gain is financially worthless. Even so they steal. They have more than enough funds to pay for themselves, and the bank account is just as fat whether they steal small sums or not. There is something else behind it, a kind of understanding that they have rights that others do not have, that they are entitled to benefits that result from power and position, and that it is ok for them to steal and lie and cheat, but not for others. They appear as heroes in the narrative of their own lives. Keltner calls it "narratives of exceptionalism" (2016, pp. 103, 130–135) – stories which people of power tell themselves, explaining away their moral shortcomings. They misattribute their own success by exaggerating their own unique giftedness, good genes and hard work, and they misinterpret the misfortune and poverty of others, attributing poverty to traits of laziness, incompetence and bad genes.

Power is manifested in the form of social position, wealth or influence in social, political and commercial contexts. People with money, higher education, higher social position, and more prestigious jobs react differently to stories about other people's misfortune than people lower on the ladder. This also can be measured by means of a brain scan. The level of empathy is generally weaker among people with successful careers and in high positions than among those lower on the success ladder.

Even on a purely psychological level - detached from material manifestations - people can place themselves in an imaginary power hierarchy. Keltner (2016) had a group of participants in an experiment where they were asked to think about people who were known to have great power in society. They were then asked to place themselves on a rung of a twelve-step ladder that represented different levels of power in society. Another group was asked to think of people and groups known to have little power, such as the unemployed and homeless, before this group was also asked to place themselves on a twelve-step ladder. It turned out that the group that had first focused on people with great power tended to place themselves lower on the power ladder than the group that had first thought about the powerless. The thought of other people standing lower on the ladder of power made them feel higher up, in possession of more power. The experiment was followed by a survey of empathy scores. The participants were shown a facial expression and asked to answer what best characterized the expression. The group that had first focused on people without power not only placed themselves higher on the power ladder, but they also immediately lost some of their capacity for empathy.

The poor and powerless spend more time during the day thinking about how others are doing, and they express greater pity for those who are in pain and difficulty. They are also inspired by hearing others tell their life stories, while people with power are more inspired by telling stories about themselves!

Managers with a lack of empathy will have problems in businesses organizations that require good relationships with employees and the ability to truly care for clients. If they lack the ability to be humble and fail to meet others on an equal level, they may have problems mobilizing people for whatever vision. At a nastier level, they contribute to the powerlessness of others. As seen from the outside, the tech-billionaire mega-project entrepreneurs mentioned in chapter 14 all seemed to belong to this group.

The brutality of Powerlessness

Successful people tend to explain the powerlessness of others as a consequence of indolence and stupid decisions: Children in poor families don't care about education and good grades! They lack the ability to "delay gratification" - the effort that is necessary even if the reward comes late! They are a burden on public funding!

Such thoughts are widespread, and they are off the mark! It represents lousy thinking, and it is caused by brain damage. The poor are not poor because they are stupid, but poverty and hopelessness can sometimes lead them to make more risky or stupid decisions (Bregman, 2018). Poor children work at least as hard as others in school. Poor parents strive at least as much as others to give their children a good upbringing, and they take whatever jobs they can get, sometimes working two shifts to take on the responsibility of providing for them. But something happens that keeps holding them back.

In the 1990s, we received research that showed that powerlessness and poverty affect people's health - even in societies with access to socialized health and social services. Social class, defined by wealth, level of education and social prestige, predicts people's health status. Chronic diseases, lower life expectancy, and problems of blood pressure, and several forms of cancer and arthritis are predominant among poor people – even when having access to medical care. Keltner says that powerlessness "gnaws" at the nervous system. It "gnaws" in the form of stresses and threats. There are bills to be paid and children who need food or clothes and money for a birthday present for a child's friend. There is a lack of money for payment for a camp school, and there are problems in keeping a job, the humiliation of standing hat in hand, the stigma of social security, malicious gossip on Facebook. There is also the burden of growing up in a depraved environment, with racism and name-calling. Powerlessness means being exposed to much more intense threats than other people, and to the bullying and direct physical threats of poor living environments. This gives rise to more or less chronic stress. And stress causes illness.

The life of the poor and powerless is an eternal struggle against injustice and contempt, uncertainty and insecurity about everything important in life. More than others, poor and powerless people struggle with most things. Anxiety, insomnia, depression and despair characterize life, and sometimes lead to alcohol abuse, drugs and crime. Good relationships, friendships, and harmonious sexual relationships become more demanding to keep up.

It also affects the children and their upbringing. What "gnaws" at the parents affects the whole family. For children it causes problems with concentration in school. Persistent stress affects the grades, and it affects the body: It wears on the entire nervous system, blood vessels and arteries, the digestive system, the immune system, and brain cells, and it can even cause changes in the DNA system. Powerlessness leads to physical and mental suffering and to a shortening of life expectancy. As Keltner says, powerlessness explains an unusual level of illness and early death among his neighbors in Kayo Drive, his childhood's Broadway. The paradox of power produces leaders with brain damage and too little ability to care. Powerlessness and poverty reveal the cost-side of the paradox.

In our time, the tempered threats of climate change, nature depletion, environmental toxins, and AGI may take a toll on many people's nerves, causing increasing depression, anxiety, and discouragement. But most of us in wealthy societies believe in technological solutions and in the chance to buy oneself out – at least for now.

The powerless in society, on the other hand, already live in environments characterized by pollution and noise, impoverished neighborhoods and inhospitable work environments. Poor and powerless people have a greater chance of being investigated by the police, being exposed to crime, or ending up in crime and prison. Poor women are more vulnerable to rape and abuse. All of them are more susceptible to humiliation and impulsive reactions from people higher on the power ladder. The "Me too" movement was everywhere a confrontation between the relatively powerless against the more powerful, as was in many ways the Black Lives Matter (BLM) movement and previous civil rights movements.

Power can cause people of means to become more impulsive and daring in their pursuit of satisfaction. Powerlessness causes people to become reserved and inhibited in their behavior. Powerlessness puts people on defensive alerts, and a hypervigilant defensive posture affects the brain physically. In a survey (Keltner, 2016, p. 146), people who had grown up in poverty were compared with others who had grown up in prosperity. Participants were shown pictures of people with different facial expressions. Just the image of an angry person was enough to put poor people's amygdala on full alert. Faced with a threatening situation, cortisol stress hormone levels skyrocketed. The body often responds with a "fight or flight" reaction or by being locked up in a "freeze". Blood pressure rises and the heart beats hard and fast. Blood goes to the muscles and to important organs. The result is another form of physical brain damage.

Mobilization from the bottom

The present chapter has presented a counterpoint to the war-economy metaphor. The world music metaphor is about inclusion and innovation, about the freedom of expression, variety of experience, and intimacy of communication, about shared values and about sharing as a value in itself. It is about emergent change – growing from below. The metaphor points to good people and the good in people, but also to corrupt leadership, the power paradox and powerlessness. The latter is about misery and destitute, and a desperate need for change. The world music metaphor does not work for people without minimal means for life sustenance. Just as men had to learn that women and girls represent half of the world's population, with half of the talents for problem-solving and development, the present generation has to learn that poor people, representing near 50 percent of the world population (at income levels below the US poverty line) also represent half of the world's talent pool.

Failing to add income and stability to the powerless will cause further disintegration. Failure to deal with the paradox of power and the brain damage caused by wealth and power will result in further waste, injustice and depletion, potentially also dystopianism.

Under the influence of Newtonian physics, social scientists (with little success) searched for empirical invariance, political "laws of nature", determinism and sureness of prediction. Present developments point in the direction of unpredictable emergence and non-linearity of development, with the potential for both tempered and ultimate threats, but also with the budding of hope and possibilism for the renewal of society – which is the topic of the next chapter.

Chapter 16

The "*dugnad*" Metaphor and the Promise of Hope

There is a Thanksgiving story told by Bob Sutton (2016) about Kurt Vonnegut (famous author of among others "Slaughterhouse V") and Joe Heller (equally famous writer of "Catch 22") who were invited to a dinner party by a billionaire on Shelter Island. Kurt Vonnegut recalled:

"I said, "Joe, how does it make you feel
to know that our host only yesterday
may have made more money
than your novel 'Catch-22'
has earned in its entire history?"
And Joe said, "I've got something he can never have."
And I said, "What on earth could that be, Joe?"
And Joe said, "The knowledge that I've got enough."
Not bad! Rest in peace!"
— Kurt Vonnegut

The story has been retold in different versions and locations – seldom attributed to Joe Heller. An anonymous version tells that there was a poor artist living by himself in a modest cottage. One day a very rich man passed by and commented to the artist. Why do you carry on with your work? I make more money in a day than you make a year. The poor man answered: "I've got something you can never have". "What might that be?" The rich man countered. "Enough!", said the painter.

The story is about the "Midas disease". It is also about Aristotle's distinction between "natural wealth" and "wealth getting" – signifying the contrast between limited needs satisfaction and the unlimited craving for more.

We have seen that the idea of unlimited material growth is morally and ecologically bankrupt. People on premium class are being well-fed to the extent that there is not enough for those on economy class. Spaceship Earth is already overburdened. It would not be able to carry all the supplies that would be needed if everybody on board were to enjoy the standard of the premium class. That makes for a supreme ethical challenge to the fundamentals of development and our definition of progress. The inequality of wealth distribution is not just a cause for philanthropy and aid to the poor, but it is a direct challenge to continued affluence growth and the existence of plutocratic "wealth pumps" in society, constantly pumping wealth and resources from a suffering economy class to a "brain damaged" class of premium passengers.

We have also recognized that science insiders are concerned that AI-driven productivity may cause mass unemployment and absolute poverty below subsistence level in decades to come, with the *absurdum* result that multitudes of humans around the world will come to constitute a "derivative

DOI: 10.1201/9781003518815-19

sector" of the economy – a sideliner to the AI-driven commercial market economy and a landing spot for "useless" and "irrelevant" blue- and white-collar labor.

Where most of what it takes to run a market economy, and where dealing with the "economic problem" of production and distribution is taken care of by machines, governments will have a choice between a universal "right to live" and universal doom. If the choice is – as we may hope – for the "right to live", some version of guaranteed minimum income for all will have to be in place. But the question still remains, what will people be doing? What will keep them occupied? For what end? And on what political-economic premises? Will the "precariat" prevail, or are there more magnanimous alternatives?

The principle of reciprocity

For the near future there will most likely be a shortage of labor in many societies. Birth rates are going down, and green transition jobs and caring for the elderly will add to job vacancies (supposedly also countering resistance to immigration). But the AI transition is not just about future productivity gains and job stealing. While seen by many as just another tool in the hands of man, it represents even at the present an ultimate version of market-capitalism as a social and societal system.

As keenly observed by Karl Polanyi in his 1944 classic "The Great Transformation", the modern market economy and market-economic rationality did away with social customs and traditions, including mechanisms of collective reciprocity. Where the market economy brought calculative attitudes and cost-benefit type assessment to the forefront even in interpersonal relationship, the social nature of work, exchanges and sharing was replaced by the price mechanism and "appraisiveness" of both things and relationships. Since industrialization, the modern economy was in a way dis-embedded from society, bringing in chronic inequality and insecurity to the poor, materialism, competitiveness, alienation and "calculativeness" to the many, and enormous riches to the select few.

The dugnad tradition

The dugnad metaphor represents the opposite of the calculative framework, reintroducing reciprocity and bringing in inclusion, security and hope. The "*dugnad*" metaphor also complements the former two metaphors of war-economy and world music. It is introduced here as a guide to societal transformation. It points historically to norms and traditions for cooperative work in an Aristotelian "natural wealth" economy. It points forward to institutional arrangements for economic decoupling of growth and for life in harmony with nature in an emerging new world society. It points toward social bonding and equitable justice, and to human life and well-being beyond AGI dominance and transhumanist experimentation, for a future logically linked to the emergence of a "derivative sector".

It is not suggested here that a dugnad economy will come to replace the market economy. On the contrary, an AI driven market economy may come to replace people from much of commercial activity, opening for widespread "uselessness" and "irrelevance", with people being referred to what I call a "derivative sector" – a sector in need of content and institutional guidelines.

We may need to remind ourselves again that most of today's advanced technologies and scientific insights were not thought of, much less anticipated, a hundred years ago. Likewise, at present, we are unable to conceptualize what will be the most important technological, societal and scientific threats and solutions a hundred years from now. It is up to the present generation, however, to decide whether future solutions will be guided by the special interests of super-rich tech billionaires, by AGI and transhumanist experimenters, for money enrichment for the few and power-grabs by unaligned robots, or by the norms of world music and survival of the friendliest.

The Norwegian concept of "*dugnad*" is old, and generally translated "voluntary work". The word [pronounced ˈdʉːgnɑd] originates from the Old Norse *dugnaðr*, referring both to help and support, and to the virtue or "being good enough" for participation – reflecting both the solidarity aspect of mutual contribution, the reciprocity element of governance, and the capability aspect of getting a job done. For the grounding of the dugnad concept, we shall also trace the principle of reciprocity to traditional societies the world over. To the *dugnad* tradition, we may add the concept of "commoning", representing a presently growing and rich variety of dugnad-like activity and related conceptualizations around the world.

The essence of a *dugnad* is the voluntary participation in cooperative work, often followed or accompanied by social events. While parallel recipes are known worldwide, they can take many different forms and serve different functions in society. Varieties of *dugnad* type institutionalization also reflect great varieties of traditions, degrees of voluntariness, scope of involvement, and economic function.

The Norwegian tradition of *dugnad* can be traced several centuries back, mainly to rural communities (Lenk, 2011; Lorentzen and Dugstad, 2011). *Dugnad* is still an important institution in Norwegian society, and while important adaptations have been made to the needs of modern society, it is still considered a key reference for reciprocity and collaboration, philanthropy, participation, inclusion, and prosocial behavior (Simon and Mobekk, 2019).

Unlike neighboring Sweden and Denmark, Norway was for centuries among the poorest countries in Europe. People were scattered in isolated small hamlets and settlements, far from governmental administrative bodies and from cities and markets. People had to depend on each other, on relatives, friends, and neighbors, when in need. General poverty, but also relative equality at a low level of income, and the fact that there was no nobility or squire tradition and very little governmental presence may have added to local self- reliance and to reciprocal dependency for expertise and work requiring engagement of several people. With little money in circulation, the contribution of work power was the main instrument of assistance in small farm and fishery communities. There was also a disciplinary element of reciprocity involved. Neighbors were dependent on each other, and those who did not participate in the *dugnad,* for other than health reasons, could not hope for equivalent assistance for themselves.

Ingredients

The main tasks supported by the traditional *dugnad* were manual jobs requiring many workers for construction or maintenance, such as launching a boat or securing a harbor, building a house or a barn, raising a new roof, building a road or a bridge – or arranging a wedding or a funeral. *Dugnads* are still in use for such purposes, more so in the countryside than in cities, and more based on family relations and personal friendship than on neighborhood participation.

Increasingly, however, from the roots of basic necessities in small and isolated communities, the modern *dugnad* is used for mobilizing members and parents of members of sport clubs, youth clubs, school music bands, community houses and the like, for lottery administration and economic support, by selling cakes or hot dogs, or whatever to support the activity. Also, voluntary organizations, religious groups, and social movements may rely on *dugnads* for financial support and member recruitment, while housing associations, kindergartens, and local schools may rely on *dugnads* for picking garbage, for painting and other maintenance, or for adding additional structures (in line with the American "work party" tradition).

At another and quite different scale, the modern concept of *dugnad* is referred to also as a national mobilization instrument. After the Second World War, Norway was in financial ruin. The post-war Prime Minister, Einar Gerhardsen, then called for a "national *dugnad*" to rebuild and modernize the country, and for people to build homes for themselves. He made a name for himself as the "*dugnad* general", as a facilitator of affordable housing loans, low-priced building sites, cooperatives formations, and for the encouragement of joint voluntary work for job creation and economic development. Local co-ops were also to some extent based on *dugnad* participation. The post-war years was overall a period of unprecedented *dugnad* activity, especially focused on family house building (Simon and Mobekk, 2019), but also on local community development, sports and youth activities.

A recent example of *dugnad*, or "the *dugnad* spirit", was the appeal to people for limiting the spread of the Covid 19 pandemic. References to the "*dugnad* spirit", meaning appeals for support of good causes, is regularly heard in the news, whether for sport events, health promotion, fundraising, environmental protection, moderation of wage demands, moderation of CEO bonuses, and for national fundraising programs, for poverty alleviation or special causes linked to "the greater good".

Dugnads seem to stimulate prosocial behavior, and *dugnad* occasions have certain motivating characteristics (Biglan et al., 2012). The voluntary and cooperative work context may itself be a stimulant – normally work is done in a good mood, with humor, jokes, and laughter, and with opportunities for cross-trade learning and local gossip. Voluntary work is known to improve your mood, lift depression and anxiety, increase one's sense of satisfaction and happiness.

At the traditional *dugnad*, it is also normal to have a social gathering following the work – with food and drink, storytelling, and music. The main reward of doing good is to feel good. This may answer some of the "what is in it for me" questions. To support their own children, take care of their own property, or even defend their own membership in a social movement, is still part of the "what is in it for me".

Beyond personal gratification the question of reciprocity remains, even in a modern setting of the modern market economy. At the opposite end of *dugnad* qualities, there is the economic issues of externalities and "free riders", and the "tragedy of the commons" problem of individualism (Hardin, 1968). Why do people engage in *dugnad*, when the alternative may be to relax or buy your way out?

It has been shown that volunteering is in itself gratifying, of benefit to the helper as well as to the helped (Wilson, 2000), with improved life-satisfaction, mental health, and occupational achievement as added bonuses. For young people voluntary work may be a defense against delinquency and a source of pride, skill development, empowerment, social interaction, positive attitudes, and overall learning (Rider and Hall, 2002). The *dugnad* version of volunteering has been shown to support the integration of immigrants and refugees (Penner, 2021), while the group dynamics of voluntary work may add to group identity, belongingness, motivation, competence, and to the overall positive experience of volunteering (Gray and Stevenson, 2020).

Voluntarism lost

The element of voluntarism may be lost, as work content and administrative tasks are institutionalized. In the Soviet Union, the first mass *Subbotnik* was initiated by Lenin in 1919 at a railway depot in Moscow, calling for communists and supporters to (voluntarily) put in an extra hour of work on workdays, and six hours of work without pay on Saturdays (Boldonova and Tsetsura, 2015). Over the years it became a weekly tradition, exported to other republics of the USSR, and later spread to Eastern Europe. In Czechoslovakia it was known as *Action Z* (*Akce Z*), from Czech word *zvelebování*, which means "improvement". Garbage removal was a typical Saturday job for university students and workers. In an article, "The Grand Initiative", Lenin called *subbotniks* "the actual beginnings of communism". Institutionalized *subbotniks* lasted till the collapse of the USSR. By then it had long since lost its meaning of being voluntary (in Russian *volonter*, or *dobrovolets*). And for students and workers it had become an added reason to hate communism, not unlike what in the West is called "civil conscription", the near opposite of a *dugnad*. Civil conscriptions (where striking is illegal) are sometimes mandated in times of war, as part of a war-economy package, or more permanently in small-community fire protection or infrastructure maintenance, but it is about command – not *dugnad*.

Beyond the gratification of voluntarism, the *dugnad* presupposes a meaningful task, openness of participation, inclusion of all (even the young and the old, the handicapped and the professionals), productive work and social activity. It builds on egalitarianism and inclusion (Lorentzen and Dugstad, 2011) but also on the disciplinary element of exclusion. Penner (2021) tells about a father who wanted to enlist his young son in the local football-club, offering to pay NOK 5,000 for not having to participate in the *dugnad* and for not having to take his turn driving the kids. The offer was rejected by the women leader of the club: "Yes, but you have given birth to children,' I said. 'You have to take care of them!" Membership was refused, and the father had to change his schedule in order to have his son registered in the club. Differences in status are linked to key competences, more than to wealth and social status. Participation in *dugnad* is both a calling and a measure of good citizenship behavior, of socializing in the community, and the integration of immigrants (Penner, 2021).

The universal experience

There is no agenda in the present text for replacement of modern markets with voluntarism and *dugnad* exchanges. But we all know that modern societies would break down immediately if families and communities didn't support themselves with voluntary engagement, based not on the price mechanism of service and labor markets, but on care and reciprocity.

For all kinds of goods and services that can easily be specified and monitored, markets work incomparably well. On the other hand, for other tasks that are less easily specified, in limited supply, or surrounded by uncertainty, agents may be expected to act opportunistically. In such cases, according to transaction-costs theory both businesses, governmental contractors and consumers may be expected to avoid the market and seek "internal" solutions. In family and local settings, the solution is normally mutual assistance and voluntarism – still not worth recording in the GDP statistics. According to estimates by AmeriCorps (https://americorps.gov/newsroom/press-release/607-million-americans-volunteered-2021-americorps-reports) and based on US Census Bureau data from in 2021:

> "Nearly 51 percent of Americans, or 124.7 million people, informally helped their neighbors at least once in the past year. Additionally, more than 23 percent of Americans, or 60.7 million people, formally volunteered with an organization, giving more than 4.1 billion hours of service with an estimated economic value of $122.9 billion."

In "The Great Transformation" Karl Polanyi (1944) observed that prior to industrialization, no economy was controlled by markets, and profits made from exchange was traditionally never a dominant marker of local production and national economies. Local markets, trade and barter, emerged in most places, without the dominating position of the modern market economy. Max Weber similarly noticed in "General Economic History" (from 1923), that commercial transactions originally was mainly between ethnic groups, that local markets did not promote the growth of a merchant class, and that money was not in use in local and near relations. The normal renumeration of labor was not based on money but on reciprocity. Before the coming of "economic man", humans were most of all "social beings" – "submerged in his social relationships", "safeguarding his social standing" by means of reciprocity (Polanyi, 1944, p. 53). Reciprocity served the maintenance of social ties through codes of honor, social prestige, generosity, and belongingness, and it also served individual interests by way of redistribution, harvest storage, sharing and provisional security.

In fact, by reference to early anthropologic and economic-history research, Polanyi identified reciprocity and redistribution as the two anchoring elements of traditional societies. Clearly, responsibility for storage and redistribution was also associated with power and privilege in the hands of heads of family, village elders or local overlords, often also tied to responsibility for defense (as under feudalism). And much of early political history can be traced to rivalry, jealousy and violence associated with the distributary role. Still, the main pillar of production was the principle of reciprocity.

The commoning and the *Buen Vivir* of the commons

In a broad sense, *dugnad* is about the promotion of common interests, and about institutional solutions where markets and political governance are absent. It works where individual and private solutions are not up to the task. It also works in many situations where "free ridership" or exploitation of collective resources might be tempting. A special case of failure to organize was reported by biologist Garrett Hardin in a much read "Science" article about "The Tragedy of the Commons" (1968). Hardin's message was that common pasture lands have tended to be over-exploited because herd owners have gained a private bonus from adding additional heads, while the totality of income for the group of cattle owners went down due to overgrazing. Thus, the common interest was sacrificed.

The article became an important reference point, warmly embraced by proponents of private ownership in general, and land ownership legislation in particular. On the other hand, it was also increasingly criticized by historians and social scientists pointing to successful historical ways of organizing to overcome the tragedy. From the rows of critical voices have also come a new wave of "commoning" research, reporting on historical solutions and on future oriented experimentation.

The commoning paradigm

It is suggested that sensible commoning may play an important role for solutions to the conundrum of environmental and social threats of our time (Bollier, 2020). Where the *dugnad* experience serves as a useful metaphor, the variety of commoning experiences serves as an important guide to social organizing experimentation for the "derivative sector".

David Bollier sees commoning as both an overriding paradigm, a discourse, an ethic, and a set of social practices with a shared focus on mutual support, conflict resolution, and the management of shared natural resources. It is also a paradigm that complements the market economy and bureaucratic government, for commoners to reclaim their "common wealth", while also defending resources of nature against commercial interests. Where Hardin's solution was privatization, commoning offer opportunities for collective agency, for preservation of local communities and traditions and for development of cultural values beyond consumerism and materialism, promoting care-giving and shared environmental responsibility. As for the varieties of forms, Elinor Ostrom in her book "Governing the Commons" (1990) documented hundreds of ways commoning was practiced, mostly in rural settings and in poor nations – earning her a Nobel Prize in economics in 2009.

The variety of models can be an obstacle in research, making theorizing and generalization difficult. But in a battle for survival and well-being in a world of threats and conundrum, variety is a definite systemic, cybernetic advantage, as described by Ross Ashby in his "law of requisite variety" (1958). As also underscored by Hirsman (Chapter 10), for a good development strategy we should account for unique experiences, local traditions, cultural values, and natural contexts..

Bollier (2020) reminds us that in the jargon of management and economics, people are referred to as "human resources". And even in UN reports from the Environmental panel, nature is referred to as "nature's services" or formally "Biodiversity and Ecosystem Services", in both cases reflecting the priority of a strictly commercial valuation. Bollier reminds us how Karl Polanyi in "The great transformation" (1944) described how the market and consumerist culture of industrialization came to replace the roles of kinship and tradition, religion, morals, and community as ordering and priority-setting principles in society. Commoning in many ways seek to contradict and reverse dominating cultural trends of modernity, also by use of modern technology: Commoning may cover the use of open value networks, crowdsourcing, buying clubs, collectively owned factories, local farming, ecological food production, stakeholder trusts, open-source programming based on shared codes, "blockchain ledger" communication, historical models of cooperative finance, marketing and exchange systems, and a huge variety of "Small is beautiful" style innovations and activities adapted to developmental, cultural and environmental contexts (Schumacher, 2011 [1973]).

A basic value of commoning is that of care – of people, communities, wildlife and the planet, of value to the caregivers and to recipients alike, adding meaning in the life of people destined by the market economy to otherwise become "irrelevant" and "useless". As mentioned, voluntary "care work" or "affective labor" (Singh, 2013), is not included in standard GDP accounting. The "care economy", fully necessary for the well-being of society, is routinely ignored in market economics, overseen as a non-economic, sustainable and self-replenishing "free resource", to be ignored or exploited, traditionally and in present society to a great extent carried out by women. We can foresee that it will make up the core of a future "derivative sector".

As a curiosity, when GDP accounting was first introduced in Norway, in the aftermath of the Second World War, the value of housewives' work in the home was included – but excluded when qualifying for Marshall aid. As cited by Bollier (2020, p. 17): "A 1980 report by the UN stated the situation with savage clarity: "Women represent 50 percent of the world adult population and one third of the official labor force, they perform nearly two thirds of all working hours, receive only one tenth of the world income and own less than 1 percent of world property." Bollier also refers to German writer Ina Praetorius (2015), who has commented on the "taken for granted" issue, suggesting that the recognition of the care economy may serve as a model for alternative thinking about commoning for the entire economy. Without the unpaid services of housewives, caring for family and up-bringing (often on top of a regular paid job), and of families and relatives, neighbors and friends, caring for the sick and the elderly, and of commoner volunteers caring for neighborhoods, sports, and youth activities, societies would break down.

To the significance of care for humans, we may add commoning stewardship in the governance of wilderness and in the honoring of the intrinsic value of nature. This applies to species protection and rewilding, and to the preservation of forest, fisheries, pastures, groundwater, and eco-systems. Such initiatives are often small-scale, but in total they add up to providing livelihood for two billion people (Bollier, 2020) – a fact more or less ignored in mainstream economic textbooks, since it represents "affective labor", not accounted for in market transactions. This all represents stupidity at a high level! On the other hand, recognizing commoning experiences around the world, and world music style innovation and development, also means to recognize robustness for survival, and a chance of meaning to life in a future AI or AGI-mastered "derivative sector".

A buen vivir experiment

For a stimulating philosophy or worldview, even in some countries recognized in the constitution (Ecuador, Bolivia), *buen vivir* means literally "good living". It is a contrast to commercialization and a complement to traditional economic institutions of reciprocity and sharing, based on solidarity among people, respect for nature for its own value, honoring of indigenous cultures, and enhancing decolonization. It points to grand societal transformations, as envisioned by Ulrich Beck in his notion of metamorphose level change (2016), but still anchored in real life, present world experience. It is about a process of transformation, with a focus on "unmaking of capitalism".

An interesting case was studied by Giuseppe Feola and colleagues (2021) at the Territorios Campesinos Agroalimentarios (TCA) del Norte de Nariño y Sur del Cauca, one among a number of agro-food farming territories supported by the peasant movement in Colombia. The idea behind TCAs was to create alternative forms of food production and at the same to protect nature and to overcome the alienation of farm workers. Marginalized indigenous and afro-descendant people were to be included in the process and given access to land. Ecosystems, threatened by agro-industrial and extractive industries, were to be restored. Bottom-up democratic processes and a great deal of political creativity was exercised, and new forms of governance emerged, among others to overcome the capital versus nature contradiction of capitalism.

The changes brought about have not come as a one-time (revolutionary) transition to a new political-economic order. Transformation has emerged from a long interactive and dynamic sequence of meetings, consultations, and confrontations, with information diffusion and consciousness-raising activities, with 600–1200 participants at *mingas* meetings (meaning "coming together for the common good"), setting the stage for decision-making, soliciting admissions and popular consent.

Participants have shared a history of community level direct political action, defending themselves against combined social and environmental injustice, such as when the Canadian Gran Colombia Gold company in 2011 wanted to exploit six thousand hectares of land for gold mining (the Mazamorras project). The peasant communities mobilized collectively, including occupation of two company encampments. Local leaders were met with death threats, but strong local opposition forced local and central authorities to reverse former decisions, and the Gran Colombia Gold company had to cease exploration.

The TCA communities of Nariño and Cauca also shared a history of "direct action", in cases where the central government failed to provide basic social, health, security, and educational services, adequate infrastructure and technical support to local communities. Community level "direct action" or "de facto action" means that local peasant communities acted autonomously to provide for needs and solve problems. In our context, we may see direct actions as expressions of Hirschman's concept of possibilism (Chapter 10). Initiators and change agents are peasants and local community inhabitants, deriving legitimacy from their history of ties to the ground and from their ability to organize collectively, deriving change agency from discussions of the "impossible" leading to "possibilism". TCA is also a vibrant expression of what is meant by *buen vivir* in practice.

Feola and colleagues (2021) defines a TCA as also a "discursive space" where visions for a just and dignified future are allowed to grow, and where motivation and autonomous change agency are allowed to flourish. Four principles are fundamental to the process: autonomy, coexistence, participation, and a profound respect for life and nature. The political-economic solution is called an *economia propia*, signifying an economy of one's own making, responding to local needs and values – contrasting predominant capitalist notions of global commercial efficiency, self-interest, cut throat competition, and threats of violence. *Economia propia* seeks to advance local sufficiency and nature protection through crop diversification and provision for local needs before selling surpluses at national and international markets. It also includes public or common control with fresh-water resources and distribution.

Where neoliberalism in government and business seek profits from the extraction of minerals for export, the TCA stand for the protection of basic food production, local needs provision and the preservation of nature and life styles tied to history and land. Where neoliberalism focuses on short term commercial gain, the TCA takes a twenty-to-thirty-year perspective, respecting the intrinsic value of belongingness to land and tradition, but also promoting progressive initiatives in areas such as women's rights, economic equality, education, and democracy. While such rights belong to a standard package of "development" and "modernization", the TCA also rejects mainstream ideas of economic development, such as work organization based on hired labor and the promotion of consumerism - "refusing what the system wants us to want".

Life itself is the central organizing principle, not profit-seeking or "wealth getting", and not the roles of people as consumers or agricultural workers. Agriculture is not seen as a form of business, or a source of profit, entrepreneurship, or employment. Agriculture is to be undertaken for its own sake, both a duty, a lifestyle, and an expression of loyalty both to forefathers and descendants. Peasants are not dispensable barriers to development and progress, to be urbanized or industrialized away in favor of modern, efficient, industrial agriculture and city life. Also, in the "discursive space" of the TCA, the slogans of neoliberal capitalist development, such as "natural resources" and "human resources" are rejected in favor of terms such as "right to life", "right to territory", "vital goods", and "ancestral heritage" – encouraging the "possibilism" of "counterhegemonic grammar", reaffirming the dignity and traditional autonomy and cultural relevance of peasant life.

The TCA experiment will go on, and similar experiments are reported from all parts of the globe. We may envision alternative possibilism formulas for change, and we may hope for success and survival. But such experiments are also vulnerable and risky, and no one knows the outcome. But what we do know is that once the global society bubbles over with bottom-up experiments, with social (and peasant) movements, new political orders may emerge, allowing for solutions that we have no chance of seeing today, including also solutions for a vibrant "derivative sector".

The transition will be significantly more comfortable when accompanied and complemented with changes from above and in a spirit of hope. Perhaps hope will be the engine we need to implement metamorphose level changes. In that case, we need to search for its secret explosive power.

Hope

Hope is generally defined as a state of mind, a mental state characterized by the expectation of something positive. The Cambridge dictionary adds that for there to be hope, there must also be a certain, even minimal, probability that what is hoped for will come true. The element of probability is the basis of hopeful expectation. But also, in order for hope to have any meaning, there must be an element of love, marking a difference between hope coming true and failure.

Unnoticed in the above definition, is that what makes hope relevant is the existence of some threat of a negative outcome. Without the possibility of something negative, hope would not be relevant. The definition of trust provides a parallel: Trust can be defined as a "state of reliance on the character, ability, strength, or truth of someone or something" (Webster), and also "to believe that someone is good and honest and will not harm you" (when harm is possible), or "that something is safe and reliable" (when it may not be safe and reliable), or "to hope and expect that something is true" (Cambridge Dictionary – when in fact there is a chance that it is not true - my additions). Without the possibility that our reliance on character, ability and truth is misplaced, there is no reason for trust. It is in situations of possible doubt and loss that trust makes a difference. Likewise, it is in situations of threats or despair, potential loss or defeat, that hope makes a difference. And the existence of threat or loss is in turn anchored in love. It is because we care about something, love someone, or feel a loss or defeat related to some outcome, that hope emerges as a contrast – as a state of mind characterized by the expectation of something positive.

With reference to the apostle Paul (1st Corinthians 13:13), hope is intimately linked to faith and love, sometimes referred to as the three theological virtues. Beyond the spiritual essence of each of hope, faith, and love, there is also a logical connection. It is love that underscores the importance of hope, and faith is an expression of expectation (or probability). To this may also be added the element of agency and action. Faith expresses itself in action and following. And, remarkably, as soon as you start to act for a good cause, out of love, you also feed into a spirit of hope. Effort and agency may have many causes (threats, greed, hatred), but love and care are still the strongest among people of good will.

In the following, we shall first consider the role of the love component, and the difference between hope and optimism, before concluding the chapter with a summary of despair elements and a round-off with causes for hope.

Hope and love

Jürgen Moltmann is best known for his book, "Theology of Hope: On the Ground and the Implications of a Christian Eschatology" (1993 [1964]), earning him reputation as a leading theologist of hope (he died in 2024, 98 years old). In his latest book, "The Spirit of Hope – Theology for a World in Peril" (2020), he first sums up some major threats to life in our time – nuclear weapons, economic and political inequality, ecological collapse and, ultimately, lack of love of life.

He says that life today is in danger - not because of mortality (we have always had to face death). But life is in danger because it is no longer loved, affirmed and accepted. He refers to Albert

Camus who, after the Second World War, wrote "this is the mystery of Europe: life is no longer loved". At the age of 96, Moltmann says that he still re-lives the horrors of war. He was thirteen years old when the Second World War began. He had dreams of studying mathematics and physics, but in 1943 he was enlisted in the German army. He survived a bombardment of Hamburg while 40,000 people were killed. In the last days of the war, he was sent to the Western Front where he surrendered to the British. Then followed three years as a prisoner of war, with transfers from one camp to another - before starting on his theological pathway.

70 years later, he says that it was his generation's destiny to experience the murderous war, and that at the time of war it was no longer about winning or losing - only about meeting death. He says that he understands Camus: a life that is no longer loved is a life that is prepared to kill, and also deserving to die. The war ended in 1945, but life had become meaningless by then.

The beginning of the 21st century brought progress and optimism, but also the dark side – feelings about the fragility and vulnerability of our new world. And at the very extreme, it brought what Moltmann refers to as "a new religion of death". It is not a religion for the masses of people, but it has produced small, fanatical communities based on the acceptance of suicide bombs. It is not about Islam, but about the cultivation of death and terror. Moltmann quotes the Taliban leader, Mullah Omar in Afghanistan saying: "Your young people love life ----- our young people love death". For our purposes, it is referred to by Moltmann as a reminder that the love of life is basic to all human endeavors (and to all wilding in nature).

On the morning of March 4, 2004, four commuter trains into Madrid were hit by ten coordinated bomb blasts. 191 people were killed and 1,460 injured. Moltmann quotes a statement from the Al-Qaeda network which was behind the attack: "You love life, we love death". He also quotes a German youth who joined the Taliban in Afghanistan: "We don't want to win: we want to kill and be killed". Perhaps killing is identified with divine power to rule over life and death. Perhaps the love of such a form of power, in a world of perceived powerlessness is the only thing they can attach to any form of hope.

Moltmann does not stop there but makes an interesting observation. The "balance of terror", masked as "peace", between the Soviet Union and the West, between the Warsaw Pact and NATO, with which we lived for 40 years after World War II, was based on mutual deterrence. But for suicide bombers and followers of a religion of death, there is no effective deterrent. They don't love life anymore, and they don't mind dying with their victims. Death is no longer a threat. As Moltmann says, all peace agreements, disarmament agreements, and non-proliferation agreements are rooted in what may seem self-evident, namely that the parties at least have a vested interest in survival and a will to live.

The logic of deterrence builds on a love-of-life prerequisite, and we may ask what happens when a party, or a whole nation, faces an enemy, and perhaps faces defeat and is gripped by a religion of death? What happens when they no longer care about life, while still having enough weapons to exterminate both themselves and much of life on earth? What happens when political or military leaders of such groups, driven into a corner, and standing without hope, find that they might as well drag the rest of the globe down with them?

Deterrence only works as long as the opposing parties have a desire to survive! Without fear of death, the balance of terror does not work. The parties to deterrence do not need sympathy for each other, as long as they are able to love themselves. Survival itself is a minimum grounding of hope. Today we have terrorist groups practicing the new religion of death. We also have young and old people around the world eyeing no lovable life prospects in our post-modern world. Some of them are captured by the religion of death.

In our context, it is not about the iron curtain, cold war and terror balance. Instead of deterrence and threats of nuclear weapons, we may envision a competition between forces of nature destruction and climate catastrophe on the one side, and the agents of mitigation and protection, on the other side. It is the love of life – love of animals and nature, of people and society, that is the driving force of the latter. To solve the problems of natural destruction, climate change and world peace, we need at least the will to live, in peace and productive community with people of all colors, ethnicities and national groupings – across borders and generations.

Hope or optimism

Terry Francis Eagleton is one of Europe's leading intellectuals, and author of "Hope without optimism" (2015). He says that philosophers have long since abandoned the topic of "hope". Hope is a factor that we have strangely chosen to overlook - "a curiously neglected notion", and he quotes the British writer and social critic, Raymond Williams (1921–1988) saying that we live with "the felt loss of a future". Instead of hope, we have a world with forecasts and predictions and varying degrees of optimism and pessimism. But it is a poor substitute. Some among us are chronic optimists, others are notorious pessimists, and most of us are somewhere in between. But subjective optimism or pessimism does not sustain a notion of future and hope that has been lost in the meantime.

As Eagleton writes, there can be many reasons to believe that things will turn out well, but the fact that some among us claim to be optimists is not one of them. A statement such as "it won't rain, because I'm going to a garden party" testifies to excessive optimism and poor understanding of natural laws and weather forecasts. In the face of the world's great challenges, both optimism and pessimism drive us to passivity. The optimist says that things will go well - we don't need to do anything - a well-known attitude among climate deniers. The pessimist says that things are going wrong, and there is no use trying to do anything about it (recall Edsel Murphy's Law – "Anything That Can Go Wrong, Will Go Wrong"). The optimist is an optimist because he is an optimist – not because of a fundamental truth about the world. The pessimist is a pessimist because he is a pessimist – not because of special insight.

Optimism and pessimism are partly hereditary and partly learned. Eagleton refers to them as a "quirk" of personal temperament, or a "gut feeling", not the result of deep reflection, nor a virtue or a moral quality. Optimism and pessimism can influence our judgments and decisions, but they do not change the mathematics behind the analyses or the arguments behind a decision.

Both optimists and pessimists may use the language of hope. The pessimist may speak about hopelessness, while the optimist may see great hope. There is still a difference between hope/hopelessness and optimism/pessimism. It is about two different scales – two different phenomena. Optimism/pessimism is rooted in personal temperament. Genuine hope is rooted in the combination of argument and analysis, faith and commitment, love and desire, virtue and the will to action. To the extent that optimism or pessimism affects hope, it is about something situational or case-specific, about different views on available data and analytical assessments. For real hope we do better by studying possibilities and probabilities, than by eliciting individual temperaments. Hope may be graded and modified by facts and analyses, regardless of whether I personally happen to be an optimist or a pessimist. And hope, stimulated by love, is a prime mover for action, for the natural world and for the human society.

Love of nature

In "Climate – A New Story" the environmentalist and wildlife expert, Charles Eisenstein (2018), says that it is our love of nature, animals, and species, and our sorrow for lost species and lost nature, that will be the main driving forces for change – more than dry numbers, scientific forecasts, and detailed climate reports. In the introductory chapter of his book, he remembers a childhood experience of watching a flock of starlings flying past, being told by his father that similar flocks of passenger pigeons used to fill the sky – but now extinct: "I still remember the event that made me into an environmentalist" – seven years old.

For myself, I remember an early morning at our forest cabin in South Norway. I was by then a grown-up person, spending a couple of days at the cabin with my aging father. In the early morning, we heard the calling signals of a majestic *tiur* grouse just outside the cabin. We looked at it for a long time and listened to his repeated calling. "He cannot find a mate", my father commented in a very sad tone. The species is not extinct, not even red-listed. But it was a reminder of vulnerability and the force of love for environmental concern. It was also a reminder of the shifting baselines problem. In my childhood, I sometimes went out early, skiing the long moorland stretches to watch the grouse-play in late winter/early spring, with their noisy and violent mating games. It is an experience that my own children and grandchildren have not shared.

In a world of rapid urbanization, baselines of nature, habitats, and wildlife, are not only shifting. For large parts of the world population, they are non-existent. They have no experience or personal awareness of life in nature. Nothing to compare to – no baseline. Factory farming and large-scale logging are just part of industrial production systems, with no attachment to nature, and no feelings to be stirred – perhaps apart from some subconscious signal reminding us of a long-lost past.

Eisenstein says that there is a double set of extinctions taking place. Species go extinct, but so do human feelings about nature love and loss. Extinctions in nature do not occur because people are evil, or in lack of willpower and character, but because we are caught in lifestyles, work environments, and cultural materialism, where the element of loss and the crying for an extinct species is not part of life. We have become addicted to lifestyle and consumption parameters, and we are unable to break the addiction, not because of lack of information or willpower, but because we lack the emotional drive of love for hope and engagement.

In contrast to what he calls "the standard narrative" (linked to BAU), and "the story of separation", that make us "numbed" to environmental depletion, and that destroys parts of ourselves, Eisenstein opts for "a new story" of real need satisfaction. It is a story of connection, relationships, beauty in life, ecological healing, intimacy, sacredness, community and relational self-development. We are being numbed to deeper feelings, engaging in competitive behaviors, sometimes even lured to go for climate strategies that in turn may worsen the natural environment, or framing green transition paths in ways that are part of the problems.

Eisenstein refers to Thich Nhat Hanh's term "interbeing", not suggesting the elimination of the self, nor a belief that "we're all one", but for an expansion of the self-concept to include relations and relation-building, making us part of something greater than ourselves, including our very "beingness" at an expanded and heightened level. We have to restore the internal ecosystem in order to save the external environment. It is done not by cultivating and repeating environmental, societal, and geopolitical threats and warnings, but by cultivating love and relationships while eliciting hope.

The standard narrative is framed in a totality of circumstances, and it is both addictive and "sticky". Eisenstein sums up some of the stickiness elements surrounding and impacting an American fracking executive: corporate culture, culture of the energy industry, performance pressures, economic pressures, attacks from environmental activists (appearing to be both misguided and ignorant), national aims for American energy independence, ideologies of growth and progress, faith in technological innovation, "Earth as a thing" perspective, childhood "success and career" programming. It is sticky, and it is easy to get stuck.

The standard narrative tells us that the quantitative achievement of more of everything will relieve us of the pains and frustration of constant want. Instead, Eisenstein reminds us that, beyond the basics, the quantitative solutions will only cause us to experience greater want. Satisfaction comes from a qualitative improvement, and it can relieve us of the want for limitless more, and from the competition that comes with it. It is about the difference between Aristotle's "natural wealth" and "wealth getting", between limited needs and unlimited greed.

First-hand experience of the blessings of nature is basic to the love of nature. It can be memories of a childhood trout-lake, or the smell in a primeval forest, the taste of wild strawberries and cherries, the sensation of swimming naked in a forest lake. For improvement, schools can add nature exposure and discovery to the syllabus. Direct participation in *dugnad* type activities for nature and animal protection, such as plastic garbage collection or the planting of trees and removal of invading species will add to the learning process and to the anchoring of baselines. Participation in rewilding will enhance the experience and the learning. Family farming may open windows of attraction to organic food, animal welfare, insect protection, flower variety, birds and wildlife. For our age of digital communication, a great number of fantastic nature movies may add to understanding and care for animals and ecosystems. And, as activists repeatedly tell me, love of nature drives action, and with action comes hope and more action.

Hope from the past

For an expansion of our bases of hope, Rebecca Solnit (2005) found "Hope in the Dark", countering despair and defeatism by reminding of the many victories, of small and larger scale initiatives and movements – all too easily forgotten at times of setbacks. The world-music metaphor, and inspiration from the pragmatism philosophy of William James, is a reminder that variety in the past is indicative of variety in the future, opening for possibilism. James' is a philosophy that may be thought of as emphasizing "fruits over roots," looking for the practical results of ideas and the difference ideas make in the real world, more than at the metaphysical arguments about the grounding of truth. In the merger between pragmatism and free will, as evidenced in "The Will to Believe" (James, 2014 [1897]) and "Pragmatism" (James, 2014 [1907]) he was concerned with the role of chance, coming before choice, and emerging from experience. Choices are based on a variety of experience, from our own past and from the observations of others, stored in memory and guiding our actions. Since such experiences also include positive outcomes, they bear witness to the possibilities for the future.

Solnit has expanded on that same pragmatist thinking by reminding us that hope doesn't mean denying negative realities. We need to face the real world, the environmental burden and the long list of threats. But we also need to remember all kinds of reasons for hope. Unfortunately many positive experiences are lost to amnesia. Hope does not imply that everything was fine in the past or will be fine in the future. But hope means to recognize possibilities and openings for visions and dreams and transformation. It means to open up the window to novel action. At a time of great geopolitical, social-political and technological uncertainty, having hope is suggestive of possibilism. Strong oligarchic forces are exploiting the shock doctrine of uncertainty, but the same uncertainty may also serve as an inspiration for people around the world to come up with ideas, and to form new movements and organizations, new alliances and bonds.

In a preview to the third edition of "Hope in the Dark", Solnit (2017) suggest the metaphor of mushrooms. Many mushrooms emerge from the ground after rainfall, as if from nowhere. Invisible to us, they are linked underground to a "fruiting body" of fungus. Our history of environmental action and social movements may appear to represent spontaneous and largely unconnected phenomena. The fungus metaphor tells us another story. We are not alone. We have connections, not always visible and conscious, but always ready to bloom after the rainfall. The destructive forces of denials, delays and obstructions will not prevail forever. Like mushrooms, mitigative and adaptive action for the planet and for the safety of society will grow spores, form stringy hyphae and root-like mycelium, finally growing primordia that develop into mushrooms. From there it continues to release more spores and grow more mushrooms, feeding planet-saving movements.

A final reflection

Hope is only relevant against a backdrop of love but also of threats to what is loved. It is because of heightened awareness of the shadows of life – as envisioned in the Introduction to this text – that the promotion of hope is urgent. Hope is fostered not by overlooking or disregarding threats, but by taking them for real, searching for solutions and engaging in mitigatory action.

We started out in Part One with a statement of physical threats and the environmental burden, to which we added problems of stickiness and green energy provision, and also the possibilities and challenges of an emerging artificial intelligence (AI) revolution. In Part Two we added psychological, social, economic, political, and technological hurdles – and explanations for problem denials and delays of mitigative action. Finally, Part Three presented metaphors for transformative change – for top-down war-economic scale political action, for bottom-up "world music" level innovative involvement from below and from the periphery, and reciprocal *dugnad* style "possibilist" alternatives to the predominant political order and the political-economic systems of the present.

Tempered and ultimate threats

For the dark backdrop, we have recognized the triple threats to the environment, related to climate, the animal world and environmental toxins, with climate approaching tipping points, bringing irreversible heating and extreme weather. We have also noticed threats to ecosystems, loss of biomass and outlooks for the sixth extinction, observing that wild animals now make up only 4 percent of biomass, against 36 percent for humans and 60 percent for husbandry animals. To the long lists of environmental concerns we also had to add the stickiness of fossil energy and the enormous challenge of green energy provision at a scale sufficient to carry production and consumption at present levels. We noticed that green electricity now accounts for 40 percent of global electricity production, but only some 12 percent is based on fast-growing solar and wind power. And since electric power adds only 20 percent to the full energy package, the green energy growth sector (of wind and solar) makes up only 2.5 percent of the total. There is a long way to go.

As for the AI turn, the announcement of global societal change at a scale hardly seen before should be taken for a fact. Incredible opportunities related to health, happiness and intelligent action are emerging, but so are also opportunities for immiseration, totalitarianism and dystopia. An AI-driven "Singularity" of artificial general intelligence (AGI) may or may not be passed – ever. But already AI is surpassing human recall, reasoning, communication and decision-making in ever more fields, potentially taking over both a sizeable share of the job market and of managerial governance. Alignment of AI development and AI governance to human values may break down – either because AI will not come with human AGI intelligence, or because humans and AI developers cannot agree about values among themselves, or because AGI-equipped robots simply decide to go for alternative values (such as maximizing the output of paper clips or eliminating humans from the food chain). As noted by Hawking (2018) AI does not need AGI, neither for posing opportunities or threats.

For the shadow side of life, the main concern of the present text is not about the ultimate "threats to humanity's long-term potential", as dramatically described by Toby Ord in "The Precipice" (2020), but rather about the "tempered" threats to the good life and the ecological health of the planet.

Denials and delays

In Part Two we sought answers to why mitigative action takes so long. For handling climate and environmental issues, and for a tempered, controlled AI development, concerted action and collective resolve is badly needed. Several subversive factors are at work to compromise international agreements and national policies for protection, solidarity, and for mitigative and adaptive action.

Under the neoliberalist political order, strong political denial of climate science has followed a long line of environmental and health problem denials, from leaded gasoline, acid rain, and cigarette smoking to ozone layer depletion and climate change. Most bluntly, President Donal Trump came to reject the full line of the green transition in his inauguration speech. An important tipping point in green transition support may come as prognoses of latent extreme weather trauma is turned into manifest damage. It is about an emerging sense of crisis or sense of urgency. That may also come into play for the regulations of AI developments, removing control from the hands of tech billionaires and totalitarian regimes, allowing for democratic controls. Much will hinge on the transition from latent to manifest environmental damage and from AI as tech dreams to competitor for jobs and governance control.

To the long list of causes of delay and inaction, we have traced the historical roots of developmental conflicts, in modernization and colonialism, immiseration, anger and *ressentiment* at national and international levels, causing disintegration and rivalry where cooperation is needed for mitigation, bringing neo-fascism and authoritarianism where democratic reform and solidarity is needed. As for the Green Transition, we have also made deeper dives into the challenges of growth and inequality and, for addition of the AI transition, into job market erosion. The logic of the problem was summarized at the end of Chapter 13, and it runs as follows.

We have seen that economic growth historically has been linked to resource depletion and environmental problems, for the last 70 years adding an enormous burden to the "net zero" target for 2050. A "net zero" means to eliminate pollution and depletion from all of present day production and consumption. That is in itself an enormous challenge. For the future of economic growth, however, a net zero goal means that all new activity must be absolutely de-coupled from the environmental burden. The absolute decoupling of economic growth stand out as a main challenge of economic policy for the present and for the next few decades. Added together, it is about achieving an absolute decoupling of all economic activity and future growth from nature destruction, for climate and for wilderness and human environments.

AI is expected to provide significant productivity growth, allowing for significant economic growth. Without economic growth, increased productivity will lead to unemployment. Various instruments are available for retraining, sharing of jobs, public works, etc., but generally, with improved productivity in the absence of economic growth there will be increased pressure on employment and wages. And with productivity growth over a long period of time, enormous economic growth will be required to maintain full employment and keep up wage levels. Huge economic growth without absolute decoupling will, however, significantly increase the environmental footprint (related to climate, nature, and pollutants), which by all means must be avoided.

In summary, we must achieve several things at once. First of all, for a net zero, there must be an absolute decoupling between environmental footprint and economic activity – both for present activity and for new, growth-related activity. It will mean the end of greenwashing, free-ridership and illusionary reporting. Secondly, we need to be realistic and honest about the possibility that the imperative of decoupling will reduce the growth potential of the economy and the productivity effects of AI, even allowing for "de-growth" or "negative growth", for higher prices and restricted consumption. Thirdly, AI development may be used for green efficiency and overall greening of the economy, but AI may also produce challenges related to the environmental burden and to "unalignment" and dystopian outlooks, for which strong regulations are needed. Fourthly, we have to be prepared for AI facilitation of robot-driven production of most things, disrupting labor markets, causing joblessness, wage levels to drop and returns on capital to increase, causing the emergence of a "derivative sector" of "irrelevant" and "useless" people. Fifthly, livable conditions and meaningful lives will have to be secured by means of some form of guaranteed minimum income and inheritance. Respecting the ageless "right to live" imperative will mean to allow the "derivative sector" to thrive, with meaningful activity absolutely decoupled from environmental destruction, and freed from the repressive norms of the commercial sector.

For a hopeful future

For those worrying about the preservation of BAU and the neoliberal political order, the story has to be told and retold that with *ressentiment* and the disintegrative forces of the 2000s, with latent environmental damage turning manifest, with all the uncertainty of AI, and with the political-economic revolution of Trump II, BAU is already past. It is time for possibilism and a breakaway. There is no chance that the political economy of the 2050s will be a prolongation of neither what was crushed by Trump nor what was installed by the tech billionaires in 2025. As little as we in 1925 could foresee the challenges and opportunities of 2025, can we today foresee the world of 2125. And the life and work of 2050 may be more different from 2025 than 2025 was from 1925. Hopefully, some values will prevail, however, such as the right to live and the desire for peace and security and for meaningful engagement.

Here is my list of bases for hope:

1. Across philosophies, religions, and cultures, nearly a hundred percent of humanity prefers life over death – at least for themselves and for family and friends. We noticed with reference to the theologist of hope, Jürgen Moltmann, that even for the workings of the cold war "terror balance" and "nuclear deterrence", the love of one's own life, and that of one's nation was mandatory for the mechanism to work. In spite of threats and cases of genocide and ethnic cleansing, there is above all a pan-humanist preference for life over death.
2. Rather than being chosen on the basis of fitness (including capacity for violence and combat), anthropologists are increasingly focusing on "survival of the friendliest" and the most empathic, as dominant traits behind the evolution of human nature. Historically people have been good at offering food and shelters to people in need, and they have generally sought to avoid war and violence. People of good will is an important basis for hope. While the paradox of power is solidly at work, the precariat does not have to prevail and expand at a time of AI dominance. A derivative sector funded by guaranteed minimum income may take survival of the friendliest to a new level of humanitarian progress.
3. In all of nature, and among all species, life preservation and expansion is a driving force. More than 70 percent of the earth's biomass has been lost over the last 60 years, but everywhere, and among all species, individuals of the natural world seek propagation and survival. Animals, birds, and all living beings are our number one allies. We can support our allies by "wilding" and "rewilding" initiatives. Rewilding is only possible because nature has the capacity for wilding. It starts with the mystery of life itself. Even without human agency and ecological project management, wilding is working in nature on its own, making desolate areas bloom again and near-extinction species rise to multiply again – if only given a chance. We can accept and respect the intrinsic value of all varieties of nature and ecosystems, even beyond their commercial contributions and their "services" to people. And we can expect all living beings to respond exuberantly, with propagation and proliferation.
4. Hope can be anchored in history. The pragmatist philosophy of William James tells us to look to personal and societal history and seek out varieties of initiatives and solutions, customs and conventions that have served us well in the past. Rebecca Solnit has more recently reminded us of "Hope in the Dark" – the importance of digging out success stories from the past in order to gain hope for the future. Pragmatism is also a call to look back to all kinds of historical technologies (such as windcatcher structures), many of them long time forgotten, that may ease our dependence on electricity and other modern technologies that come with "stickiness" and additions environmental burden.
5. Hope can also be anchored in the future, in technology and society and in prospects for future innovation. For all of our green transition challenges and AI potentials, there are solutions to be discovered, entirely new scientific insights to emerge, for both technical improvement and societal progress. Principles for solidarity and peace, and technologies for environmental protection and climate stability are waiting to be discovered. The "Russian roulette" gamble with the planet, even at the point of "threats to humanity's long-term potential", still suggest a 5/6 chance of survival. While the probability of less dramatic suffering is considerably higher than the chance of ultimate collapse, it would still be irrational not to have hope.
6. Finally, the existence of hope is itself a solid footing for further hope. Hope for human and animal life on earth works like a self-fulfilling prophecy. Hope rooted in moral conviction and faith in people, in trust mpathy can breed empathy. We don't have to seek escapes on another planet or in trans-human life forms. The American sociologist, Robert K. Merton introduced the term "self-fulfilling prophecy" to the social sciences (1948), suggesting that beliefs and

expectations, whether well-founded or ill-conceived, would impact the outcome. Self-fulfilling prophecies can be about positive or negative outcomes, at personal and collective levels, acting as positive feedback loops, causing wars (First world war!) or miracles, bank runs and social movements, based always on predictions contributing to their accomplishment. Hope is a more narrow category, serving self-fulfilling prophecies that the belief that something positive will come true, anchored in love and concern, simply because we hang on to hope.

About nature, there is a probing question I have sometimes asked friends: What if there were no people on Earth, only an enormous biodiversity of animals, plants, and other creatures, with no markets, no traders, and no monetary or barter value at all. Would that mean that all of nature, and the Earth itself would be entirely worthless? Under capitalism we act as if it would. The question belongs to our reasoning about love and hope. Can we love nature, beyond the commercial value of occupation and extraction? I believe so, and I envision hope. While in the English language (unlike most other languages) material wealth is referred to in personal quality terms ("he is worth $250 billion"), the truth is that we love people, not only for their value in labor markets or financial markets. We value an Olympic medal, not for its value by metal weight, but for the respect it bestows on the laureates. Love of nature and people, for something more than their Midas value, is a starting point for hope and protective environmental action.

As individuals, we have varying capacities and freedoms for impacting politics in defense of the planet, where politicians are bestowed responsibility for execution. But as individuals, we are also "world music" style carriers of innovative action, compassionate care, artistic and creative potential, beyond the right to vote and influence politics. We can opt for mutual benefit, and for those excluded from gainful employment we may develop an emergent social-political "derivative sector". Instead of "uselessness" and "irrelevance" the derivative sector carries the potential to free people from the straight jacket of materialist competition, conspicuous consumption, egocentrism and greed, and from the paradox of power and the Midas disease of money. Instead, it carries hope for fulfilling lives, supporting endless needs of care, inclusion, compassion and artistic creativity.

The fact that computers are superior chess players has not stopped humans from enjoying the game. Life may be enjoyed, even when AI and AGI play most games and perform most operations better. The "derivative sector" may still come to offer more rewarding life experiences than those offered by the traditional market sector, or by a future AI-framed commercial sector – more in line with a good *dugnad* event.

References

Abbott, R., and Bogenschneider, B. (2018). Should robots pay taxes: Tax policy in the age of automation. *Harvard Law & Policy review*, 12: 145.

Acemoglu, D. (2024). Don't Believe the AI Hype. *Project Syndicate*. Retrieved May 21, 2024, from https://www.project-syndicate.org/commentary/ai-productivity-boom-forecasts-countered-by-theory-and-data-by-daron-acemoglu-2024-05.

Acemoglu, D., and Restrepo, P. (2018). Artificial intelligence, automation, and work. In *The economics of artificial intelligence: An agenda* (pp. 197–236). University of Chicago Press.

Acemoglu, D., and Restrepo, P. (2020). Robots and jobs: Evidence from US labor markets. *Journal of Political Economy*, 128(6): 2188–2244.

Achakulwisut, P., Lazarus, M., Asvanon, R., Almeida, P. C., Fauzi, D., Ghosh, E. et al. (2023). The Production Gap Report 2023: Phasing down or phasing up? Top fossil fuel producers plan even more extraction despite climate promises.

Ackeman, F. (2007). Debating Climate Economics: The Stern Review vs. Its Critics. Report to Friends of the Earth. In.

Advisors, D., de Wit, W., and Bigaud, N. (2019). No Plastic in Nature: Assessing plastic ingestion from nature to people. *WWF. Gland, Switzerland. Acedido a*, 15.

Afanasyev, V. G. (1971). *The Scientific Management of Society*. Progress Publishers.

Agarwal, A., Baskaran, S., Parekh, N., Cho, C.-L., Henkel, R., Vij, S. et al. (2021). Male infertility. *The Lancet*, 397(10271): 319–333.

Aghion, P., Antonin, C., Bunel, S., and Jaravel, X. (2020). What are the labor and product market effects of automation? New evidence from France. *CEPR Discussion Paper No.* DP14443.

Ahmadi, M. H., Ghazvini, M., Sadeghzadeh, M., Alhuyi Nazari, M., Kumar, R., Naeimi, A. et al. (2018). Solar power technology for electricity generation: A critical review. *Energy Science & Engineering*, 6(5): 340–361.

Åhman, H. (2013). Social sustainability–society at the intersection of development and maintenance. *Local Environment*, 18(10): 1153–1166.

Ahmed, N., and Wahed, M. (2020). The De-democratization of AI: Deep learning and the compute divide in artificial intelligence research. *arXiv preprint arXiv:*2010.15581.

Ahmed, S., Darwish, A. S., and Farrell, P. (2022). Benchmarking sustainability of megaprojects: A review. Sustainable Energy Development and Innovation: Selected Papers from the World Renewable Energy Congress (WREC) 2020.

Aldhous, P. (2024). At what point does driving an electric vehicle get you a carbon emissions payoff? *Proceeding of the National Academy of Sciences*.

Alexander, M. (2011). The new jim crow. *Ohio State Journal of Criminal Law*, 9, 7.

Allen, D., Spoelman, N., Linsley, C., and Johl, A. (2024). The Fraud of Plastic Recycling: How Big Oil and the plastics industry deceived the public for decades and caused the plastic waste crisis. *Center for Climate Integrity. United States of America.*

Allen, P., and Greaves, M. (2011). The singularity isn't near. *Technology Review*, 12(2011): 7-8.

Alvaredo, F., Atkinson, A. B., Piketty, T., and Saez, E. (2013). The top 1 percent in international and historical perspective. *Journal of Economic Perspectives*, 27(3): 3–20.

Anastas, P., and Eghbali, N. (2010). Green chemistry: principles and practice. *Chemical Society Reviews*, 39(1): 301–312.

Andersen, K. (2020). *Evil geniuses: The unmaking of America: A recent history*. Ebury Press Penguin Random House.

Anderson, S., and Alperstein, O. (2023). Wall Street Bonuses Decline But Still Dwarf Worker Pay Increases Since 2008 Crash.

Anderson, T. R., Hawkins, E., and Jones, P. D. (2016). CO_2, the greenhouse effect and global warming: from the pioneering work of Arrhenius and Callendar to today's Earth System Models. *Endeavour*, 40(3): 178–187.

Anthony, M. A., Bender, S. F., and van der Heijden, M. G. (2023). Enumerating soil biodiversity. *Proceedings of the National Academy of Sciences*, 120(33): e2304663120.
Applebaum, A. (2024). *Autocracy, Inc. - The Dictators Who Want to Run the World*. Penguin Random House UK.
Arendt, H. (2006 [1963]). *Eichmann in Jerusalem - A Report on the Banality of Evil*. Penguin Books.
Arendt, H. (2017 [1951]). *The Origins of Totalitarianism*. Penguin Classics.
Arens, M., Åhman, M., and Vogl, V. (2021). Which countries are prepared to green their coal-based steel industry with electricity?-Reviewing climate and energy policy as well as the implementation of renewable electricity. *Renewable and sustainable energy reviews*, 143: 110938.
Aronson, S., and Barak, B. (2023, April 27th). Five Worlds of AI (a joint post with Boz Barak).
Ashby, W. R. (1956). *An Introduction to Cybernetics*. Champan and Hall.
Ashby, W. R. (1958). Requisite Variety and its implications for the control of complex systems. *Cybernetica (Namur)*, 1(2).
Assi, L., Carter, K., Deaver, E. E., Anay, R., and Ziehl, P. (2018). Sustainable concrete: Building a greener future. *Journal of Cleaner Production*, 198: 1641–1651.
Atkinson, A. B., and Piketty, T. (2007). *Top incomes over the twentieth century: a contrast between continental european and english-speaking countries*. Oxford University Press.
Autor, D. H. (2015). Why are there still so many jobs? The history and future of workplace automation. *Journal of Economic Perspectives*, 29(3): 3–30.
Autor, D. H., and Dorn, D. (2013). The growth of low skill service jobs and the polarization of the US labor market. *The American Economic Review*, 103(5): 1553–1597.
Autor, D. H., Levy, F., and Murnane, R. J. (2003). The skill content of recent technological change: An empirical exploration. *The Quarterly Journal of Economics*, 118(4): 1279–1333.
Auty, R. M. (1995). Economic development and the resource curse thesis. In *Economic and political reform in developing countries* (pp. 58–80). Palgrave Macmillan.
AXA. (2024). *AXA Future Risks Report 2024*. AXA. https://www.axa.com/en/news/2024-future-risks-report.
Bailey, K. L. (2009). *An evaluation of the impact of Hurricane Katrina on crime in New Orleans, Louisiana* Texas State University - San Marcos]. San Marcos.
Baily, M. N., and Kane, A. (2024). *How will AI affect Productivity*. Brookings Institution.
Baiman, R., Clarke, S., Elsworth, C., Field, L., MacCracken, M., Macdonald, J. et al. (2024). Addressing the urgent need for direct climate cooling: Rationale and options. *Oxford Open Climate Change*, 4(1): kgae014.
Bandura, A. (1989). Human agency in social cognitive theory. *American Psychologist*, 44(9): 1175.
Bandura, A. (2001). Social cognitive theory: An agentic perspective. *Annual Review of Psychology*, 52(1): 1–26.
Banerjee, S. B., Jermier, J. M., Peredo, A. M., Perey, R., and Reichel, A. (2021). Theoretical perspectives on organizations and organizing in a post-growth era. *Organization*, 28(3): 337–357.
Banuri, S., and Nguyen, H. (2023). Borrowing to keep up (with the Joneses): Inequality, debt, and conspicuous consumption. *Journal of Economic Behavior & Organization*, 206: 222–242.
Bar-On, Y. M., Phillips, R., and Milo, R. (2018). The biomass distribution on Earth. *Proceedings of the National Academy of Sciences*, 115(25): 6506–6511.
Barnett, C. K., and Pratt, M. G. (2000). From threat-rigidity to flexibility-Toward a learning model of autogenic crisis in organizations. *Journal of Organizational Change Management*, 13(1): 74–88.
Barrett, A. M., and Baum, S. D. (2017). A model of pathways to artificial superintelligence catastrophe for risk and decision analysis. *Journal of Experimental & Theoretical Artificial Intelligence*, 29(2): 397–414.
Bartels, L. M. (2009). Economic inequality and political representation. *The unsustainable American state*, 167–196.
Barten, O., and Meindertsma, J. (2023, July 20). An AI Pause Is Humanity's Best Bet For Preventing Extinction. *Time magazine*.
Bassin, M. (2016). *The Gumilev mystique: Biopolitics, Eurasianism, and the construction of community in modern Russia*. Cornell University Press.
Bates, T. R. (1975). Gramsci and the Theory of Hegemony. *Journal of the History of Ideas*, 36(2): 351–366.
Battistoni, A. (2017). The false promise of universal basic income. *Dissent*, 64(2): 51–62.
Baum, S. (2017). A survey of artificial general intelligence projects for ethics, risk, and policy. *Global Catastrophic Risk Institute Working Paper*, 17–11.
Baum, S. D., Goertzel, B., and Goertzel, T. G. (2011). How long until human-level AI? Results from an expert assessment. *Technological Forecasting and Social Change*, 78(1): 185–195.
Baumberg Geiger, B. (2016). The stigma of claiming benefits: a quantitative study. *Journal of Social Policy*, 45(2): 181–199.
Baumberg Geiger, B., Bell, K., and Gaffney, D. (2012). *Benefits stigma in Britain*.
Beaudry, P., Green, D. A., and Sand, B. M. (2016). The Great Reversal in the Demand for Skill and Cognitive Tasks. *Journal of Labor Economics*, 34(S1): 199–247.

Beck, U. (2005). *Power in the global age: A new global political economy*. Polity.
Beck, U. (2016). *The metamorphosis of the world: How climate change is transforming our concept of the world*. John Wiley & Sons.
Bekar, C., Carlaw, K., and Lipsey, R. (2018). General purpose technologies in theory, application and controversy: a review. *Journal of Evolutionary Economics*, 28: 1005–1033.
Bendell, J. (2018). *Deep adaptation: A map for navigating climate tragedy* [Occasional Papers].
Berg, A., Buffie, E. F., and Zanna, L.-F. (2018). Should we fear the robot revolution?(The correct answer is yes). *Journal of Monetary Economics*, 97: 117–148.
Berman, T., and Taft, N. (2021). Global Oil Companies Have Committed to 'Net Zero'Emissions. *It'sa Sham*.
Berners-Lee, M. (2021). *There Is No Planet B: A Handbook for the Make Or Break Years-Updated Edition*. Cambridge University Press.
Bianchini, S., Damioli, G., and Ghisetti, C. (2023). The environmental effects of the "twin" green and digital transition in European regions. *Environmental and Resource Economics*, 84(4): 877–918.
Biggs, D., Biggs, R., Dakos, V., Scholes, R. J., and Schoon, M. (2011). Are we entering an era of concatenated global crises? *Ecology and Society*, 16(2).
Biglan, A., Flay, B. R., Embry, D. D., and Sandler, I. N. (2012). The critical role of nurturing environments for promoting human well-being. *American Psychologist*, 67(4): 257.
Blasco, J. L., King, A., and Jayaram, S. (2018). *How to report on the SDGs: What good looks like and why it matters*
Blinder, A. S. (2009). How many US jobs might be offshorable? *World Economics*, 10(2): 41.
Blinder, A. S., and Watson, M. W. (2016). Presidents and the US economy: An econometric exploration. *American Economic Review*, 106(4): 1015–1045.
Blum, C., Bunke, D., Hungsberg, M., Roelofs, E., Joas, A., Joas, R. et al. (2017). The concept of sustainable chemistry: Key drivers for the transition towards sustainable development. *Sustainable Chemistry and Pharmacy*, 5: 94–104.
Blumenthal, J., Diamond, M. L., Hoffmann, M., and Wang, Z. (2022). Time to break the "lock-in" impediments to chemicals management. *Environmental science & technology*, 56(7): 3863–3870.
Blyth, M., and Fraccaroli, N. (2025). *Inflation - A Guide for Users and Losers*. W. W. Norton & Company.
Blythe, C., and Jepson, P. (2020). *Rewilding: The radical new science of ecological recovery* (Vol. 14). Icon Books.
Boers, N. (2021). Observation-based early-warning signals for a collapse of the Atlantic Meridional Overturning Circulation. *Nature Climate Change*, 11(8): 680–688.
Bofinger, P., and Scheuermeyer, P. (2019). Income Distribution and Aggregate Saving: A Non-Monotonic Relationship. *Review of income and wealth*, 65(4): 872–907.
Bohle, D., and Greskovits, B. (2009). Varieties of Capitalism and Capitalism «tout court». *European Journal of Sociology/Archives Européennes de Sociologie*, 50(3): 355–386.
Boldonova, I., and Tsetsura, K. (2015). Volunteering movement in Russia: Regional presence and future perspectives. *Styles of Communication*, 7(1): 7–22.
Bollier, D. (2020). Commoning as a transformative social paradigm. In *The new systems reader* (pp. 348–361). Routledge.
Boretti, A. (2023). How close is a hydrogen hypersonic commercial aircraft? *International Journal of Hydrogen Energy*, 48(28): 10684–10691.
Bork, R. H. (1978). *The Antitrust Paradox: A Policy at War with Itself*. Basic Books.
Boston, J., and Lempp, F. (2011). Climate change: explaining and solving the mismatch between scientific urgency and political inertia. *Accounting, Auditing & Accountability Journal*, 24(8): 1000–1021.
Bostrom, N. (2005). A history of transhumanist thought. *Journal of evolution and technology*, 14(1).
Bostrom, N. (2017). *Superintelligence*. Oxford University Press.
Bostrom, N., Douglas, T., and Sandberg, A. (2016). The Unilateralist's Curse and the Case for a Principle of Conformity. *Social epistemology*, 30(4): 350–371.
Boulding, K. E. (2013). The economics of the coming spaceship earth. *In*: Boulding, K. (Ed.). *Environmental quality in a growing economy* (pp. 3-14). Resources for the Future - RFF Press.
Bowen, M., and Dabbar, P. (2022). Reducing Russian involvement in western nuclear power markets. *Columbia University, Center on Global Energy Policy URL: https://www. energypolicy. columbia. edu/publications/reducing-russian-involvementwestern-nuclear-power-markets (accessed: 29.04. 2023)*.
Bower, J. L., and Christensen, C. M. (1995). Disruptive technologies: Catching the wave. *Harvard Business Review*, *73*(1): 43–53.
Boyle, A. D., Leggat, G., Morikawa, L., Pappas, Y., and Stephens, J. C. (2021). Green New Deal proposals: Comparing emerging transformational climate policies at multiple scales. *Energy Research & Social Science*, 81: 102259.
Brand, C. (2024). Confronting mobesity is vital for the global electrification of transport. *Nature Energy*, *9*. https://www.nature.com/articles/s41560-024-01559-x.epdf?sharing_token=TP5da_G3UnaQczzOdWvcS

9RgN0jAjWel9jnR3ZoTv0PwWr-LMfRlrkSbEYsJ_4EshM0w7SneQoipS4Xkxo7Yys66V876_WefxmC_Ea7Du6PkJiV82ETjpYYvqtdvzedNQqXErPlo3747PyDCkjhp_zz4O5idcpOW9OBUKeOrQbc%3D.
Bregman, R. (2018). *Utopia for realists*. Bloomsbury Publishing.
Bregman, R. (2020). *Humankind: A hopeful history*. Bloomsbury Publishing.
Brooks, D. R., Hoberg, E. P., Boeger, W. A., and Trivellone, V. (2022). Emerging infectious disease: an underappreciated area of strategic concern for food security. *Transboundary and Emerging Diseases*, 69(2): 254–267.
Brooks, R. C. (1915). Metropolitan free cities. *Political Science Quarterly*, 30(2): 222–234.
Brozović, D. (2023). Societal collapse: A literature review. *Futures*, 145: 103075.
Brulle, R. J. (2023). Advocating inaction: a historical analysis of the Global Climate Coalition. *Environmental Politics*, 32(2): 185–206.
Brynjolfsson, E. (2022). The turing trap: The promise & peril of human-like artificial intelligence. *Daedalus*, 151(2): 272–287.
Bryson, B. (2004). *A short history of nearly everything*. Crown.
Burke, M., Hsiang, S. M., and Miguel, E. (2015). Global non-linear effect of temperature on economic production. *Nature*, 527(7577): 235–239.
Butlin, P. (2021). Ai alignment and human reward. Proceedings of the 2021 AAAI/ACM Conference on AI, Ethics, and Society.
Büntgen, U., Myglan, V. S., Ljungqvist, F. C., McCormick, M., Di Cosmo, N., Sigl, M. et al. (2016). Cooling and societal change during the Late Antique Little Ice Age from 536 to around 660 AD. *Nature Geoscience*, 9(3): 231–236.
Byrne, M. P., and O'Gorman, P. A. (2018). Trends in continental temperature and humidity directly linked to ocean warming. *Proceedings of the National Academy of Sciences*, 115(19): 4863–4868.
Cabigiosu, A., and Cabigiosu, A. (2020). An overview of the luxury fashion industry. *In*: Cabigiosu, A. (Ed.). *Digitalization in the Luxury Fashion Industry: Strategic Branding for Millennial Consumers* (pp. 9–31). Palgrave Macmillan, Cham.
Caesar, L., McCarthy, G. D., Thornalley, D. J. R., Cahill, N., and Rahmstorf, S. (2021). Current Atlantic Meridional Overturning Circulation weakest in last millennium. *Nature geoscience*, 14(3): 118–120. https://doi.org/10.1038/s41561-021-00699-z.
Calderón-Garcidueñas, L., Gónzalez-Maciel, A., Reynoso-Robles, R., Delgado-Chávez, R., Mukherjee, P. S., Kulesza, R. J. et al. (2018). Hallmarks of Alzheimer disease are evolving relentlessly in Metropolitan Mexico City infants, children and young adults. APOE4 carriers have higher suicide risk and higher odds of reaching NFT stage V at≤ 40 years of age. *Environmental Research*, 164: 475–487.
Callendar, G. S. (1961). Temperature fluctuations and trends over the earth. *Quarterly Journal of the Royal Meteorological Society*, 87(371): 1–12.
Cancino, J., Soto, K., Tapia, J., Muñoz-Quezada, M. T., Lucero, B., Contreras, C. et al. (2023). Occupational exposure to pesticides and symptoms of depression in agricultural workers. A systematic review. *Environmental Research*, (231): 116190.
Carr, L., Iacoboni, M., Dubeau, M.-C., Mazziotta, J. C., and Lenzi, G. L. (2003). Neural mechanisms of empathy in humans: a relay from neural systems for imitation to limbic areas. *Proceedings of the National Academy of Sciences*, 100(9): 5497–5502.
Carrington, D. (2025, April 3). Climate crisis on track to destroy capitlism, warns top insurer. *The Guardian*.
Carroll, C. D., Slacalek, J., and Sommer, M. (2011). International evidence on sticky consumption growth. *Review of Economics and Statistics*, 93(4): 1135–1145.
Cartwright, J., George, A., Howgego, J., Clark, S., and Jones, N. (2024). Five climate megaprojects that might just save the world. *New Scientist*.
Case, A., and Deaton, A. (2020). *Deaths of Despair and the Future of Capitalism*. Princeton University Press.
Cellan-Jones, R. (2014). Hawking: AI could end human race. *BBC News*, 2.
Chafkin, M. (2021). *The Contrarian: Peter Thiel and Silicon Valley's Pursuit of Power*. Bloomsbury Publishing.
Chalmers, D. J. (2016). The singularity: A philosophical analysis. *Science fiction and philosophy: From time travel to superintelligence*, 171–224.
Chanza, N., and Musakwa, W. (2021). Indigenous practices of ecosystem management in a changing climate: Prospects for ecosystem-based adaptation. *Environmental Science & Policy*, 126: 142–151.
Chayes, S. (2020). *Everybody Knows: Corruption in America*. C. Hurst & Co.
Chernoff, A., and Jagtiani, J. (2024). *Beneath the Crypto Currents: The Hidden Effect of Crypto "Whales"*. F. R. B. Philadelphia.
Cho, J., Puspitasari, F. D., Zheng, S., Zheng, J., Lee, L.-H., Kim, T.-H. et al. (2024). Sora as an AGI World Model? A Complete Survey on Text-to-Video Generation. *arXiv preprint arXiv:2403.05131*.

Chu, S. (2009). Carbon capture and sequestration. In (Vol. 325, pp. 1599-1599): American Association for the Advancement of Science.
Churchman, C. (1967). West, "Wicked Problems,". *Management Science*, 14(4).
Climate, A. (2023). Production Gap Report 2023.
Clynes, M. E., and Kline, N. S. (1960). Cyborgs and space. *Astronautics*, 14(9): 26–27.
Cobb, C., Halstead, T., and Rowe, J. (1995). *The genuine progress indicator: summary of data and methodology* (Vol. 15). Redefining Progress San Francisco.
Cobb, J., and Daly, H. (1989). For the common good, redirecting the economy toward community, the environment and a sustainable future. In: Boston: Beacon Press.
Coeckelbergh, M. (2021). AI for climate: freedom, justice, and other ethical and political challenges. *AI and Ethics*, 1(1): 67–72.
Collins, J. (2005). Level 5 leadership: The triumph of humility and fierce resolve. *Harvard Business Review*, 83(7): 136.
Collins, J., and Porras, J. I. (1994). *Built to Last - Successful Habits of Visionary Companies*. Harper Collins.
Collins, R. (2008). *Violence: A micro-sociological theory*. Princeton University Press.
Congleton, R. D. (1992). Political institutions and pollution control. *The review of economics and statistics*, 412–421.
Connolly, R., Connolly, M., Carter, R. M., and Soon, W. (2020). How much human-caused global warming should we expect with business-as-usual (BAU) climate policies? A semi-empirical assessment. *Energies*, 13(6): 1365.
Conti, G. O., Ferrante, M., Banni, M., Favara, C., Nicolosi, I., Cristaldi, A. et al. (2020). Micro-and nano-plastics in edible fruit and vegetables. The first diet risks assessment for the general population. *Environmental Research*, 187: 109677.
Cook, B. (2023). *Energy Storage: 4 alternatives to lithium-ion batteries currently exciting investors*. M. Company.
Cook, I., and Cook, I. (2020). The climate change denial industry. *The Politics of the Final Hundred Years of Humanity (2030-2130)*, 65–84.
Cook, J. (2016). Countering climate science denial and communicating scientific consensus. In *Oxford Research Encyclopedia of Climate Science*.
Corbyn, Z. (2021, March 28). "Shanna Swan: Most couples may have to use assisted reproduction by 2045". *The Observer*.
Costanza, R., Kubiszewski, I., Giovannini, E., Lovins, H., McGlade, J., Pickett, K. E. et al. (2014). Development: Time to leave GDP behind. *Nature*, 505(7483): 283–285.
Cowls, J., Tsamados, A., Taddeo, M., and Floridi, L. (2021). The AI gambit: leveraging artificial intelligence to combat climate change—opportunities, challenges, and recommendations. *AI & SOCIETY*, 1–25.
Cox, K. D., Covernton, G. A., Davies, H. L., Dower, J. F., Juanes, F., and Dudas, S. E. (2019). Human consumption of microplastics. *Environmental science & technology*, *53*(12): 7068–7074.
Crafts, N. (2021). Artificial intelligence as a general-purpose technology: an historical perspective. *Oxford Review of Economic Policy*, 37(3): 521–536.
Cremer, H., and Pestieau, P. (2011). The tax treatment of intergenerational wealth transfers. *CESifo Economic Studies*, 57(2): 365–401.
Croeser, E. (2020). *Ecosocialism and climate justice: An ecological neo-Gramscian analysis*. Routledge.
D'Orlando, F. (2020). Technological unemployment and the resurgence of political economy. *American Review of Political Economy*, 15(1).
Daly, H. E. (1991). *Steady-state economics: with new essays*. Island press.
Damario, M. A. (2014). General aspects of fertility and infertility. *Human Fertility: Methods and Protocols*, 3–23.
Dauth, W., Findeisen, S., Suedekum, J., and Woessner, N. (2018). Adjusting to robots: Worker-level evidence. *Opportunity and Inclusive Growth Institute Working Papers*, 13.
Davenport, T. H., and Mittal, N. (2023). *All-in on AI: How smart companies win big with artificial intelligence*. Harvard Business Press.
Davies, H., McKernan, B., and Sabbagh, D. (2023, December 1). The Gospel': how Israel uses AI to select bombing targets in Gaza. *The Guardian*.
Dawkins, R. (2016 [1976]). *The selfish gene*. Oxford University Press.
Dayan, N., Twitto, M., Rochman, Y., Beitler, U., Zion, I. B., Bortnikov, E. et al. (2021). The end of Moore's law and the rise of the data processor. *Proceedings of the VLDB Endowment*, 14(12): 2932–2944.
De Loecker, J., Eeckhout, J., and Unger, G. (2020). The rise of market power and the macroeconomic implications. *The Quarterly Journal of Economics*, 135(2): 561–644.
De Mesquita, B. B., and Smith, A. (2011). *The dictator's handbook: why bad behavior is almost always good politics*. Hachette UK.

de Oliveira-Souza, R., Hare, R. D., Bramati, I. E., Garrido, G. J., Ignácio, F. A., Tovar-Moll, F. et al. (2008). Psychopathy as a disorder of the moral brain: fronto-temporo-limbic grey matter reductions demonstrated by voxel-based morphometry. *Neuroimage*, 40(3): 1202–1213.

Deepinder, F., and Braunstein, G. D. (2011). Gynecomastia: incidence, causes and treatment. *Expert Review of Endocrinology & Metabolism*, 6(5): 723–730.

Delannoy, L., Longaretti, P.-Y., Murphy, D. J., and Prados, E. (2021). Peak oil and the low-carbon energy transition: A net-energy perspective. *Applied energy*, 304: 117843.

Dempsey, N., Bramley, G., Power, S., and Brown, C. (2011). The social dimension of sustainable development: Defining urban social sustainability. *Sustainable Development*, 19(5): 289–300.

Desideri, U., Zepparelli, F., Morettini, V., and Garroni, E. (2013). Comparative analysis of concentrating solar power and photovoltaic technologies: Technical and environmental evaluations. *Applied Energy*, 102: 765–784.

Dillman, K. J., Árnadóttir, Á., Heinonen, J., Czepkiewicz, M., and Davíðsdóttir, B. (2020). Review and meta-analysis of EVs: embodied emissions and environmental breakeven. *Sustainability*, 12(22): 9390.

Dinneen, J. (2024). Gaseous gold. *New Scientist*, 261(3476): 32–35.

Dinse, G. E., Parks, C. G., Weinberg, C. R., Co, C. A., Wilkerson, J., Zeldin, D. C. et al. (2022). Increasing prevalence of antinuclear antibodies in the United States. *Arthritis & Rheumatology*, 74(12): 2032–2041.

Ditlevsen, P., and Ditlevsen, S. (2023). Warning of a forthcoming collapse of the Atlantic meridional overturning circulation. *Nature communications*, 14(1): 2032–2041. https://doi.org/10.1038/s41467-023-39810-w.

Doran, P. T., and Zimmerman, M. K. (2009). Examining the scientific consensus on climate change. *Eos, Transactions American Geophysical Union*, 90(3): 22–23.

Drexler, K. E. (1986). *Engines of Creation: The Coming Era of Nanotechnology*. Anchor Books.

Dryzek, J. S. (2022). *The politics of the earth: Environmental discourses*. Oxford University Press.

Dyke, J. G., Knorr, W., and Watson, R. (2021). 4. Why Net Zero Policies Do More Harm than Good. *Negotiating Climate Change in Crisis*, 39.

Eagleton, T. (2015). *Hope without optimism*. University of Virginia Press.

Economist, T. (2010). Cutting out the middle men. *The Economist* (4th November).

Eeckhout, J. (2022). *The Profit Paradox: How Thriving Firms Threaten the Future of Work*. Princeton University Press.

Egan, M. (2020). *Stocks perform better when a Democrat is in the White House. History proves it* C. Business. https://edition.cnn.com/2020/09/23/investing/stock-market-election-trump-biden/index.html.

Eisenstein, C. (2018). *Climate: A new story*. North Atlantic Books.

Eizenberg, E., and Jabareen, Y. (2017). Social sustainability: A new conceptual framework. *Sustainability*, 9(1): 68.

Ellis, E. (2011). The planet of no return: Human resilience on an artificial Earth. *Breakthrough Journal*, 2(Fall), 37–44.

Ember. (2023). Wind and solar reach a record 12% of global electricity in 2022. *Ember's fourth annual Global Electricity Review*.

Emmenegger, P., and Marx, P. (2019). The politics of inequality as organised spectacle: why the Swiss do not want to tax the rich. *New political economy*, 24(1): 103–124.

Enevoldsen, P., Permien, F.-H., Bakhtaoui, I., von Krauland, A.-K., Jacobson, M. Z., Xydis, G. et al. (2019). How much wind power potential does europe have? Examining european wind power potential with an enhanced socio-technical atlas. *Energy Policy*, 132: 1092–1100.

England, M. R., and Polvani, L. M. (2023). The Montreal Protocol is delaying the occurrence of the first ice-free Arctic summer. *Proceedings of the National Academy of Sciences*, 120(22): e2211432120.

Eres, R. (2016). Scanning for empathy. *Australasian Science*, 37(2): 30–31.

Eriksen, S. H., Simpson, N. P., Glavovic, B., Ley, D., Carr, E. R., Fernández-Carril, L. et al. (2024). Pathways for urgent action towards climate resilient development. *Nature Climate Change*, 1–4.

Fei, T., and Shuang-Qing, X. (2012). Definition of business as usual and its impacts on assessment of mitigation efforts. *Advances in Climate Change Research*, 3(4): 212–219.

Felten, E. W., Raj, M., and Seamans, R. (2018). A method to link advances in artificial intelligence to occupational abilities. AEA Papers and Proceedings,

Feola, G., Vincent, O., and Moore, D. (2021). (Un) making in sustainability transformation beyond capitalism. *Global Environmental Change*, 69: 102290.

Ferguson, T., and Rogers, J. (1986). *Right turn: The decline of the Democrats and the future of American politics*. Macmillan.

Festinger, L. (1957). *A Theory of Cognitive dissonance*. Stanford University Press.

Festinger, L., Riecken, H., and Schachter, S. (1956). *When Prophecy Fails: A Social and Psychological Study of A Modern Group that Predicted the Destruction of the World*. Harper-Torchbooks.

Floyd, J., Alexander, S., Lenzen, M., Moriarty, P., Palmer, G., Chandra-Shekeran, S. et al. (2020). Energy descent as a post-carbon transition scenario: How 'knowledge humility'reshapes energy futures for post-normal times. *Futures*, 122: 102565.
Flyn, C. (2022). *Islands of abandonment: Nature rebounding in the post-human landscape*. Penguin.
Flynn, C., Jardon, S. T., Dibden, J., Grant, Z., Kaya, R., Kirby, M. et al. (2024). *Peoples' Climate Vote 2024*. U. N. D. P. (UNDP). chrome-extension://efaidnbmnnnibpcajpcglclefindmkaj/https://peoplesclimate.vote/document/Peoples_Climate_Vote_Report_2024.pdf.
Flyvbjerg, B., and Gardner, D. (2023). *How Big Things Get Done: The Surprising Factors that Determine the Fate of Every Project, from Home Renovations to Space Exploration and Everything in Between*. Signal.
Flyvbjerg, B., and Turner, J. R. (2018). Do classics exist in megaproject management? In (Vol. 36, pp. 334–341): Elsevier.
Folke, C., Polasky, S., Rockström, J., Galaz, V., Westley, F., Lamont, M. et al. (2021). Our future in the Anthropocene biosphere. *Ambio*, 50: 834–869.
Fossen, F., and Sorgner, A. (2019). Mapping the future of occupations: transformative and destructive effects of new digital technologies on jobs. Форсайт, *13*(2 (eng)), 10–18.
Fraga-Lamas, P., Lopes, S. I., and Fernández-Caramés, T. M. (2021). Green IoT and edge AI as key technological enablers for a sustainable digital transition towards a smart circular economy: An industry 5.0 use case. *Sensors*, 21(17): 5745.
Fressoz, J.-B. (2024). *More and More and More - An All-Consuming History of Energy*. Allen Lane - Penguin Random House Ireland.
Frey, C. B., and Osborne, M. A. (2017). The future of employment: How susceptible are jobs to computerisation? *Technological Forecasting and Social Change*, 114: 254–280.
Friedlingstein, P., O'sullivan, M., Jones, M. W., Andrew, R. M., Hauck, J., Olsen, A. et al. (2020). Global carbon budget 2020. *Earth System Science Data Discussions*, 2020, 1–3.
Friedman, M. (1962). *Capitalism and freedom*. The University of Chicago Press.
Friedman, M., Friedman, R. D., and Friedman, R. D. (1990 [1980]). *Free to choose*. Free to Choose Enterprise.
Fry, D. P. (2013). *War, peace, and human nature: The convergence of evolutionary and cultural views*. Oxford University Press.
Fursin, G. (2020). Enabling reproducible ML and Systems research: the good, the bad, and the ugly. *August. https:// doi. org/10.5281/ZENODO, 4005773*.
Galbraith, J. K. (1967). *The new industrial state*. Houghton Millin Company.
Galbraith, J. K. (1982). Recession economics. *New York Review of Books*, 29(1): 34–34.
Garcia, J. M., and Robertson, M. L. (2017). The future of plastics recycling. *Science*, 358(6365): 870–872.
Gardiner, S. M. (2009). Saved by Disaster? Abrupt Climate Change, Political Inertia, and the Possibility of an Intergenerational Arms Race. *Journal of Social Philosophy* 40.
Gardner, H. E. (2000). *Intelligence reframed: Multiple intelligences for the 21st century*. Hachette Uk.
Garner, A. J. (2023). Observed increases in North Atlantic tropical cyclone peak intensification rates. *Scientific Reports*, 13(1): 16299.
Gault, M. (2021). Billionaires See VR as a Way to Avoid Radical Social Change. *Wired magazine*. https://www.wired.com /story/billionaires-use-vr-avoid-social-change/.
Gawdat, M. (2021). *Scary smart: the future of Artificial Intelligence and how you can save our world*. Pan Macmillan.
Gawdat, M. (2022). *That Little Voice in Your Head: Adjust the Code that Runs Your Brain*. Pan Macmillan.
Gellert, P. K., and Lynch, B. D. (2003). Mega-projects as displacements. *International Social Science Journal*, *55*(175): 15–25.
Georgieff, A., and Milanez, A. (2021). What happened to jobs at high risk of automation?
Gerstle, G. (2022). *The rise and fall of the neoliberal order: America and the world in the free market era*. Oxford University Press.
Gibb, B. C. (2019). Plastics are forever. *Nature chemistry*, 11(5): 394–395.
Gibson, S. (2013). Milgram's obedience experiments: A rhetorical analysis. *British Journal of Social Psychology*, *52*(2): 290–309.
Gilens, M. (2005). Inequality and democratic responsiveness. *Public Opinion Quarterly*, 69(5): 778–796.
Gilens, M., and Page, B. I. (2014). Testing theories of American politics: Elites, interest groups, and average citizens. *Perspectives on politics*, 12(3): 564–581.
Gillespie, S. (2019). *Climate crisis and consciousness: Re-imagining our world and ourselves*. Routledge.
Gladwell, M. (2006). *The tipping point: How little things can make a big difference*. Little, Brown.
Goessling, H. F., Rackow, T., and Jung, T. (2025). Recent global temperature surge intensified by record-low planetary albedo. *Science*, 387(6729): 68–73.
Gold, M. (2021). Green Intelligence - AI could boost efforts to fight climate change. *The Economist*

Goldblatt, C., and Watson, A. J. (2012). The runaway greenhouse: implications for future climate change, geoengineering and planetary atmospheres. *Philosophical Transactions of the Royal Society A: Mathematical, Physical and Engineering Sciences*, 370(1974): 4197–4216.

Goldsmith, D., and Rees, M. (2022). *The end of astronauts: Why robots are the future of exploration*. Harvard University Press.

Goldstein, J. E., Neimark, B., Garvey, B., and Phelps, J. (2023). Unlocking "lock-in" and path dependency: A review across disciplines and socio-environmental contexts. *World Development*, 161: 106116.

Goldstone, J. A. (1986). State breakdown in the English revolution: A new synthesis. *American Journal of Sociology*, 92(2): 257–322.

Goldstone, J. A. (1988). East and West in the seventeenth century: political crises in Stuart England, Ottoman Turkey, and Ming China. *Comparative Studies in Society and History*, 30(1): 103–142.

Goldstone, J. A. (2016 [1991]). *Revolution and rebellion in the early modern world: population change and state breakdown in England, France, Turkey, and China, 1600-1850*. Routledge.

Graeber, D. (2011). *Debt: The first five thousand years*. Melville House.

Graeber, D. (2018). *Bullshit jobs: a theory*. Simon & Schuster.

Graeber, D., and Wengrow, D. (2021). *The dawn of everything: A new history of humanity*. Penguin UK.

Grant, L., Vanderkelen, I., Gudmundsson, L., Fischer, E., Seneviratne, S. I., and Thiery, W. (2025). Global emergence of unprecedented lifetime exposure to climate extremes. *Nature*, 641(8062): 374–379.

Gray, D., and Stevenson, C. (2020). How can 'we'help? Exploring the role of shared social identity in the experiences and benefits of volunteering. *Journal of Community & Applied Social Psychology*, 30(4): 341–353.

Greenfield, P. (2023). Revealed: more than 90% of rainforest carbon offsets by biggest certifier are worthless, analysis shows. *The Guardian*, 18.

Greenspan, A. (2002). *Testimony of Chairman Alan Greenspan: Federal Reserve Board's Semiannual Monetary Policy Report to the Congress before the Committee on Banking, Housing, and Urban Affairs*. Washington, DC

Gregory, S., Simmons, A., Kumari, V., Howard, M., Hodgins, S., and Blackwood, N. (2012). The antisocial brain: psychopathy matters: a structural MRI investigation of antisocial male violent offenders. *Archives of General Psychiatry*, 69(9): 962–972.

Groom, B., Drupp, M. A., Freeman, M. C., and Nesje, F. (2022). The future, now: A review of social discounting. *Annual Review of Resource Economics*, 14: 467–491.

Grote, G. (1924). *A History of Greece* (Vol. 3). J.M. Dent & Co/E.P. Dutton & Co.

Grullon, G., Larkin, Y., and Michaely, R. (2019). Are US industries becoming more concentrated? *Review of Finance*, 23(4): 697–743.

Grüne-Yanoff, T., and Verreault-Julien, P. (2021). How-possibly explanations in economics: anything goes? *Journal of economic methodology*, 28(1): 114–123.

Guizar-Coutiño, A., Jones, J. P., Balmford, A., Carmenta, R., and Coomes, D. A. (2022). A global evaluation of the effectiveness of voluntary REDD+ projects at reducing deforestation and degradation in the moist tropics. *Conservation Biology*, 36(6): e13970.

Guliyev, H. (2023). Artificial intelligence and unemployment in high-tech developed countries: New insights from dynamic panel data model. *Research in Globalization*, 7, 100140.

Gutiérrez-Negrín, L. C. (2024). Evolution of worldwide geothermal power 2020–2023. *Geothermal Energy*, 12(1): 14.

Hacker, J. S., and Pierson, P. (2010). Winner-take-all politics: Public policy, political organization, and the precipitous rise of top incomes in the United States. *Politics & Society*, 38(2): 152–204.

Hajer, M. A. (1995). *The politics of environmental discourse: Ecological modernization and the policy process*. Clarendon Press.

Halacy, D. S. (1965). *Cyborg: Evolution of the superman*. Harper & Row.

Halal, W. E. (2014). Global Consciousness: Coming of Age for the Planet. *World Future Review*, 6(3): 261–272.

Hamilton, C. (2016). The theodicy of the "Good Anthropocene". *Environmental Humanities*, 7(1), 233–238.

Haney, C., Banks, C., and Zimbardo, P. (1973). *Interpersonal dynamics in a simulated prison*.

Hansen, J. (2005). Is there still time to avoid 'dangerous anthropogenic interference'with global climate. *American Geophysical Union. http://www. columbia. edu/~ jeh1/keeling_talk_and_slides. pdf.*

Hansen, N., Toscani, F., and Zhou, J. (2023). Europe's Inflation Outlook Depends on How Corporate Profits Absorb Wage Gains. *IMF Blog*, 26.

Hanson, R. (2008). Economics of the singularity. *iEEE SpEctrum*, 45(6): 45–50.

Hanson, R. (2016). *The age of Em: Work, love, and life when robots rule the earth*. Oxford University Press.

Haque, M. K., Azad, M. A. K., Hossain, M. Y., Ahmed, T., Uddin, M., and Hossain, M. M. (2021). Wildfire in Australia during 2019–2020, Its impact on health, biodiversity and environment with some proposals for risk management: a review. *Journal of Environmental Protection*, 12(6): 391–414.

Harari, Y. N. (2014). *Sapiens: A brief history of humankind*. Random House.
Harari, Y. N. (2017). *Homo deus - A Brief History of Tomorrow*. Vintage.
Harari, Y. N. (2020). *Read Yuval Harari's blistering warning to Davos in full*. https://www.weforum.org/agenda/2020/01/yuval-hararis-warning-davos-speech-future-predications/).
Hardin, G. (1968). The tragedy of the commons: the population problem has no technical solution; it requires a fundamental extension in morality. *Science*, 162(3859): 1243–1248.
Hart-Landsberg, M. (2023). Planning an Ecologically Sustainable and Democratic Economy. *Monthly Review*(July-August).
Haslam, S. A., Reicher, S. D., and Birney, M. E. (2014). Nothing by mere authority: Evidence that in an experimental analogue of the Milgram paradigm participants are motivated not by orders but by appeals to science. *Journal of Social Issues*, 70(3): 473–488.
Hausfather, Z., and Peters, G. P. (2020). Emissions–the 'business as usual'story is misleading. *Nature*.
Hawking, S. (2018). *Brief answers to the big questions*. Bantam Books.
Haworth, A. (2013). Why have young people in Japan stopped having sex. *The Guardian*, 2013.
Hayek, F. A. (1960). *The Constitution of Liberty*. Routledge and Kegan Paul.
Hayek, F. A. (2007 [1944]). *The Road to Serfdom: Text and Documents*. University of Chicago Press.
Hayek, F. A. (2012). *Hayek on Hayek: An autobiographical dialogue*. University of Chicago Press.
Hebdige, D. (1994). *Hiding in the Light*. Routledge.
Helmers, E., Dietz, J., and Weiss, M. (2020). Sensitivity analysis in the life-cycle assessment of electric vs. combustion engine cars under approximate real-world conditions. *Sustainability*, 12(3): 1241.
Hendriks, M., Burger, M., and Commandeur, H. (2023). The influence of CEO compensation on employee engagement. *Review of Managerial Science*, 17(2): 607–633.
Herrmann, U. (2022). *Das Ende des Kapitalismus: Warum Wachstum und Klimaschutz nicht vereinbar sind–und wie wir in Zukunft leben werden*. Kiepenheuer & Witsch.
Hertwich, E. G., Gibon, T., Bouman, E. A., Arvesen, A., Suh, S., Heath, G. A. et al. (2015). Integrated life-cycle assessment of electricity-supply scenarios confirms global environmental benefit of low-carbon technologies. *Proceedings of the National Academy of Sciences*, 112(20): 6277–6282.
Hicks, J. (1946). *Value and Capital*. Clarendon.
Hirschman, A. O. (1971). *A bias for hope*. Yale university press.
Hirschman, A. O. (2013). 16 essays. In J. Adelman (Ed.), *The Essential Hirschman*. Princeton University Press.
Hochkirch, A., Bilz, M., Ferreira, C. C., Danielczak, A., Allen, D., Nieto, A. et al. (2023). A multi-taxon analysis of European Red Lists reveals major threats to biodiversity. *Plos one*, 18(11).
Hofstede, G. (1984). *Culture's consequences: International differences in work-related values*. Sage.
Hooker, S. (2021). The hardware lottery. *Communications of the ACM*, 64(12): 58–65.
Hope, D., and Limberg, J. (2022). The economic consequences of major tax cuts for the rich. *Socio-Economic Review*, 20(2): 539–559.
Hou, Q., Zhen, M., Qian, H., Nie, Y., Bai, X., Xia, T. et al. (2021). Upcycling and catalytic degradation of plastic wastes. *Cell Reports Physical Science*, 2(8).
Howard, L. (1833). *The climate of London: deduced from meteorological observations made in the metropolis and at various places around it* (Vol. 3). Harvey and Darton, J. and A. Arch, Longman, Hatchard, S. Highley [and] R. Hunter.
Hu, W., and Deng, Z. (2023). A review of dynamic analysis on space solar power station. *Astrodynamics*, 7(2): 115–130.
Huang, X., and Song, J. (2023). Urban moisture and dry islands: spatiotemporal variation patterns and mechanisms of urban air humidity changes across the globe. *Environmental Research Letters*, 18(10): 103003.
Huber, E., Petrova, B., and Stephens, J. D. (2022). Financialization, labor market institutions and inequality. *Review of international political economy*, 29(2): 425–452.
IEA. (2018). *Technology Roadmap - Low-Carbon Transition in the Cement Industry*. IEA. https://www.iea.org/reports/technology-roadmap-low-carbon-transition-in-the-cement-industry.
IEA. (2021). The Role of Critical Minerals in Clean Energy Transitions.
Ilievski, F., Oltramari, A., Ma, K., Zhang, B., McGuinness, D. L., and Szekely, P. (2021). Dimensions of commonsense knowledge. *Knowledge-Based Systems*, 229: 107347.
ILO. (2018). *World employment and social outlook 2018: Greening with jobs*. International Labour Organisation (ILO)
Inglehart, R., Foa, R., Peterson, C., and Welzel, C. (2008). Development, freedom, and rising happiness: A global perspective (1981–2007). *Perspectives on Psychological Science*, 3(4): 264–285.
IPCC. (2023). *Climate Change 2023: Synthesis Report. Contribution of Working Groups I, II and III to the Sixth Assessment Report of the Intergovernmental Panel on Climate Change* IPCC.

Jackson, T. (2013). *Prosperity without growth*. Routledge.
Jackson, T. (2017). *Prosperity without growth: Foundations for the economy of tomorrow*. Routledge.
Jackson, T. (2021). *Post growth: Life after capitalism*. John Wiley & Sons.
Jacobson, M. Z., Delucchi, M. A., Bauer, Z. A., Goodman, S. C., Chapman, W. E., Cameron, M. A., Bozonnat, C., Chobadi, L., Clonts, H. A., and Enevoldsen, P. (2017). 100% clean and renewable wind, water, and sunlight all-sector energy roadmaps for 139 countries of the world. *Joule*, 1(1): 108–121.
James, W. (1911). The Moral Equivalent of War. In *Lecture 11 in Memories and Studies.* (pp. 267–296). Longman Green and Co
James, W. (2014 [1897]). The Will to Believe And Other Essays in Popular Philosophy. In. Cambridge University Press.
James, W. (2014 [1907]). *Pragmatism-A New Name for Some Old Ways of Thinking*. Cambridge University Press.
Janes, R. R. (2022). The value of museums in averting societal collapse. *Curator: The Museum Journal*, 65(4): 729–745.
Johnsen, H. C. G. (2025). *Friedrich Hayek - Betingelsene for et liberalt samfunn*. Cappelen Damm Akademisk.
Johnsson, F., Karlsson, I., Rootzén, J., Ahlbäck, A., and Gustavsson, M. (2020). The framing of a sustainable development goals assessment in decarbonizing the construction industry–Avoiding "Greenwashing". *Renewable and sustainable energy reviews*, 131: 110029.
Jones, E. L. (2003). *The European miracle: environments, economies and geopolitics in the history of Europe and Asia*. Cambridge University Press.
Jones, K. E., Patel, N. G., Levy, M. A., Storeygard, A., Balk, D., Gittleman, J. L. et al. (2008). Global trends in emerging infectious diseases. *Nature*, 451(7181): 990–993.
Jones, N. (2024). The new car batteries that could power the electric vehicle revolution. *Nature*, 626(7998): 248–251.
Joppa, L., Luers, A., Willmott, E., Friedmann, S. J., Hamburg, S. P., and Broze, R. (2021). *Microsoft's million-tonne CO2-removal purchase—lessons for net zero*. Nature Publishing Group.
Judge, T. A., Bono, J. E., Ilies, R., and Gerhardt, M. W. (2002). Personality and leadership: a qualitative and quantitative review. *Journal of Applied Psychology*, 87(4): 765.
Kahneman, D. (2011). *Thinking fast and slow*. Allen Lane and Penguin Books.
Kahneman, D., and Tversky, A. (1984). Choices, values, and frames. *American Psychologist*, 39(4): 341.
Kasting, J. F. (1988). Runaway and moist greenhouse atmospheres and the evolution of Earth and Venus. *Icarus*, 74(3): 472–494.
Katz, L. F., and Margo, R. A. (2014). Technical Change and the Relative Demand for Skilled Labor: The United States in Historical Perspective. *In*: Boustan,L. P., C. Frydman, and R. A. Margo (Eds.). *Human Capital in History: The American Record* (pp. 15-57). Univesity of Chicago Press.
Keltner, D. (2016). *The power paradox: How we gain and lose influence*. Penguin.
Kemp, L., Xu, C., Depledge, J., Ebi, K. L., Gibbins, G., Kohler, T. A. et al. (2022). Climate Endgame: Exploring catastrophic climate change scenarios. *Proceedings of the National Academy of Sciences*, 119(34).
Kennedy, R. (1968). *Remarks At The University Of Kansas*. John F. Kennedy Presidential Library and Museum.
Kenner, D. (2019). *Carbon inequality: The role of the richest in climate change*. Routledge.
Kenny, C. (2005). Why are we worried about income? Nearly everything that matters is converging. *World Development*, 33(1): 1–19.
Kerr, S., and Parker, C. (2019). Making climate real: Climate consciousness, culture and music. *King's Law Journal*, 30(2): 185–193.
Keynes, J. M. (1930). Economic possibilities for our grandchildren. In *Essays in persuasion* (pp. 321–332). Springer.
Keynes, J. M. (1939). Paying for the War. *The Times*, 14: 41–51.
Keynes, J. M. (1940). *How to Pay for the War: A Radical Plan for the Chancellor of the Exchequer*. Macmillan.
Keys, B. J., and Mulder, P. (2024). *Property Insurance and Disaster Risk: New Evidence from Mortgage Escrow Data.*
Kiatpongsan, S., and Norton, M. I. (2014). How much (more) should CEOs make? A universal desire for more equal pay. *Perspectives on Psychological Science*, 9(6): 587–593.
Kimberley, J. R. (1981). Managerial Innovation. *In*: Nystrom, P., and W. Starbuck (Eds.). *Handbook of Organizational Design* (Vol. 1, pp. 84–104). Oxford University Press.
Klein, N. (2007). *The Shock Doctrine*. Penguin Books.
Klein, N. (2023, May 8). AI machines aren't 'hallucinating'. But their makers are. *The Guardian*.
Klotz, L. (2021). *Subtract: The untapped science of less*. Flatiron Books.
Knudsen, H. (2021). Equal Pay for a Green Future. *In*: Örtenblad, A. (Ed.). *Debating Equal Pay for All – Economy, Practicability and Ethics* (pp. 183–200). Palgrave Macmillan.

Knudsen, H., and Holbek, J. (2007). Diversity - Creativity - Leadership: A Journey from Ancient Greece to Modern Society. *In*: April, K. and M. Shockley (Eds.). *Diversity - New Realities in a Changing World* (pp. 226–247). Palgrave MacMillan.

Koch, M., Manuylov, I., and Smolka, M. (2021). Robots and firms. *The Economic Journal*, 131(638): 2553–2584.

Kopatz, V., Wen, K., Kovács, T., Keimowitz, A. S., Pichler, V., Widder, J. et al. (2023). Micro-and Nanoplastics Breach the Blood–Brain Barrier (BBB): Biomolecular Corona's Role Revealed. *Nanomaterials*, 13(8): 1404.

Kornieieva, Y. (2020). Non-financial reporting challenges in monitoring SDGs achievement: investment aspects for transition economy. *International Journal of Economics and Business Administration VIII*(1): 62–71.

Kotter, J. P. (1996). *Leading Change*. Harvard Business School Press.

Kotter, J. P. (2008). *A sense of urgency*. Harvard Business Press.

Kotz, M., Kuik, F., Lis, E., and Nickel, C. (2024). Global warming and heat extremes to enhance inflationary pressures. *Communications Earth & Environment*, 5(1): 116.

Krznaric, R. (2024). *History for tomorrow: inspiration from the past for the future of humanity*. WH Allen, Ebury Publishing.

Kubiszewski, I., Costanza, R., Franco, C., Lawn, P., Talberth, J., Jackson, T. et al. (2013). Beyond GDP: Measuring and achieving global genuine progress. *Ecological Economics*, 93: 57–68.

Kurzweil, R. (1999). *The age of spiritual machines: how we will live, work and think in the new age of intelligent machines*. Orion.

Kuznets, S. (1955). Economic Growth and Income Inequality. *American Economic Review*, 45(March), 1–28.

Lamb, H. H. (1965). The early medieval warm epoch and its sequel. *Palaeogeography, Palaeoclimatology, Palaeoecology*, 1: 13–37.

Landau, C., Bradley, K., Burns, E., Flessner, M., Gage, K., Hager, A. et al. (2023). The silver bullet that wasn't: Rapid agronomic weed adaptations to glyphosate in North America. *PNAS Nexus*, 2(12): pgad338.

Landgrebe, J., and Smith, B. (2022). *Why machines will never rule the world: artificial intelligence without fear*. Routledge.

Landrigan, P. J. (2024). Plastics, Fossil Carbon, and the Heart. *The New England journal of medicine*, 390(10): 948–950.

Lappé, F. M. (2011). *Diet for a small planet: The book that started a revolution in the way Americans eat*. Ballantine Books.

Lappé, F. M. (2013). Beyond the scarcity scare: reframing the discourse of hunger with an eco-mind. *The Journal of Peasant Studies*, 40(1): 219–238.

Latouche, S. (2009). *Farewell to growth*. Polity.

Laux, J., Wachter, S., and Mittelstadt, B. (2024). Trustworthy artificial intelligence and the European Union AI act: On the conflation of trustworthiness and acceptability of risk. *Regulation & Governance*, 18(1): 3–32.

Lawrence, N. D. (2024). *The atomic human: Understanding ourselves in the age of AI*. Penguin Random House.

Layard, R. (2005). Rethinking public economics: The implications of rivalry and habit. *Economics and happiness*, 1(1): 147–170.

Lazarus, R. J. (2008). Super wicked problems and climate change: Restraining the present to liberate the future. *Cornell Law Review*, 94: 1153–1234.

Le Bon, G. (1896). *Psychologie des foules*. Alcan.

Le Page, M. (2019). Destruction of nature is as big a threat to humanity as climate change. *New Scientist*, *6*(3229). https://www.newscientist.com/.

Lebling, K., Leslie-Bole, H., Byrum, Z., and Bridgwater, L. (2022). *6 Things to Know About Direct Air Capture*. W. R. Institute.

Ledari, M. B., Khajehpour, H., Akbarnavasi, H., and Edalati, S. (2023). Greening steel industry by hydrogen: Lessons learned for the developing world. *International Journal of Hydrogen Energy*, 48(94): 36623–36649.

Ledford, H. (2024). Human body's ageing'clock'ticks faster after heat stress. *Nature*. https://doi.org/Nature. 2024 Dec 9. doi: 10.1038/d41586-024-04007-8. Online ahead of print.

Lee, K.-F. (2018). *AI superpowers: China, Silicon Valley, and the new world order*. Houghton Mifflin.

Leivestad, E. H. (2021, November 11). Utenfor glasshuset – Hvorfor opplevde Fjodor Dostojevskij det berømte krystallpalasset i London som en trussel? *Klassekampen*, 20-21.

Lenk, C. (2011). *Human tissue research: A European perspective on the ethical and legal challenges*. Oxford University Press Inc.

Leontief, W. (1983). Technological advance, economic growth, and the distribution of income. *Population and Development Review*, 9(3): 403–410.

Lepori, M. (2015). There is no Anthropocene: Climate change, species-talk, and political economy. *Telos*, *172*, 103-124.

Lesseur, C., Pirrotte, P., Pathak, K. V., Manservisi, F., Mandrioli, D., Belpoggi, F. et al. (2021). Maternal urinary levels of glyphosate during pregnancy and anogenital distance in newborns in a US multicenter pregnancy cohort. *Environmental Pollution*, 280(July): 117002.

Levine, H., Jørgensen, N., Martino-Andrade, A., Mendiola, J., Weksler-Derri, D., Mindlis, I. et al. (2017). Temporal trends in sperm count: a systematic review and meta-regression analysis. *Human Reproduction Update*, 23(6): 646–659.

Levrel, H., and Martinet, V. (2021). Ecological economists: the good, the bad, and the ugly? *Ecological Economics*, 179: 106694.

Levy, F., and Murnane, R. J. (2004). *The new division of labor: How computers are creating the next job market*. Princeton University Press.

Light, M. T., He, J., and Robey, J. P. (2020). Comparing crime rates between undocumented immigrants, legal immigrants, and native-born US citizens in Texas. *Proceedings of the National Academy of Sciences*, 117(51): 32340–32347.

Liu, Y., Zhang, K., Li, Y., Yan, Z., Gao, C., Chen, R. et al. (2024). Sora: A Review on Background, Technology, Limitations, and Opportunities of Large Vision Models. *arXiv preprint arXiv:*2402.17177.

Lohmann, S. (1998). An information rationale for the power of special interests. *American Political Science Review*, 92(4): 809–827.

Lorentzen, H., and Dugstad, L. (2011). *Den norske dugnaden: Historie, kultur og fellesskap*. Høyskoleforlaget.

Lovins, L. H., Von Weizsacker, E., Lovins, A. B., and Milani, B. (1998). Factor four: doubling wealth, halving resource use. *Alternatives Journal*, 24(2): 33.

Luther, K. M. (1968). *Where Do We Go from Here: Chaos or Community?* Bantam Books.

Lutter, M., and Allen, N. (2024). Building Freedom Cities. *City Journal*. https://www.city-journal.org/article/building-freedom-cities

Lüscher, M. (1961). Air-conditioned termite nests. *Scientific American*, 205(1): 138–147.

Ma, R., Lin, Y., and Lin, B. (2023). Does digitalization support green transition in Chinese cities? Perspective from Metcalfe's Law. *Journal of Cleaner Production*, 425(November): 138769.

Maihali, M. A. (2023, August 31). *Development and Debt Crisis: Measures to Reduce the Debt in Third World Countries* African Scholar Publications and Research International Proceedings of the 23rd Academic Conference on Transformation Agenda for Third World Communities: Multidisciplinary Approach., Nasarawa State University, Keffi, Nasarawa State, Nigeria.

Malmodin, J., and Lundén, D. (2018). The energy and carbon footprint of the global ICT and E&M sectors 2010–2015. *Sustainability*, 10(9): 3027.

Malmodin, J., Lövehagen, N., Bergmark, P., and Lundén, D. (2024). ICT sector electricity consumption and greenhouse gas emissions–2020 outcome. *Telecommunications Policy*, 102701. https://ssrn.com/abstract=4424264 or http://dx.doi.org/10.2139/ssrn.4424264.

Mankiw, N. G., and Reis, R. (2006). Pervasive stickiness. *American Economic Review*, 96(2): 164–169.

Mann, M. E. (2012). *The hockey stick and the climate wars: Dispatches from the front lines*. Columbia University Press.

Mann, M. E., Bradley, R. S., and Hughes, M. K. (1998). Global-scale temperature patterns and climate forcing over the past six centuries. *Nature*, 392(6678): 779–787.

Mann, M. E., Bradley, R. S., and Hughes, M. K. (1999). Northern hemisphere temperatures during the past millennium: Inferences, uncertainties, and limitations. *Geophysical Research Letters*, 26(6): 759–762.

Marfella, R., Prattichizzo, F., Sardu, C., Fulgenzi, G., Graciotti, L., Spadoni, T. et al. (2024). Microplastics and nanoplastics in atheromas and cardiovascular events. *New England Journal of Medicine*, 390(10): 900–910.

Marshall, A. (2009 [1890]). *Principles of economics: unabridged eighth edition*. Cosimo, Inc.

Matlin, S. A., Mehta, G., Hopf, H., and Krief, A. (2016). One-world chemistry and systems thinking. *Nature chemistry*, 8(5): 393–398.

McCarthy, J. (2007). *What is Artificial Intelligence?*. http://www-formal.stanford.edu/jmc/.

McHarg, I. (1969). *Design With Nature* The Natural History Press.

McLean, S., Read, G. J., Thompson, J., Baber, C., Stanton, N. A., and Salmon, P. M. (2023). The risks associated with Artificial General Intelligence: A systematic review. *Journal of Experimental & Theoretical Artificial Intelligence*, 35(5): 649–663.

McPeak, A. (2018). Disruptive Technology and the Ethical Lawyer. *University of Toledo Law Review*, 50: 457.

Meadows, D., H, Meadows, D., L, Randers, J., and Behrens, W. W. (1972). *The limits to growth*. Potomac Associates.

Meckling, J., and Allan, B. B. (2020). The evolution of ideas in global climate policy. *Nature Climate Change*, 10(5): 434–438.

Merton, R. K. (1938). Science and the social order. *Philosophy of Science*, 5(3): 321–337.

Merton, R. K. (1948). The self-fulfilling prophecy. *The Antioch Review*, 8(2): 193–210.

Meys, R., Kätelhön, A., Bachmann, M., Winter, B., Zibunas, C., Suh, S. et al. (2021). Achieving net-zero greenhouse gas emission plastics by a circular carbon economy. *Science*, 374(6563): 71–76.

Michaux, S. P. (2021). *Restructuring the Circular Economy into the Resource Balanced Economy* (9522174122). G.-G. S. o. Finland.

Michel, S. L., Swingedouw, D., Ortega, P., Gastineau, G., Mignot, J., McCarthy, G. et al. (2022). Early warning signal for a tipping point suggested by a millennial Atlantic Multidecadal Variability reconstruction. *Nature communications*, 13(1): 5176.

Mill, J. S. (1871). *Principles of political economy: with some of their applications to social philosophy* (Vol. 1). Longmans, Green, Reader, and Dyer.

Mills, J. H., and Waite, T. A. (2009). Economic prosperity, biodiversity conservation, and the environmental Kuznets curve. *Ecological Economics*, 68(7): 2087–2095.

Milman, O. (2024, December 5). How climate risks are driving up insurance permiums around the US – visualized. *The Guardian*.

Mincer, J., and Danninger, S. (2000). *Technology, Unemployment, and Inflation* National Bureau of Economic Research.

Mishra, P. (2017). *Age of anger: A history of the present*. Macmillan.

Mohanasundaram, S., Varadharajan, V., Selvaraju, M., Manikandan, S., Ramasamy, S., Jayakumar, M. et al. (2024). Green ammonia as peerless entity for realm of clean-energy carrier toward zero carbon emission: Purviews, neoteric tendencies, potentialities and downsides. *Fuel*, 365(June): 131118.

Molina, M. J., and Rowland, F. S. (1974). Stratospheric sink for chlorofluoromethanes: chlorine atom-catalysed destruction of ozone. *Nature*, 249(5460): 810–812.

Moltmann, J. (1993). *Theology of hope: On the ground and the implications of a Christian eschatology*. Fortress Press.

Moltmann, J. (2020). *The spirit of hope: Theology for a world in peril*. Westminster John Knox Press.

Moore, G. E. (2006). Cramming more components onto integrated circuits, Reprinted from Electronics, volume 38, number 8, April 19, 1965, pp. 114 ff. *IEEE Solid-State Circuits Society Newsletter*, 11(3): 33–35.

Moravec, H. (1988). Mind Children: The Future of Robot and Human Intelligence. *Harvard University Press*.

Murphy, D. (2009). The Net Hubbert Curve: what does it mean. *The oil drum*(June 22), 806-807.

Murphy, D. J. (2014). The implications of the declining energy return on investment of oil production. *Philosophical Transactions of the Royal Society A: Mathematical, Physical and Engineering Sciences*, 372(2006), 20130126.

Murray, C. (2012). *Coming apart: The state of white America, 1960–2010*. Crown Forum.

Murray, C., and Phillips, M. (2001). The British underclass: Ten years later/Commentary. *Public Interest*, *145*(25).

Nakajima, S., Hayashi, Y.-Y., and Abe, Y. (1992). A study on the "runaway greenhouse effect" with a one-dimensional radiative–convective equilibrium model. *Journal of Atmospheric Sciences*, 49(23): 2256–2266.

Nedland, S. (2023). *Kultur - Retur. Musikkens reiser i historien og i geografien*. Svein Sandnes Bokforlag.

Nersisyan, Y., and Wray, L. R. (2021). Can we afford the green new deal? *Journal of Post Keynesian Economics*, 44(1): 68–88.

Ngo, R., Chan, L., and Mindermann, S. (2022). *The alignment problem from a deep learning perspective* arXiv preprint arXiv:2209.00626.

Nikolova, E., and Polansky, J. (2021). Conversionary protestants do not cause democracy. *British Journal of Political Science*, 51(4): 1723–1733.

Nordhaus, W. D. (1993). Reflections on the economics of climate change. *Journal of Economic Perspectives*, 7(4): 11–25.

Nordhaus, W. D. (1997). Discounting in economics and climate change; an editorial comment. *Climatic Change*, 37(2): 315–328.

Nordhaus, W. D. (2007). A review of the Stern review on the economics of climate change. *Journal of Economic Literature*, 45(3): 686–702.

Nordhaus, W. D., and Tobin, J. (1972). Is growth obsolete? In *Economic research: Retrospect and prospect, volume 5, economic growth* (pp. 1-80). NBER.

Noto La Diega, G., and Walden, I. (2016). Contracting for the 'Internet of Things': Looking into the Nest. *Queen Mary School of Law Legal Studies Research Paper* (219).

Nozick, R. (1974). *Anarchy, State, and Utopia*. Basic Books.

O'Brien, I. (2024, September 15). Data center emissions probably 662% higher than big tech claims. Can it keep up the ruse? *The Guardian*.

O'Donovan, C., Gurakan, S., and Wu, X. (2025). *Visions, values, voices: a survey of artificial intelligence researchers*.

OECD, O. f. E. C.-o. a. D. (2022). *Global Plastics Outlook: Policy Scenarios to 2060*. OECD Publishing.

Offer, J., and Pinker, R. (2017). *Social Policy and Welfare Pluralism: Selected Writings of Robert Pinker*. Policy Press.

Ollinaho, O. I. (2023). What is 'business as usual'? Towards a theory of cumulative sociomaterial change. *Globalizations*, 20(4): 611–627.

Open Markets Institute. (2023). *Monopoly by the numbers, Anti-Monopoly Basics*. Open Markets Institute. Retrieved June 28, 2023 from https://www.openmarketsinstitute.org/learn/monopoly-by-the-numbers.

Ord, T. (2020). *The precipice: Existential risk and the future of humanity*. Hachette Books.

Oreskes, N., and Conway, E. M. (2011). *Merchants of doubt: How a handful of scientists obscured the truth on issues from tobacco smoke to global warming*. Bloomsbury Publishing.

Oruka, H. O. (1990). *Sage philosophy: Indigenous thinkers and modern debate on African philosophy* (Vol. 4). Brill.

Ostrom, E. (1990). *Governing the commons: The evolution of institutions for collective action*. Cambridge University Press.

Ouyang, L., Wu, J., Jiang, X., Almeida, D., Wainwright, C., Mishkin, P. et al. (2022). Training language models to follow instructions with human feedback. *Advances in Neural Information Processing Systems*, 35: 27730–27744.

Paddison, L. (2021). *How the rich are driving climate change - The world's wealthiest people make a huge contribution to climate change through carbon-hungry activities. How can we reduce emissions from the rich?*

Pan, X., Wang, H., Wang, Z., Lin, L., Zhang, Q., Zheng, X. et al. (2019). Carbon Palma Ratio: A new indicator for measuring the distribution inequality of carbon emissions among individuals. *Journal of Cleaner Production*, 241: 118418.

Parrique, T., Barth, J., Briens, F., Kerschner, C., Kraus-Polk, A., Kuokkanen, A. et al. (2019). *Decoupling Debunked - Evidence and Arguments Against Green Growth as a Sole Strategy for Sustainability.*

Paul, K. C., Krolewski, R. C., Lucumi Moreno, E., Blank, J., Holton, K. M., Ahfeldt, T. et al. (2023). A pesticide and iPSC dopaminergic neuron screen identifies and classifies Parkinson-relevant pesticides. *Nature Communications*, 14(1): 2803.

Pearson, H. (2024). The rise of eco-anxiety: scientists wake up to the mental-health toll of climate change. *Nature*, *628*(8007): 256–258.

Penner, A. (2021). «Den gode borgeren» og «den flinke flyktningen»: fortellinger om dugnad, ideer om homogenitet og rasialisering. In *Fortellinger om integrering i norske lokalsamfunn* (pp. 134–155). Universitetsforlaget.

Perry, G. (2013). *Behind the shock machine: The untold story of the notorious Milgram psychology experiments*. Scribe Publications Pty Ltd.

Perry, G. (2018). *The lost boys: inside Muzafer Sherif's Robbers Cave experiment*. Scribe Publications Pty Ltd.

Persson, L., Carney Almroth, B. M., Collins, C. D., Cornell, S., De Wit, C. A., Diamond, M. L. et al. (2022). Outside the safe operating space of the planetary boundary for novel entities. *Environmental Science & Technology*, 56(3): 1510–1521.

Phillips, K. D. (2013). Empathy for psychopaths: Using fMRI brain scans to plea for leniency in death penalty cases. *The Law & Psychology Review* 37, 1.

Pianta, S., and Rettl, P. (2023). *Global Harms, Local Profits: How the Uneven Costs of Natural Disasters Affect Support for Green Political Platforms*. H. B. School.

Piketty, T. (2014). *Capital in the twenty-first century*. Harvard University Press.

Piketty, T. (2020). *Capital and ideology*. Harvard University Press.

Piketty, T. (2022). *A brief history of equality*. Harvard University Press.

Piketty, T., Saez, E., and Stantcheva, S. (2014). Optimal taxation of top labor incomes: A tale of three elasticities. *American Economic Journal: Economic Policy*, 6(1): 230–271.

Pinker, R. (2006). From Gift Relationships to Quasi-markets: An Odyssey along the Policy Paths of Altruism and Egoism. *Social Policy & Administration*, 40(1): 10–25.

Pinker, S. (2011). *The better angels of our nature: The decline of violence in history and its causes*. Penguin.

Pinker, S. (2018). *Enlightenment now: The case for reason, science, humanism, and progress*. Penguin UK.

Pitesa, M., and Thau, S. (2013). Masters of the universe: How power and accountability influence self-serving decisions under moral hazard. *Journal of Applied Psychology*, 98(3): 550.

Plous, S. (1993). *The psychology of judgment and decision making*. Mcgraw-Hill Book Company.

Polanyi, K. (1944). *The great transformation*. Beacon Press.

Popp, M., Schmidt, H., and Marotzke, J. (2016). Transition to a moist greenhouse with CO_2 and solar forcing. *Nature communications*, 7(1): 10627.

Postrel, V. (2008). Inconspicuous consumption. *The Atlantic*.

Powell, L. (1971). Memorandum, to: Eugene Snydor, Education Committee, U. S. Chamber of Commerse. *In*: U. S. C. o. Commerse (Ed.).

Power, M. C., Adar, S. D., Yanosky, J. D., and Weuve, J. (2016). Exposure to air pollution as a potential contributor to cognitive function, cognitive decline, brain imaging, and dementia: a systematic review of epidemiologic research. *Neurotoxicology*, 56: 235–253.

Praetorius, I. (2015). *The Care-Centered Economy: Rediscovering what has been taken for granted.* Heinrich-Böll-Stiftung.
Putnam, R. D. (2000). *Bowling alone: The collapse and revival of American community.* Simon and schuster.
Qian, N., Gao, X., Lang, X., Deng, H., Bratu, T. M., Chen, Q. et al. (2024). Rapid single-particle chemical imaging of nanoplastics by SRS microscopy. *Proceedings of the National Academy of Sciences*, 121(3): e2300582121.
Rabaia, M. K. H., Abdelkareem, M. A., Sayed, E. T., Elsaid, K., Chae, K.-J., Wilberforce, T. et al. (2021). Environmental impacts of solar energy systems: A review. *Science of The Total Environment*, 754: 141989.
Rand, A. (1943). *The Fountainhead.* Bobbs-Merrill.
Rand, A. (1957 [1992]). *Atlas Shrugged.* Random House.
Raugei, M., Sgouridis, S., Murphy, D., Fthenakis, V., Frischknecht, R., Breyer, C. et al. (2017). Energy Return on Energy Invested (ERoEI) for photovoltaic solar systems in regions of moderate insolation: A comprehensive response. *Energy Policy*, 102: 377–384.
Rausch, H. (1982). *Theoria: Von ihrer sakralen zur philosophischen Bedeutung.* Wilhelm Fink Verlag.
Raworth, K. (2012). *A safe and just space for humanity: can we live within the doughnut?* Oxfam.
Raworth, K. (2017). *Doughnut economics: seven ways to think like a 21st-century economist.* Chelsea Green Publishing.
Reeve, J. R., Hoagland, L. A., Villalba, J. J., Carr, P. M., Atucha, A., Cambardella, C. et al. (2016). Organic farming, soil health, and food quality: considering possible links. *Advances in Agronomy*, 137: 319–367.
Reganold, J. P., and Wachter, J. M. (2016). Organic agriculture in the twenty-first century. *Nature Plants*, 2(2): 1–8.
Reid, W. V., Chen, D., Goldfarb, L., Hackmann, H., Lee, Y.-T., Mokhele, K. et al. (2010). Earth system science for global sustainability: grand challenges. *Science*, *330*(6006): 916–917.
Richie, H., and Roser, M. (2024). *Land Use.*
Rider, C., and Hall, C. (2002). Inclusive volunteering: Benefits to participants and community. *Therapeutic Recreation Journal*, 36(3): 247–259.
Riess, H. (2017). The science of empathy. *Journal of Patient Experience*, 4(2): 74–77.
Ripple, W. J., Wolf, C., Gregg, J. W., Levin, K., Rockström, J., Newsome, T. M., Betts, M. G., Huq, S., Law, B. E., and Kemp, L. (2022). *World scientists' warning of a climate emergency 2022.* Oxford University Press.
Ripple, W. J., Wolf, C., Gregg, J. W., Rockström, J., Mann, M. E., Oreskes, N. et al. (2024). The 2024 state of the climate report: Perilous times on planet Earth. *BioScience*, biae087.
Ripple, W. J., Wolf, C. N., and TM, G. (2017). 15,364 scientist signatories from 184 countries; World Scientists' Warning to Humanity: A Second Notice. *BioScience*, 67(12): 1026–1028.
Rittel, H. W., and Webber, M. M. (1973). Dilemmas in a general theory of planning. *Policy Sciences*, 4(2): 155–169.
Robeyns, I. (2024). *Limitarianism: The case against extreme wealth.* Random House.
Robinson, F. S. (2013). The human future: upgrade or replacement? *The Humanist*, 73(4): 22–26.
Robinson, K. S. (2020). *The ministry for the future*. Orbit.
Robock, A., Oman, L., and Stenchikov, G. L. (2007). Nuclear winter revisited with a modern climate model and current nuclear arsenals: Still catastrophic consequences. *Journal of Geophysical Research: Atmospheres*, 112(D13).
Rockström, J., Steffen, W., Noone, K., Persson, Å., Chapin, F. S., Lambin, E. F. et al. (2009). A safe operating space for humanity. *Nature*, 461(7263): 472–475.
Roelen, K. (2020). Receiving social assistance in low-and middle-income countries: Negating shame or producing stigma? *Journal of Social Policy*, 49(4): 705–723.
Rogers-Dillon, R. (1995). The dynamics of welfare stigma. *Qualitative sociology*, 18(4): 439–456.
Roine, J., Vlachos, J., and Waldenström, D. (2009). The long-run determinants of inequality: What can we learn from top income data? *Journal of public economics*, 93(7-8): 974–988.
Rolnick, D., Donti, P. L., Kaack, L. H., Kochanski, K., Lacoste, A., Sankaran, K. et al. (2022). Tackling climate change with machine learning. *ACM Computing Surveys (CSUR)*, 55(2): 1-96.
Roose, K. (2019, January 25). The hidden automation agenda of the Davos elite. *The New York Times.*
Rosen, J. (2024). *The Pursuit of Happiness: How Classical Writers on Virtue Inspired the Liv es of the Founders and Defined America.* Simon & Schuster.
Rosling, H. (2023). *Factfulness.* Flammarion.
Rothbard, M. N. (1978). *For a new liberty: The libertarian manifesto.* Ludwig von Mises Institute.
Rushkoff, D. (2018). How tech's richest plan to save themselves after the apocalypse. *The Guardian.* https://www.theguardian.com/technology/2018/jul/23/tech-industry-wealth-futurism-transhumanism-singularity
Rushkoff, D. (2022). *Survival of the richest: Escape fantasies of the tech billionaires.* WW Norton & Company.
Sachs, J. D., and Warner, A. (1995). Natural resource abundance and economic growth. In *NBER Working Paper (5398).* National bureau of economic research Cambridge, Mass., USA.

Salamanca, F., Georgescu, M., Mahalov, A., Moustaoui, M., and Wang, M. (2014). Anthropogenic heating of the urban environment due to air conditioning. *Journal of Geophysical Research: Atmospheres*, 119(10): 5949–5965.

Salmon, P. M., Hulme, A., Walker, G. H., Waterson, P., Berber, E., and Stanton, N. A. (2020). The big picture on accident causation: A review, synthesis and meta-analysis of AcciMap studies. *Safety Science*, 126: 104650.

Sandkühler, F., Stewart, H., Korzekwa, R. C., and Brauner, J. (2023). *Expert Survey on Progress in AI* (AI Impacts, Issue. Retrieved from osf.io/vmdny.

Sankaran, K. (2019). Carbon emission and plastic pollution: how circular economy, blockchain, and artificial intelligence support energy transition? *Journal of Innovation Management*, 7(4): 7–13.

Sankaran, S., Müller, R., and Drouin, N. (2020). Creating a 'sustainability sublime'to enable megaprojects to meet the United Nations sustainable development goals. *Systems Research and Behavioral Science*, 37(5): 813–826.

Sayed, E. T., Olabi, A. G., Alami, A. H., Radwan, A., Mdallal, A., Rezk, A. et al. (2023). Renewable energy and energy storage systems. *Energies*, 16(3): 1415.

Sbordone, A. M. (2002). Prices and unit labor costs: a new test of price stickiness. *Journal of Monetary Economics*, 49(2): 265–292.

Scheffer, M., van Nes, E. H., Kemp, L., Kohler, T. A., Lenton, T. M., and Xu, C. (2023). The vulnerability of aging states: A survival analysis across premodern societies. *Proceedings of the National Academy of Sciences*, 120(48): e2218834120.

Schillinger, D., and Kearns, C. (2017). Do sugar-sweetened beverages cause obesity and diabetes? *Annals of Internal Medicine*, 167(1): 72–73.

Schmidt, I. (2024). Billionaires' Survivalist Bunkers Go Absolutely Bonkers With Fiery Moats and Water Cannons. *The Hollywood Reporter*. https://www.yahoo.com/entertainment/billionaires-survivalist-bunkers-absolutely-bonkers-211655155.html?guccounter=1&guce_referrer=aHR0cHM6Ly93d3cuZ29vZ2xlLmNvbS8&guce_referrer_sig=AQAAAIrRE2a6Qfg84zByu9jGaC066YES-i6Ovm014zLFkp1qZnNbtwN2a0TTKQwPoXIV0V-WoQi8QOQwwky5SUWZz8cwtaUhrm_HHZrcQI_m36UYwGgQca0EE_MpaSLrnznp85E5OAc6zdiGpbR0a0rx2hT0R_O4FmotRTJAWjeyLY7U.

Schmitt-Grohé, S., and Uribe, M. (2006). *Comparing two variants of Calvo-type wage stickiness* (NBER Working Paper Series, Issue.

Schröder, E., and Storm, S. (2020). Economic growth and carbon emissions: The road to "hothouse earth" is paved with good intentions. *International Journal of Political Economy*, 49(2): 153–173.

Schulz, A.-C., and Flickinger, M. (2020). Does CEO (over) compensation influence corporate reputation? *Review of Managerial Science*, 14(4): 903–927.

Schumacher, E. F. (2011 [1973]). *Small is beautiful: A study of economics as if people mattered*. Random House.

Schumpeter, J. A. (1994 [1942]). *Capitalism, Socialism and Democracy*. Routledge.

Schwaller, F. (2025). Will AI improve your life? Here's what 4,000 researchers think. *Nature*.

Schön, D. A. (1979). Generative metaphor: A perspective on problem-setting in social policy. pp. 137–163. *In*: Ortony, A. (Ed.). *Metaphor and thought* (2nd ed.). Cambridge University Press.

Schöngart, S., Nicholls, Z., Hoffmann, R., Pelz, S., and Schleussner, C.-F. (2025). High-income groups disproportionately contribute to climate extremes worldwide. *Nature Climate Change*, 1–7.

Schaake, M. (2024). *The Tech Coup: How to Save Democracy from Silicon Valley*. Princeton University Press.

Scott, L., and McGill, A. (2018). SDG Reporting Challenge 2018-From promise to reality: Doe s business really care about the SDGs. *And what needs to happen to turn words int o action. PWC.*

Senge, P. M. (1990). *The fifth discipline: The Art & Practice of the Learning Organization* (Vol. 1). Century.

Shafik, N. (1994). Economic development and environmental quality: an econometric analysis. *Oxford economic papers*, 46(Supplement_1): 757–773.

Shen, M., Huang, W., Chen, M., Song, B., Zeng, G., and Zhang, Y. (2020). (Micro) plastic crisis: un-ignorable contribution to global greenhouse gas emissions and climate change. *Journal of Cleaner Production*, 254: 120138.

Sherif, M. (2015). *Group conflict and co-operation: Their social psychology*. Psychology Press.

Shingikai, K. (1974). *Measuring net national welfare of Japan: Report of the NNW Measurement Committee*. Printing Bureau, Ministry of Finance.

Shue, H. (2023). Unseen urgency: delay as the new denial. *Wiley Interdisciplinary Reviews: Climate Change*, 14(1): e809.

Shulman, C., and Sandberg, A. (2010). Implications of a Software-Limited Singularity. European Conference of Computing and Philosophy - ECAP10: VIII, Shulman, Carl, and Anders Sandberg. 2010. "Implications of a Software-Limited Singularity." Mainzer.

Silvertown, J. (2024). *Selfish Genes to Social Beings: A Cooperative History of Life*. Oxford University Press.

Simon, C., and Mobekk, H. (2019). Dugnad: A fact and a narrative of Norwegian prosocial behavior. *Perspectives on behavior Science*, 42: 815–834.

Singer, P. (1975). *Animal Liberation: A New Ethic for Our Treatment of Animals*. New York Review.

Singh, N. M. (2013). The affective labor of growing forests and the becoming of environmental subjects: Rethinking environmentality in Odisha, India. *Geoforum*, *47*, 189-198.

Sirror, H. (2024). Innovative Approaches to Windcatcher Design: A Review on Balancing Tradition Sustainability and Modern Technologies for Enhanced Performance. *Energies*, 17(22): 5770.

Sisson, E. O. (1924). THE SENSE OF CRISIS. *Texas Review*, 9(3): 221–227.

Sivakrishna, A., Adesina, A., Awoyera, P., and Kumar, K. R. (2020). Green concrete: A review of recent developments. *Materials Today: Proceedings*, 27: 54–58.

Smalley, R. E. (2001). Of chemistry, love and nanobots. *Scientific American*, 285(3): 76–77.

Smil, V. (2014). *Materials and dematerialization - Making the modern world:* John Wiley & Sons.

Smil, V. (2019). Energy (r) evolutions take time. *World Energy*, 44(10-14): 10–15.

Smil, V. (2022). *How the world really works: The scientist's Guide to Our Past, Present and Future*. Penguin Books.

Soga, M., and Gaston, K. J. (2018). Shifting baseline syndrome: causes, consequences, and implications. *Frontiers in Ecology and the Environment*, 16(4): 222–230.

Solnit, R. (2005). *Hope in the Dark - Untold Histories Wild Possibilities*. Canongate Books Ltd.

Solnit, R. (2017). Grounds for hope. *Tikkun*, 32(1): 30–39.

Song, Y., Johnson, D., Peng, R., Hensley, D. K., Bonnesen, P. V., Liang, L. et al. (2018). A physical catalyst for the electrolysis of nitrogen to ammonia. *Science advances*, 4(4): e1700336.

Sotos, M. E. (2015). *GHG protocol scope 2 guidance*.

Sovacool, B. K. (2010). A critical evaluation of nuclear power and renewable electricity in Asia. *Journal of Contemporary Asia*, 40(3): 369–400.

Spash, C. (1999). The development of environmental thinking in economics. *Environmental Values*, 8(4): 413–435.

Spash, C. L. (2009). The new environmental pragmatists, pluralism and sustainability. *Environmental Values*, 18(3): 253–256.

Spash, C. L. (2011). Social ecological economics: Understanding the past to see the future. *American Journal of Economics and Sociology*, 70(2): 340–375.

Spash, C. L. (2017). Social ecological economics. In *Routledge Handbook of Ecological Economics* (pp. 3-16). Routledge.

Spiering, D. J. (2006). Rewilding North America, A Vision for Conservation in the 21st Century. *Natural Areas Journal*, 26(2): 223–225.

Spitzer, M. (2021). *The Musical Human: A History of Life on Earth–A BBC Radio 4'Book of the Week'*. Bloomsbury Publishing.

Springmann, M., Van Dingenen, R., Vandyck, T., Latka, C., Witzke, P., and Leip, A. (2023). The global and regional air quality impacts of dietary change. *Nature Communications*, 14(1): 6227.

Stalker, L., Talukder, A., Strand, J., Josh, M., and Faiz, M. (2022). Gold (hydrogen) rush: risks and uncertainties in exploring for naturally occurring hydrogen. *The APPEA Journal*, 62(1): 361–380.

Standing, G. (2014). The precariat. *Contexts*, 13(4): 10–12.

Stanley, T. J., Danko, W. D., and Stanley, S. (1996). The millionaire next door: The surprising secrets of America's Wealthy. In. Atlanta: Longstreet Press.

Stanton, E. A. (2007). *Inequality and The Human Development Index* [Doctoral, University of Massachusetts]. Amherst.

Staw, B. M., Sandelands, L. E., and Dutton, J. E. (1981). Threat rigidity effects in organizational behavior: A multilevel analysis. *Administrative Science Quarterly*, 26(4): 501–524.

Steffen, W., Richardson, K., Rockström, J., Cornell, S. E., Fetzer, I., Bennett, E. M. et al. (2015). Planetary boundaries: Guiding human development on a changing planet. *Science*, 347(6223): 1259855.

Stegmann, P., Daioglou, V., Londo, M., van Vuuren, D. P., and Junginger, M. (2022). Plastic futures and their CO_2 emissions. *Nature*, 612(7939): 272–276.

Stein-Perlman, Z., Weinstein-Raun, B., and Grace, K. (2022). Expert survey on progress in AI. *AI Impacts. Available online at: https://aiimpacts. org/2022-expert-survey-on-progress-in-ai (accessed December 7, 2022)*.

Stern, N. H. (2007). *The economics of climate change: the Stern review*. Cambridge University Press.

Stevenson, H. (2019). Contemporary discourses of green political economy: AQ method analysis. *Journal of Environmental Policy & Planning*, 21(5): 533–548.

Stiglitz, J. E. (2017). The overselling of globalization. *Business Economics*, 52: 129–137.

Stigliz, J. (2002). Globalization and its Discontents, 2002. In. New Delhi: Penguin Books India.

Stockman, D. (2013). *The triumph of politics: Why the Reagan revolution failed*. Public Affairs.

Sun, Y., and Gong, H. (2023). Firm financialization and cost stickness behavior. *Finance Research Letters*, 57: 104236.

Supran, G., and Oreskes, N. (2017). Assessing ExxonMobil's climate change communications (1977–2014). *Environmental Research Letters*, 12(8): 084019.

Suri, V., and Chapman, D. (1998). Economic growth, trade and energy: implications for the environmental Kuznets curve. *Ecological Economics*, 25(2): 195–208.

Sutton, B. (2016, November 23, 2016). "The knowledge that I have got enough" Kurt Vonnegut, Joe Heller, and a Thanksgiving Message. https://medium.com/@bobsutton/kurt-vonnegut-joe-heller-and-a-thanksgiving-message-8a31ca397888.

Sverdrup-Thygeson, A. (2020). *Extraordinary Insects: The Fabulous, Indispensable Creatures Who Run Our World.* Simon & Schuster.

Swan, S. H., and Colino, S. (1920). *Count Down: How Our Modern World Is Threatening Sperm Counts, Altering Male and Female Reproductive Development, and Imperiling the Future of the Human Race*. Simon & Schuster.

Swerdloff, R. S., and Ng, C. M. (2023). Gynecomastia: etiology, diagnosis, and treatment. In Feingold K. R., Anawalt B., and e. a. Blackman M. R. (Eds.), *Endotext [Internet]*. MDText.com.

Sæther, A. (2017). *Samuel Pufendorf – The Grandfather of Modern Political Economy?* [Doctoral Dissertation, Norwegian School of Economics - NHH]. Bergen.

Tainter, J. (1988). *The collapse of complex societies*. Cambridge university press.

Taplin, J. (2023). *The End of Reality: How four billionaires are selling out our future*. Penguin Random House.

Tassava, C. J. (2008). The American Economy during World War II. *FH. Net Encyclopedia, editied by Robert Whaples. URL http://eh. net/encyclopedia/the_american_ economy_during_world_war_ii.*

Tendera, P., and Rubiś, W. (2016). World Music: A Transcultural Phenomenon. New Music Concepts: 2nd International Conference ICNMC.

Theobald, R. (1965). *Free men and free markets*. C. N. Potter.

Thompson, N. C., Greenewald, K., Lee, K., and Manso, G. F. (2020). The computational limits of deep learning. *arXiv preprint arXiv:2007.05558.*

Thompson, W. (1824). *An inquiry into the principles of the distribution of wealth most conducive to human happiness: applied to the newly proposed system of voluntary equality of wealth.* Longsman, Hurst, Ross, Orme, Brown and Green.

Tica, J., Globan, T., and Arčabić, V. (2022). Managing the impact of globalization and technology on inequality. *Economic research-Ekonomska istraživanja*, 35(1): 1035–1060.

Tierney, J. (1998, November 30). The Big City; Rich and Poor, Consumed By Consuming. *The New York Times.*

Tonn, B., and Stiefel, D. (2014). Human extinction risk and uncertainty: Assessing conditions for action. *Futures*, 63: 134–144.

Torres, E., García-Fernández, A., Iñigo, D., Lara-Romero, C., Morente-López, J., Prieto-Benítez, S. et al. (2023). Facilitated adaptation as a conservation tool in the present climate change context: a methodological guide. *Plants*, 12(6): 1258.

Tracker, N. Z. (2023). *Net Zero Stocktake 2023: NewClimate Institute*. Oxford Net Zero, Energy and Climate Intelligence Unit and Data-Driven EnviroLab.

Trainer, T. (2010). *The Transition: To a Sustainable and Just World.* Envirobook.

Trainer, T. (2012). De-growth: Do you realise what it means? *Futures*, 44(6): 590–599.

Turchin, P. (2018). Historical dynamics: Why states rise and fall.

Turchin, P. (2023). *End Times - lElites, Counter-Elites and the Path of Political Disintegration*. Penguin Random House UK.

Turchin, P., Whitehouse, H., Gavrilets, S., Hoyer, D., François, P., Bennett, J. S. et al. (2022). Disentangling the evolutionary drivers of social complexity: A comprehensive test of hypotheses. *Science Advances*, 8(25): eabn3517.

Tyner Sr, G., Costanza, R., and Fowler, R. G. (1988). The net-energy yield of nuclear power. *Energy*, 13(1): 73–81.

Tyson, A., Kennedy, B., and Funk, C. (2021). Gen Z, Millennials stand out for climate change activism, social media engagement with issue. *Pew Research Center*, 26.

Ugai, T., Sasamoto, N., Lee, H.-Y., Ando, M., Song, M., Tamimi, R. M. et al. (2022). Is early-onset cancer an emerging global epidemic? Current evidence and future implications. *Nature Reviews Clinical Oncology*, 19(10): 656–673.

Uno, R. (1989). Economic Growth and Environmental Change in Japan; Net National Welfare and Beyond. pp. 307–332. *In*: Archibugi, F. and P. Nijkamp (Eds.). *Economy and Ecology: Towards Sustainable Development.* Springer.

Vadén, T., Lähde, V., Majava, A., Järvensivu, P., Toivanen, T., Hakala, E. et al. (2020). Decoupling for ecological sustainability: A categorisation and review of research literature. *Environmental Science & Policy*, 112: 236–244.

Van den Bergh, J. C. (2009). The GDP paradox. *Journal of Economic Psychology*, 30(2): 117–135.

Van der Waal, J. W., and Thijssens, T. (2020). Corporate involvement in sustainable development goals: Exploring the territory. *Journal of Cleaner Production*, 252: 119625.

van Deventer, J. S., White, C. E., and Myers, R. J. (2021). A roadmap for production of cement and concrete with low-CO 2 emissions. *Waste and Biomass Valorization*, 12: 4745–4775.

van Hassel, F., and Bovenkerk, B. (2023). How Should We Help Wild Animals Cope with Climate Change? The Case of the Iberian Lynx. *Animals*, 13(3): 453.

Van Hootegem, A., Niesen, W., and De Witte, H. (2019). Does job insecurity hinder innovative work behaviour? A threat rigidity perspective. *Creativity and Innovation Management*, 28(1): 19–29.

van Wijngaarden, A., Moore, J. C., Alfthan, B., Kurvits, T., and Kullerud, L. (2024). A survey of interventions to actively conserve the frozen North. *Climatic Change*, 177(4): 58.

Veblen, T. (2017). *The theory of the leisure class*. Routledge.

Vinuesa, R., Azizpour, H., Leite, I., Balaam, M., Dignum, V., Domisch, S. et al. (2020). The role of artificial intelligence in achieving the Sustainable Development Goals. *Nature Communications*, 11(1): 233.

Vogl, V., Åhman, M., and Nilsson, L. J. (2018). Assessment of hydrogen direct reduction for fossil-free steelmaking. *Journal of Cleaner Production*, 203: 736–745.

Volscho, T. W., and Kelly, N. J. (2012). The rise of the super-rich: Power resources, taxes, financial markets, and the dynamics of the top 1 percent, 1949 to 2008. *American Sociological Review*, 77(5): 679–699.

von Weizsacker, E. U. (2013). *Factor four: doubling wealth, halving resource use-A report to the Club of Rome*. Routledge.

Wade, R. H. (2023). *The world development report 2022: finance for an equitable recovery in the context of the international debt crisis* (0012-155X). (Development and Change, Issue. I. o. S. S. T. Hague.

Wallace-Wells, D. (2017). The Uninhabitable Earth: Famine, economic collapse, a sun that cooks us: What climate change could wreak—sooner than you think. *New York Magazine*, 9.

Wallace, J., and McGrattan, A. (2011). *The trumpet*. Yale University Press.

Wallich, P. (2008). Who's Who in the Singularity - A Guide to the Singularity True Believers, Atheists, and Agnostics. https://spectrum.ieee.org/who-is-who-in-the-singularity.

Wang, C., Zhang, L., Chang, Y., and Pang, M. (2021). Energy return on investment (EROI) of biomass conversion systems in China: Meta-analysis focused on system boundary unification. *Renewable and Sustainable Energy Reviews*, 137: 110652.

Ward, J. D., Sutton, P. C., Werner, A. D., Costanza, R., Mohr, S. H., and Simmons, C. T. (2016). Is decoupling GDP growth from environmental impact possible? *Plos one*, 11(10): e0164733.

Ward, M. H., Madrigal, J. M., Jones, R. R., Friesen, M. C., Falk, R. T., Koebel, D. et al. (2023). Glyphosate in house dust and risk of childhood acute lymphoblastic leukemia in California. *Environment International*, 172: 107777.

Washington, H., and Cook, J. (2011). *Climate change denial: Heads in the sand*. Routledge.

Wason, P. C. (1968). Reasoning about a rule. *Quarterly Journal of Experimental Psychology*, 20(3): 273–281.

Watson, J. E., Venter, O., Lee, J., Jones, K. R., Robinson, J. G., Possingham, H. P. et al. (2018). Protect the last of the wild. In: Nature Publishing Group.

Webb, M. (2019). *The impact of Artificial Intelligence on the Labor Market*.

Weick, K. E. (1995). *Sensemaking in organizations* (Vol. 3). Sage.

West, P. (2004). *Conspicuous compassion: Why sometimes it really is cruel to be kind*. Coronet Books.

West, T. A., Börner, J., Sills, E. O., and Kontoleon, A. (2020). Overstated carbon emission reductions from voluntary REDD+ projects in the Brazilian Amazon. *Proceedings of the National Academy of Sciences*, 117(39): 24188–24194.

West, T. A., Wunder, S., Sills, E. O., Börner, J., Rifai, S. W., Neidermeier, A. N. et al. (2023). Action needed to make carbon offsets from tropical forest conservation work for climate change mitigation. *Science*, 381(6660): 873–877.

Westrum, R. (1982). Social Intelligence A bout Hidden Events: Its Significance for Scientific Research and Social Policy. *Knowledge*, 3(3): 381–400.

Whitehead, A. N. (1925). *Science and the Modern World*. Cambridge University Pres.

Whiting, M. E., and Watts, D. J. (2024). A framework for quantifying individual and collective common sense. *Proceedings of the National Academy of Sciences*, 121(4): e2309535121.

Whyte, D. (2015). *How corrupt is Britain?* Pluto Press London.

Wilkinson, R., and Pickett, K. (2009). *The spirit level. Why more equal societies almost always do better*. Bloomsbury Press.

Willett, W., Rockström, J., Loken, B., Springmann, M., Lang, T., Vermeulen, S. et al. (2019). Food in the Anthropocene: the EAT–Lancet Commission on healthy diets from sustainable food systems. *The Lancet*, 393(10170): 447–492.

Wilson, E. O. (2016). *Half-earth: our planet's fight for life*. WW Norton & Company.

Wilson, J. (2000). Volunteering. *Annual Review of Sociology*, 26(1): 215–240.

Wilson, T. F. (1992). *The power" to coin" money: the exercise of monetary powers by the Congress*. ME Sharpe Inc.
Wong, C. (2023). Extreme heat harms health-what is the human body's limit? *Nature*.
Woodberry, R. D. (2010). Weber through the back door: Protestant competition, elite power dispersion and the global spread of democracy. APSA 2010 Annual Meeting Paper.
Woodberry, R. D. (2012). The missionary roots of liberal democracy. *American Political Science Review*, *106*(2), 244–274.
Woollings, T. (2020). *Jet stream: A journey through our changing climate*. Oxford University Press.
Woskov, P. (2017). Millimeter-wave directed energy deep boreholes. AIG drilling for geology II conference. Brisbane, Australia.
Wright, C., and Nyberg, D. (2017). An inconvenient truth: How organizations translate climate change into business as usual. *Academy of Management Journal*, 60(5): 1633–1661.
Wu, X., and Chen, G. (2018). Coal use embodied in globalized world economy: From source to sink through supply chain. *Renewable and Sustainable Energy Reviews*, 81: 978–993.
Wu, X., Zhou, L., Shi, J., Cheng, C. Y., and Sun, F. (2022). Multiomics analysis of male infertility. *Biology of Reproduction*, 107(1): 118–134.
WWF. (2024). *Living Planet Report 2024 – A System in Peril*.
Xu, C., Kohler, T. A., Lenton, T. M., Svenning, J.-C., and Scheffer, M. (2020). Future of the human climate niche. *Proceedings of the National Academy of Sciences*, 117(21): 11350–11355.
Yack, B. (1990). Natural right and Aristotle's understanding of justice. *Political Theory*, 18(2): 216–237.
Yasin, I., Ahmad, N., and Chaudhary, M. A. (2021). The impact of financial development, political institutions, and urbanization on environmental degradation: evidence from 59 less-developed economies. *Environment, Development and Sustainability*, 23: 6698–6721.
Yermoldina, G., Suimenbayev, B., Sysoev, V., and Suimenbayeva, Z. B. (2019). Features of space solar power station control system. *Acta Astronautica*, 158: 111–120.
Yew-Kwang, N. (1987). Diamonds are a government's best friend: burden-free taxes on goods valued for their values. *The American Economic Review*, 77(1): 186–191.
Yin, H.-T., Wen, J., and Chang, C.-P. (2023). Going green with artificial intelligence: The path of technological change towards the renewable energy transition. *Oeconomia Copernicana*, 14(4): 1059–1095.
York, R., and Bell, S. E. (2019). Energy transitions or additions?: Why a transition from fossil fuels requires more than the growth of renewable energy. *Energy Research & Social Science*, 51: 40–43.
Yudkowsky, E. (2007). The Power of Intelligence. https://intelligence.org/2007/07/10/the-power-of-intelligence/.
Zaltman, G., Duncan, R. B., and Holbek, J. (1973). *Innovation in Organizations*. John Wiley and Sons.
Zhang, D. D., Lee, H. F., Wang, C., Li, B., Pei, Q., Zhang, J., and An, Y. (2011). The causality analysis of climate change and large-scale human crisis. *Proceedings of the National Academy of Sciences*, 108(42): 17296–17301.
Zhao, J., Xu, L., Sun, J., Song, M., Wang, L., Yuan, S. et al. (2023). The Global Trends in Incidence, Death, Burden and Risk Factors of Early-onset Cancer From 1990 to 2019. *BMJ Oncology*, 2: e000049.
Zheng, J., and Suh, S. (2019). Strategies to reduce the global carbon footprint of plastics. *Nature Climate Change*, 9(5): 374–378.
Zidar, O. (2019). Tax cuts for whom? Heterogeneous effects of income tax changes on growth and employment. *Journal of Political Economy*, 127(3): 1437–1472.
Zimbardo, P. (2011). *The Lucifer effect: How good people turn evil*. Random house.
Zolōtas, X. E. (1981). *Economic growth and declining social welfare*. New York University Press.
Zuboff, S. (2023). The age of surveillance capitalism. pp. 203–213. *In*: Longhofer, W. and D. Winchester (Eds.). *Social Theory Re-Wired - new connections to classical and contemparary perspectives*. Routledge.

Index

D

E

F

G

H

N

O

P

R

S

T

For Product Safety Concerns and Information please contact our EU
representative GPSR@taylorandfrancis.com
Taylor & Francis Verlag GmbH, Kaufingerstraße 24, 80331 München, Germany

www.ingramcontent.com/pod-product-compliance
Ingram Content Group UK Ltd.
Pitfield, Milton Keynes, MK11 3LW, UK
UKHW052051230926
475702UK00025B/1094
9781032844800